ZnO Nanowires: Growth, Properties, and Energy Applications

ZnO Nanowires: Growth, Properties, and Energy Applications

Editor

Vincent Consonni

MDPI

Basel • Beijing • Wuhan • Barcelona • Belgrade • Novi Sad • Cluj • Manchester

Editor
Vincent Consonni
Université Grenoble Alpes,
CNRS, Grenoble INP, LMGP
Grenoble, France

Editorial Office
MDPI
St. Alban-Anlage 66
4052 Basel, Switzerland

This is a reprint of articles from the Special Issue published online in the open access journal *Nanomaterials* (ISSN 2079-4991) (available at: https://www.mdpi.com/journal/nanomaterials/special_issues/ZnO_nanowires).

For citation purposes, cite each article independently as indicated on the article page online and as indicated below:

Lastname, A.A.; Lastname, B.B. Article Title. *Journal Name* **Year**, *Volume Number*, Page Range.

ISBN 978-3-0365-9919-9 (Hbk)
ISBN 978-3-0365-9920-5 (PDF)
doi.org/10.3390/books978-3-0365-9920-5

Contents

About the Editor

Vincent Consonni

Vincent Consonni (PhD) received his PhD in Materials Science and Engineering in 2008 from Grenoble INP. He then spent post-doctoral stays in Paul-Drude-Institut für Festkörperelektronik, Berlin, and in Laboratoire des Technologies de la Microélectronique, Grenoble. He joined Laboratoire des Matériaux et du Génie Physique, Grenoble, as a permanent CNRS Research Scientist in 2011, received his "Habilitation à Diriger des Recherches" in 2016, and was appointed CNRS Research Director in 2023. Vincent Consonni is the Head of NanoMAT Team and he has been supervising the research line devoted to Semiconducting Nanowires and Nanostructures since 2013. His research focuses on the physical-chemistry of micro- and nano-structures involving compound semiconductors (ZnO, Ga_2O_3, TiO_2, Sb_2S_3, ...) with a special emphasis on their chemical synthesis using wet chemistry, MOCVD, and ALD. He is currently working on ZnO nanowires and related heterostructures for optoelectronic and piezoelectric devices as well as on Ga_2O_3 thin films for power electronics.

nanomaterials

Editorial

ZnO Nanowires: Growth, Properties, and Energy Applications

Vincent Consonni

Université Grenoble Alpes, CNRS, Grenoble INP, LMGP, F-38016 Grenoble, France;
vincent.consonni@grenoble-inp.fr

As a biocompatible semiconductor composed of abundant elements, ZnO, in the form of nanowires, exhibits remarkable properties, mainly originating from its wurtzite structure and correlated with its high aspect ratio at nanoscale dimensions. ZnO nanowires have thus received increasing interest in the community and have specifically emerged as a potential building block for a wide variety of devices in the field of energy conversion. Among the different energy conversion applications, ZnO nanowires have, to name just two examples, been integrated into nanostructured solar cells and piezoelectric devices. Despite the vast number of publications in the field, there is still a significant need to explore the growth of ZnO nanowires, to more precisely elucidate and control their fundamental properties, and to improve their integration into real-world engineering devices.

In terms of growth, one of the most difficult challenges consists in integrating ZnO nanowires over dedicated localized areas and dedicated substrates, both of these being relevant for the targeted devices. Durbach et al. report a nano-second laser irradiation process for generating Au catalysts over dedicated areas, thus controlling the position of ZnO nanowires grown by chemical vapor deposition through a selective growth approach compatible with large surfaces [1]. Schaper et al. show the formation of ZnO nanowires over single-walled carbon nanotubes and graphene using a full chemical vapor deposition approach, further achieving selective growth over dedicated areas [2]. Another challenge consists in developing innovative heterostructures made of ZnO nanowires combined with a selected semiconductor. Jin et al. develop the growth of semiconducting shells (i.e., ZnS and Ag_2S) deposited by successive ionic layer adsorption and reaction on ZnO nanowires, further revealing their UV-sensing properties [3]. Zhang et al. report on the fabrication of heterostructures made from ZnO nanorods covered with TiO_{2-x} mesoporous spheres, revealing their properties for photocatalytic hydrogen production [4].

The integration of ZnO nanowires into nanostructured solar cells as an electron-transporting material is driven by the expected benefits of light-trapping phenomena and efficient charge carrier management. However, beyond the proof-concept of ZnO nanowire-based solar cells, the need to carefully optimize the dimensional parameters suffers from technological challenges. Sekar et al. report on the optimization of the dimensions of ZnO nanowires and their impact on the photovoltaic properties of $FACsPb(IBr)_3$ perovskite solar cells, further exploring the use of a carbazole-based hole-transporting material [5]. Hector et al. investigate the effect of the thickness of the Sb_2S_3 shell over the photovoltaic properties of extremely thin absorber solar cells, revealing the dimensional trade-off required [6].

ZnO nanowires, with their growth direction oriented along the piezoelectric and polar *c*-axis, act as the active layer in piezoelectric devices, which are largely developed using a vertically integrated configuration. By combining finite-element method calculations with experimental data available in the literature, Lopez-Garcia et al. report on dimensional roadmap and optimization guidelines, showing that the range of optimal radius, that fully deplete ZnO nanowires in terms of charge carriers, depends on the growth technique [7]. Lopez-Garcia et al. reveal the fabrication of gravure-printed ZnO seed layers as an alternative process in order to subsequently form ZnO nanowires over flexible polymer

Citation: Consonni, V. ZnO Nanowires: Growth, Properties, and Energy Applications. *Nanomaterials* **2023**, *13*, 2519. https://doi.org/10.3390/nano13182519

Received: 24 August 2023
Accepted: 25 August 2023
Published: 8 September 2023

substrates, further characterizing their piezoelectric properties using piezoresponse force microscopy [8]. Tlemcani et al. show the integration of ZnO nanowires into flexible piezoelectric nanogenerators and compare their performance using two seed layer structures (i.e., Au/ZnO vs. ITO/ZnO) [9]. Zhai et al. report on the combination of ZnO nanowires with a cellulose nanofiber film, further revealing their electromechanical and UV-sensing properties [10].

Ultimately, assembling ZnO nanowires into hierarchical structures represents a promising approach for further increasing their integration into engineering devices. Di Mari et al. report on the formation of ZnO nanostars made of agglomerated nanowires and explore their properties as pseudo-capacitors for energy storage [11].

In summary, this Special Issue brings together more than 80 authors from different countries, who submitted 11 original research articles conveying their foundational research dedicated to ZnO nanowires. Overall, if the present Special Issue cannot fully reflect the high diversity rapidly developing in the community of ZnO nanowires, it will certainly contribute to research interest in the field.

Acknowledgments: The author is grateful to all the authors for submitting their investigations to the present Special Issue and for its successful completion. He deeply acknowledges the Nanomaterials reviewers for enhancing the quality and impact of all the submitted papers. Finally, he sincerely thanks the editors, for their continuous support during the development and publication of this Special Issue.

Conflicts of Interest: The author declares no conflict of interest.

References

1. Durbach, S.; Schniedermeyer, L.; Marx, A.; Hampp, N. Laser-Induced Au Catalyst Generation for Tailored ZnO Nanostructure Growth. *Nanomaterials* **2023**, *13*, 1258. [CrossRef] [PubMed]
2. Schaper, N.; Alameri, D.; Kim, Y.; Thomas, B.; McCormack, K.; Chan, M.; Divan, R.; Gosztola, D.J.; Liu, Y.; Kuljanishvili, I. Controlled Fabrication of Quality ZnO NWs/CNTs and ZnO NWs/Gr Heterostructures via Direct Two-Step CVD Method. *Nanomaterials* **2021**, *11*, 1836. [CrossRef] [PubMed]
3. Jin, Y.; Jiao, S.; Wang, D.; Gao, S.; Wang, J. Enhanced UV Photoresponsivity of ZnO Nanorods Decorated with Ag_2S/ZnS Nanoparticles by Successive Ionic Layer Adsorption and Reaction Method. *Nanomaterials* **2021**, *11*, 461. [CrossRef] [PubMed]
4. Zhang, B.; Li, Q.; Wang, D.; Wang, J.; Jiang, B.; Jiao, S.; Liu, D.; Zeng, Z.; Zhao, C.; Liu, Y.; et al. Efficient Photocatalytic Hydrogen Evolution over TiO_{2-x} Mesoporous Spheres-ZnO Nanorods Heterojunction. *Nanomaterials* **2020**, *10*, 2096. [CrossRef] [PubMed]
5. Sekar, K.; Nakar, R.; Bouclé, J.; Doineau, R.; Nadaud, K.; Schmaltz, B.; Poulin-Vittrant, G. Low-Temperature Hydrothermal Growth of ZnO Nanowires on AZO Substrates for $FACsPb(IBr)_3$ Perovskite Solar Cells. *Nanomaterials* **2022**, *12*, 2093. [CrossRef] [PubMed]
6. Hector, G.; Eensalu, J.S.; Katerski, A.; Roussel, H.; Chaix-Pluchery, O.; Appert, E.; Donatini, F.; Acik, I.O.; Kärber, E.; Consonni, V. Optimization of the Sb_2S_3 Shell Thickness in ZnO Nanowire-Based Extremely Thin Absorber Solar Cells. *Nanomaterials* **2022**, *12*, 198. [CrossRef] [PubMed]
7. Lopez Garcia, A.J.; Mouis, M.; Consonni, V.; Ardila, G. Dimensional Roadmap for Maximizing the Piezoelectrical Response of ZnO Nanowire-Based Transducers: Impact of Growth Method. *Nanomaterials* **2021**, *11*, 941. [CrossRef] [PubMed]
8. Garcia, A.J.; Sico, G.; Montanino, M.; Defoor, V.; Pusty, M.; Mescot, X.; Loffredo, F.; Villani, F.; Nenna, G.; Ardila, G. Low-Temperature Growth of ZnO Nanowires from Gravure-Printed ZnO Nanoparticle Seed Layers for Flexible Piezoelectric Devices. *Nanomaterials* **2021**, *11*, 1430. [CrossRef] [PubMed]
9. Slimani Tlemcani, T.; Justeau, C.; Nadaud, K.; Alquier, D.; Poulin-Vittrant, G. Fabrication of Piezoelectric ZnO Nanowires Energy Harvester on Flexible Substrate Coated with Various Seed Layer Structures. *Nanomaterials* **2021**, *11*, 1433. [CrossRef] [PubMed]
10. Zhai, L.; Kim, H.-C.; Muthoka, R.M.; Latif, M.; Alrobei, H.; Malik, R.A.; Kim, J. Environment-Friendly Zinc Oxide Nanorods-Grown Cellulose Nanofiber Nanocomposite and Its Electromechanical and UV Sensing Behaviors. *Nanomaterials* **2021**, *11*, 1419. [CrossRef] [PubMed]
11. Di Mari, G.M.; Mineo, G.; Franzò, G.; Mirabella, S.; Bruno, E.; Strano, V. Low-Cost, High-Yield ZnO Nanostars Synthesis for Pseudocapacitor Applications. *Nanomaterials* **2022**, *12*, 2588. [CrossRef] [PubMed]

nanomaterials

Article

Efficient Photocatalytic Hydrogen Evolution over TiO_{2-X} Mesoporous Spheres-ZnO Nanorods Heterojunction

BingKe Zhang [1,2], Qi Li [3], Dongbo Wang [1,2,*], Jinzhong Wang [1,2,*], Baojiang Jiang [3,*], Shujie Jiao [1,2,*], DongHao Liu [1,2], Zhi Zeng [1,2], ChenChen Zhao [1,2], YaXin Liu [1,2], ZhiKun Xun [4,5,*], Xuan Fang [6,*], ShiYong Gao [1,2], Yong Zhang [1,2] and LianCheng Zhao [1,2]

1 National Key Laboratory for Precision Hot Processing of Metals, Harbin Institute of Technology, Harbin 150001, China; zhangbingke007@163.com (B.Z.); 18245019907@163.com (D.L.); 1144420106@hit.edu.cn (Z.Z.); zhaochenstu@163.com (C.Z.); lyx15545582475@163.com (Y.L.); gaoshiyong@hit.edu.cn (S.G.); yongzhang@hit.edu.cn (Y.Z.); lczhao@hit.edu.cn (L.Z.)

2 Department of Optoelectronic Information Science, School of Materials Science and Engineering, Harbin Institute of Technology, Harbin 150001, China

3 Key Laboratory of Functional Inorganic Material Chemistry, School of Chemistry and Materials Science, Heilongjiang University, Ministry of Education of the People's Republic of China, Harbin 150080, China; liqchem@sina.com

4 College of Science, Guangdong University of Petrochemical Technology, Guandu Road No. 139, Maoming 525000, China

5 Key Laboratory for Photonic and Electronic Bandgap Materials, Ministry of Education, School of Physics and Electronic Engineering, Harbin Normal University, Harbin 150025, China

6 State Key Laboratory of High Power Semiconductor Lasers, School of Science, Changchun University of Science and Technology, 7089 Wei-Xing Road, Changchun 130022, China

* Correspondence: wangdongbo@hit.edu.cn (D.W.); jinzhong_wang@hit.edu.cn (J.W.); shujiejiao@hit.edu.cn (B.J.); jbj@hlju.edu.cn (S.J.); xuzhikunnano@163.com (Z.X.); fangx@cust.edu.cn (X.F.)

Received: 18 September 2020; Accepted: 20 October 2020; Published: 22 October 2020

Abstract: Photocatalytic water splitting into hydrogen is regarded as one of the key solutions to the deterioration of the global environment and energy. Due to the significantly reduced grain boundaries, ZnO nanorods facilitate a fast electron transfer through their smooth tunnels and are well suited as a photocatalyst. However, the photocatalytic hydrogen evolution performance of pristine ZnO nanorods is still low due to the high recombination rate of photogenerated electron-hole pairs and the less light absorption. Here, a novel structure about black ZnO nanorods (NRs)/TiO_{2-X} mesoporous spheres (MSs) heterojunction has been prepared and the photocatalytic hydrogen evolution performance has been explored. The photocatalytic activity test results showed that ZnO NRs/TiO_{2-X} MSs exhibited higher catalytic activity than ZnO NRs for hydrogen production. Compared to the pure ZnO NRs photoanode, the photocurrent of ZnO NRs/TiO_{2-X} MSs heterojunction photoanode could reach 0.41 mA/cm^2 in view of the expanding spectral response region and effective inhibition of e^-/h^+ recombination at the same condition. Using a relatively integrated experimental investigation and mechanism analysis, we scrutinized that after being treated with $NaBH_4$, TiO_2 MSs introduce oxygen vacancies expanding the photocatalytic activity of pure TiO_2, and improving conductivity and charge transport capabilities through coating on ZnO NRs. More importantly, the results provide a promising approach in the NRs/MSs composite structure serving as photoanodes for photocatalytic hydrogen production.

Keywords: ZnO NRs/TiO_{2-X} MSs heterojunction; photocatalytic hydrogen production; oxygen vacancies; efficiency; spectral response region

1. Introduction

Currently, ZnO nanostructures-based photocatalysts have attracted great attention owing to their non-toxicity, optical transparency, high photoconversion efficiency, low cost, etc. [1–7]. Due to the significantly reduced grain boundaries, one-dimensional (1D) nanostructures especially nanorods facilitate a fast electron transfer through their smooth tunnels [8–10]. Nevertheless, the photocatalytic H_2 evolution performance of pristine ZnO nanorods is still low due to the high recombination rate of photogenerated e^-/h^+ pairs and the less light absorption (only in the ultraviolet region accounting for 3–5% of solar energy) [11–15]. To overcome the above disadvantages, the heterojunction structure based on ZnO nanorods provides a new insight for suppressing charge recombination and enhancing light absorption.

Many studies on the coupling of ZnO and TiO_2 have shown an improved photocatalytic performance in dye degradation and hydrogen evolution [16–20]. Although the proper band edge position of TiO_2 relative to ZnO can assist the effective interfacial charge transfer, the fabrication of the ZnO/TiO_2 heterojunction cannot broaden the light absorption range, which is still too narrow to improve the efficiency of the photocatalyst. Recently, Ti^{3+} self-doped black TiO_2 got supreme importance in the field of light harvesting, which can be formed from conventional white TiO_2 to regulate the electronic energy band structure of TiO_2 with the expansion of its optical response from ultraviolet light to a visible and infrared region [21–28]. It can be seen from previous reports that TiO_2 mesoporous microspheres have an ultrahigh specific surface area and strong light scattering ability, which is achieved by combining the advantages of primary nanocrystalline composition and sub-micron structure [29–32]. However, up to now, the investigation on the photocatalytic performance of black TiO_{2-x} mesoporous microspheres/ZnO nanorods nanostructure heterojunctions is very rare.

In this research, a novel structure about black TiO_{2-x} mesoporous spheres/ZnO nanorods heterojunction has been prepared and the photocatalytic hydrogen evolution performance was explored. The light absorption region of the heterostructure covers the ultraviolet to infrared region nearly over the entire spectral range and exhibits excellent photocatalytic hydrogen production performance compared to the pure ZnO and ZnO/TiO_2. Additionally, as confirmed from electrochemical, electronic, and spectroscopy characterizations, the significantly improved hydrogen production performance is found in the designed ZnO/TiO_{2-x} heterojunction structure, which is rooted from the potential difference generated on the ZnO-TiO_{2-x} interface and increased light absorption from the reduction of TiO_2. The experimental results of the present work indicate a novel strategy for developing the ZnO nanostructure-based composite materials with excellent photocatalytic properties.

2. Materials and Methods

2.1. Preparation of ZnO Nanorods (NRs)

The ZnO seed layer was deposited on FTO glass (15 Ω square) by the frequency (RF) magnetron sputtering system. The FTO substrates were ultrasonically cleaned with acetone solutions, ethanol, and deionized water in an ultrasound bath for 20 min, respectively. A high-purity ZnO target (99%) with a 60.0 diameter was used for continuous sputtering at 20 W RF power for 5 min, and the working pressure was maintained at 1.0 Pa with the flow ratio of oxygen to argon 18 to 42 sccm. Then, the ZnO seed layer was annealed at 450 °C for 60 min in air.

The ZnO nanorods were grown onto the seed substrates by a simple hydrothermal method. The ZnO seed layer was immersed in a mixed solution containing 30 mL of 0.03 mmol of $Zn(NO_3)_2{\cdot}6H_2O$, 0.06 mmol of HTM in the Teflon-lined stainless steel autoclave, and then the autoclave was heated in a constant temperature electric oven at 90 °C for 6 h.

2.2. Preparation of ZnO Nanorods (NRs)/TiO_{2-x} Mesoporous Spheres (MSs) Composites

The TiO_2 MSs were synthesized under hydrothermal conditions as previously reported [33]. $NaBH_4$ and TiO_2 powders were mixed in an equal ratio, and the mixture was ground for 20 min completely. Then, the mixture was heated at 300 °C for 60 min under an argon atmosphere. The mixture was repeatedly filtered and washed with water and ethanol to remove $NaBH_4$. The prepared TiO_{2-X} MSs (1.0 g) was added to ethanol (80 mL) containing α-terpineol (4.34 mL) and ethyl cellulose (0.5 g) and stirred vigorously for 1 h, and then the solvent was removed with a rotary evaporator at 10 h to prepare the TiO_{2-X} paste. The spin coating solution was obtained by repeatedly stirring the slurry thirty times, and spin-coating on the ZnO nanorods at 3500 rpm for 20 s, followed by sintering at 450 °C for 30 min of preparation of the unreduced TiO_2 and ZnO composite material followed by the same pattern.

2.3. Characterizations

The morphology and crystallinity of the obtained material were performed by scanning electron microscopy (SEM, Carl Zeiss, Merlin Compact, Jena, Germany). The transmission electron microscope (TEM, FEI, Hillsboro, USA) and high-resolution TEM (HRTEM, FEI, Hillsboro, USA) observations were performed with a TEI Tecnai G2 F30 (FEI, Hillsboro, USA) microscope operating at 300 KV, while energy-dispersive X-ray spectroscopy (EDS) and elemental mapping were performed on the transmission electron microscope. On the spectrometer (ESCALAB 250Xi, Thermo Scientific Escalab, Waltham, USA), the X-ray photoelectron spectroscopy (XPS) measurement was carried out. The X-ray power diffraction (XRD) analysis was characterized by an X-ray diffractometer (Empyrean, Panalytical, Malvern, UK) equipped with Cu Kα radiation (λ = 1.5418 Å). Raman spectra were obtained using a 532.8 nm excitation laser on the LabRAM HR EV0 of Horiba Jobin Yvon, Langjoux, France. The optical response of all the samples was obtained by photoluminescence (PL) measurements excited by a 325 nm He-Cd laser. At room temperature, the ultraviolet visible diffuse reflectance spectra (DRS) was recorded on the Shimadzu UV1700-visible spectrophotometer.

2.4. Photoelectrochemical (PEC) Measurements

PEC measurements were conducted on the Metrohm electrochemical workstation (Autolab-PGSTAT302N) in a three-electrode cell, the obtained sample, platinum foil, and Ag/AgCl electrode as the working, counter, and reference electrodes, respectively. A 1 M KOH solution was used as an electrolyte without the additive. A model LS1000-4S-AM1.5G-1000W solar simulator (Solar Light Company, Glenside, PA, USA) equipped with a metal mesh was employed to give an irradiance of 100 mW/cm^2. Under an AM 1.5 G light source, linear sweep voltammetry (LSV) measurements were performed with a sweep rate of 1 mV/s. The photocatalytic H_2 evolution experiments were performed in the photocatalytic system evaluation device (AuLight, Beijing, CEL-SPH2N) equipped with an online gas chromatograph (SP7800, TCD, 5 molecular sieve, Beijing Keruida Limited), and nitrogen as the carrier gas. AM 1.5 was used as the light source to trigger the photocatalytic reaction. Under stirring conditions, a certain 0.05 g catalyst was dispersed into a mixed solution composed of 100-mL methanol aqueous solution (15 vol%) and methanol as a sacrificial reagent. A certain amount of H_2PtC_{16} solution was added dropwise to load 1 wt% Pt on the surface of the catalyst by in situ photo deposition. A gaseous sample was collected and analyzed every hour.

3. Results

Morphological and structural characterizations of the pure ZnO NRs and the ZnO-NRs/TiO_2 MSs composite samples are presented in Figure 1. From the top view images of Figure 1a, it can be found that the ZnO nanorods grow upward on the FTO substrate with a high diffracted intensity and each ZnO nanorod has a typical hexagonal structure cross section. The average diameter of ZnO nanorods is about 130 nm, as shown in Figure 1b. Compared with pure ZnO nanorods, no obvious morphological changes are observed on the surface of the nanorods coated with TiO_2 or TiO_{2-X}. Obviously, TiO_2 and TiO_{2-X} MSs are observed on the surface of the ZnO nanorods in Figure 1c–f. It is proven that the TiO_2 and TiO_{2-X} coating solutions deposit MSs on the NRs surface after the spin-coating process. The detailed nanostructures of ZnO-TiO_{2-X} are further performed by transmission electron microscope (TEM) and high-resolution TEM (HRTEM). It can be clearly observed from Figure 2a that the TiO_{2-X} MSs adhere on the surface of the coarsened ZnO NRs and the TiO_{2-X} MSs/ZnO NRs still keep a three-dimensional morphology. The clear lattice fringes in the HRTEM image indicate that the formation of a ZnO-TiO_{2-X} heterostructure is well-crystallized. Two different types of lattice images are obtained with d spaces of 0.350 and 0.260 nm, corresponding to the (101) plane of typical anatase TiO_2 and the (002) plane of hexagonal ZnO, respectively [34]. The hybrid photocatalyst has clear and continuous stripes, indicating that there is an intense mutual attraction between ZnO NRs and TiO_{2-X} due to suitable lattice parameters [35]. Furthermore, this also proves that the reduced TiO_2 nanocrystals are still highly crystalline, and the Ti^{3+} introduced by the reduction does not make the crystal lattice disordered. The corresponding EDS elemental mapping images of the ZnO NRs coated with TiO_{2-X} MSs are presented in Figure S1 (Supplementary Materials) (ESI†).

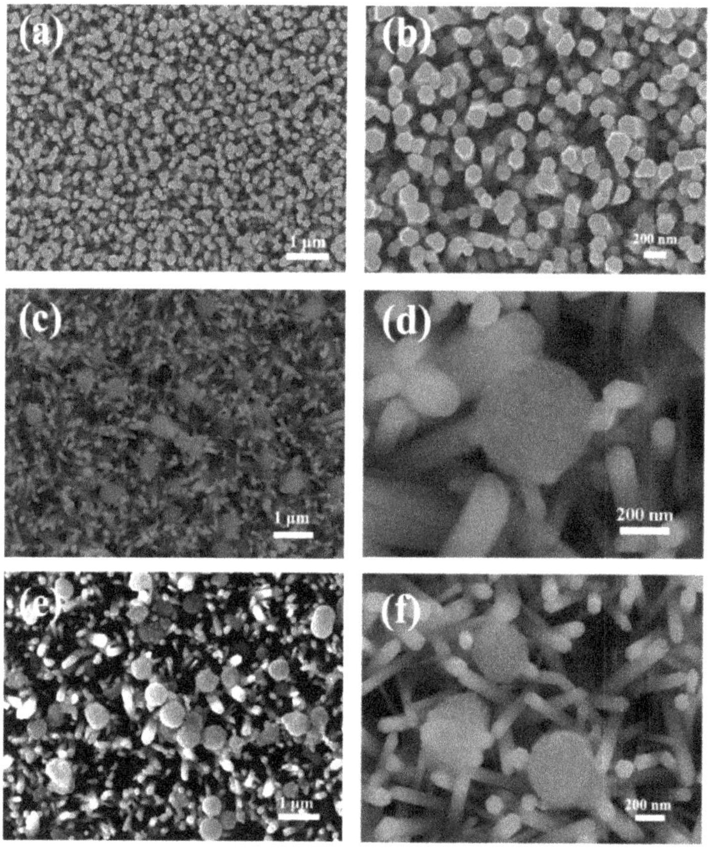

Figure 1. Typical top view SEM images of pure ZnO (**a**,**b**), ZnO/TiO_2 (**c**,**d**), ZnO/TiO_{2-X} (**e**,**f**) samples on the FTO substrate.

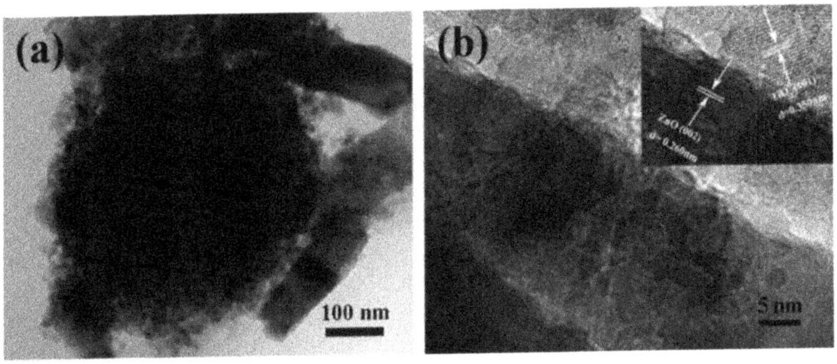

Figure 2. Structural characterizations of ZnO NRs/TiO_{2-X} MSs: (**a**) TEM image of ZnO NRs/TiO_{2-X} MSs; (**b**) HRTEM image of ZnO NRs/TiO_{2-X} MSs.

Figure 3 shows the X-ray diffraction (XRD) patterns of the fabricated ZnO, ZnO/TiO_2, ZnO/TiO_{2-X}, and FTO substrate for comparison. The XRD pattern of pure ZnO nanorods displays diffraction peaks around 31.77, 34.42, 36.25, and 47.54°, which could be well indexed to the characteristic peaks (100), (002), (101), and (102) planes of the hexagonal ZnO with a wurtzite structure (JCPDS file no. 36-1451). In addition, the main peaks with a higher intensity depict the ZnO nanorods epitaxially with a very uniform orientation grown along the C-axis on the ZnO seed layer. After spin coating TiO_2 and TiO_{2-X}, the ZnO NRs also exhibit a strong wurtzite structure diffraction, but a close observation of the XRD pattern in the range of 20 to 30° reveals a broad peak centered at 25.2° (inset of Figure 3). The peaks can be assigned to the (101) of the TiO_2 anatase phase (JCPDS 21-1272). Raman is also powerful to identify the structural changes, including determination of the presence of heterojunction surface and oxygen vacancies. Figure 4 shows the Raman spectra of the samples, where pure ZnO Raman modes that are located at 98.32 and 435.72 cm^{-1} correspond to the two typical vibration modes E_2 (low) and E_2 (high) of the hexagonal wurtzite structure zinc oxide crystals, respectively [36]. Obviously, four more peaks are observed in the Raman spectrum of the ZnO nanorods coated with TiO_2. The peaks appearing at 398.94 cm^{-1}, 518.31 cm^{-1} and 144.30 cm^{-1}, 637.54 cm^{-1} are assigned respectively to the B_{1g} mode, a doublet of the A_{1g} and B_{1g} modes and E_g mode of anatase phase TiO_2, which indicates that the ZnO nanorods are coated with TiO_2 and TiO_{2-X} successfully [37–39]. This result is in agreement with those from the XRD and SEM measurements. Nevertheless, the strongest E_g mode area at 144.30 amplified by TiO_2 shows a slight shift accompanied by a larger linewidth by the $NaBH_4$ reduction treatment. This proves that local lattice defects related to surface oxygen vacancies will be introduced after the reduction, and a blue shift will be caused due to phonon confinement or non-stoichiometry. In addition, the broadening of the peak position is also related to the decrease of crystal quality [40–42]. To further perform the surface composition and chemical state of ZnO/TiO_{2-X}, the XPS results are shown in Figure S3 (Supplementary Materials), ESI†. Full elemental X-ray photo electron spectroscopy proves the successful composite of ZnO with TiO_{2-X}. A detailed analysis of the inner Ti 2p orbital electrons shows the Ti^{4+} 2p3/2 and 2p1/2 spin-orbit splitting peaks at 457.89 and 463.59 eV, respectively. The additional small peaks in the black TiO_2 at 457.34 and 463.04 eV can be attributed to the 2p1/2 and 2p3/2 peaks of Ti^{3+}.

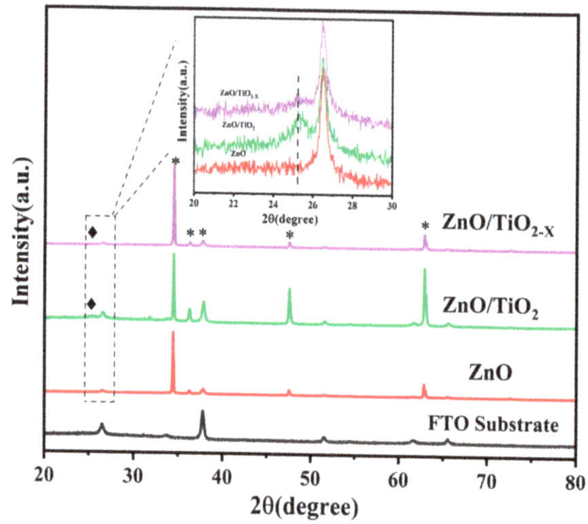

Figure 3. XRD patterns of ZnO, ZnO/TiO_2, ZnO/TiO_{2-X}, and FTO substrate samples (* represents the peak position of the hexagonal ZnO with a wurtzite structure and ♦ the peak position of TiO_2 anatase phase).

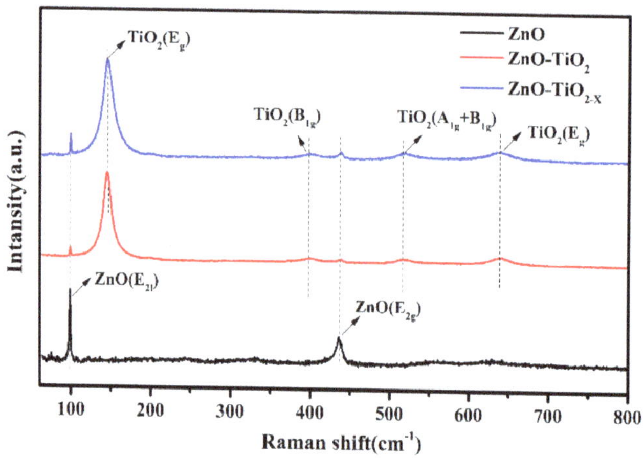

Figure 4. Raman spectra of TiO_2 NRs, ZnO/TiO_2, ZnO/TiO_{2-X}.

Figure 5 shows the absorption spectrum of the ZnO NRs, ZnO NRs/TiO_2 MSs, and ZnO NRs/TiO_{2-X} MSs heterojunction thin films grown on glass. All samples show a strong absorption of ultraviolet light, which can be attributed to the electronic transition from the valence band to the conduction band. It is indicated that the absorption onset of the ZnO/TiO_{2-X} is at ~400 nm, which has an obvious red shift compared with the ZnO and ZnO/TiO_2 composite material, which may be correlated with the amount of Ti^{3+} in the shell. The optical band gaps of the samples can be determined according to the Kubelka-Munk equation: $(\alpha h\nu)$ n = K $(h\nu - E_g)$, where "hν" and "α" are photonic energy and the optical absorption coefficient, respectively. The bandgap of the ZnO NRs, ZnO NRs/TiO_2 MSs, and ZnO NRs/TiO_{2-X} MSs can be assessed as ~3.21, ~3.10, ~3.02 eV, respectively by measuring the linear extrapolation to the hν-axis (Figure S4 (Supplementary Materials), ESI†) [43]. More importantly, the absorption of the sample combined with the TiO_{2-X} mesoporous sphere is markedly increased at a 400–600 nm region, compared with pure ZnO the enhancement of sunlight harvesting can be attributed to the lattice disorder introduced in the formation of black TiO_2 (the generation of mid gap energy levels within the band gap) [44].

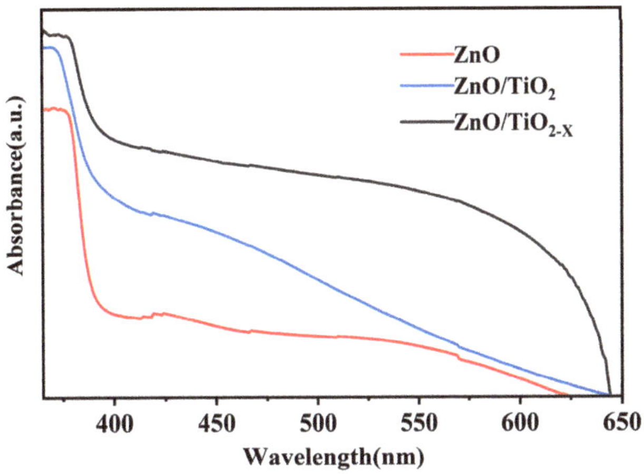

Figure 5. UV-Vis diffuse reflectance spectra of the pure ZnO, ZnO/TiO_2, $ZnO/TiO_{2\text{-}X}$ samples.

To judge the quality of ZnO crystals and perform the mechanism of photoinduced charge carriers and interfacial charge transfer over the $ZnO/TiO_{2\text{-}X}$ heterostructures, the room temperature PL spectra of the three synthesized samples are shown in Figure 6. The pure ZnO nanorods show a strong and sharp ultraviolet luminescence peak at ~372 nm, originating from the near band gap excitonic emission, and the other is located at ~738 nm, which is attributed to the second-order diffraction peak emitted at the ultraviolet band [45]. No obvious defect emission is seen in the yellow-green wavelength band of visible light, indicating that the ZnO nanorods prepared in this way have good crystallinity and fewer surface defects [46,47]. After coating with TiO_2, the emission intensity is significantly reduced, which indicated that the composite materials reduce the recombination probability of photogenerated e^-/h^+ pair. It is worth noting that the black $TiO_{2\text{-}X}$ mesoporous spheres/ZnO nanorods heterojunction has the lowest electron-hole recombination rate. Therefore, the unique heterostructure photocatalyst of the $ZnO/TiO_{2\text{-}X}$ has a positive effect on the effective separation of photogenerated carriers during the photocatalytic reactions charge carriers. This considering the fact that the number of oxygen vacancies in titanium dioxide is too low to be disclosed, it can be easily explained that the significantly defect emission peak by oxygen vacancies in the PL was not observed. It is worth noting that the emission peak of ZnO nanorods shifted to the right after coating with $TiO_{2\text{-}X}$. This behavior may be due to the quantum confinement stark effect [48–50]. Therefore, the type II band structure of $ZnO/TiO_{2\text{-}X}$ promotes the excited carriers, which confines the electron-hole recombination, and results in the emission intensity decreases. From the electron dynamics point of view, the observed lifetime shortenings obviously indicate an improvement in the charge separation for the $ZnO/TiO_{2\text{-}X}$, leading to an increase in photocatalytic performance (as below).

To perform the effect of $ZnO/TiO_{2\text{-}X}$ heterojunction on photoelectrochemical properties, the linear sweep voltammetry (LSV) behavior of these samples is recorded under a simulated AM 1.5 sunlight illumination (Figure 7a). The photocurrent starting potential of $ZnO/TiO_{2\text{-}X}$ is −0.637 vs. Ag/AgCl, which shows a slight shift compared with −0.618 vs. Ag/AgCl of ZnO/TiO_2 and the fastest growth of anodic current at a higher applied voltage. It is probably related to the best electron transportation and improved charge transfer compared to the pure ZnO electrode and ZnO/TiO_2. In addition, a set of linear scans are collected in the dark, as shown in Figure S5 (Supplementary Materials), ESI†. All three photoelectrodes show a very low dark current. The transient photocurrent response can also predict the photocatalytic activity of the electrode. The transient photocurrent response analysis is employed under illumination with several 20 s light on/off cycles at 0 vs. Ag/AgCl in Figure 7b. Transient photocurrent response experiments are explored to prove the enhancement in separation and transport efficiency of charge carriers in ZnO/TiO_{2-X} nanostructures. It can be observed that the photocurrent signal can be repeatedly switched from the "on" state to the "off" state by periodically turning the light source on and off for all electrodes, which shows that all electrodes have a good stability. It is worth noting that after coating with TiO_{2-x} the photocurrent has been significantly enhanced compared with pure ZnO nanorods. Simultaneously, the photocurrent induced by the simulated solar light irradiation of ZnO/TiO_{2-X} (0.41 mA/cm^2) at 0 vs. Ag/AgCl, is about eight times larger than that of ZnO (0.05 mA/cm^2). The pure ZnO exhibited a much lower photocurrent than ZnO/TiO_{2-X}, which may be attributed to the high recombination rate of photo-induced charges and weak light absorption. Under light, the narrow band gap of TiO_{2-X} can capture low-energy photons. The narrow band gap of TiO_{2-X} can increase the light absorption rate and the internal electrostatic field in the ZnO/TiO_{2-X} junction causes the recombination rate of photogenerated e^-/h^+ to decrease.

Using methanol as the electron donor and Pt as the co-catalyst, the photocatalytic hydrogen evolution experiment is carried out under AM 1.5. Figure 8 shows the results of hydrogen evolution experiments of ZnO and ZnO/TiO_{2-X} materials. After 5 h of xenon lamp irradiation, ZnO/TiO_{2-X} nanocomposites exhibited significant photocatalytic activities towards H_2 production performance, and its hydrogen evolution amount reaches 243.4 μmol·g^{-1} better than the previous reports [51,52]. The addition of TiO_{2-X} has a significant effect on the photocatalytic activity of ZnO, indicating that the TiO_{2-X} hollow structures could promote more effective photons absorption and prolong the lifetime of the photo-induced electrons to a certain extent.

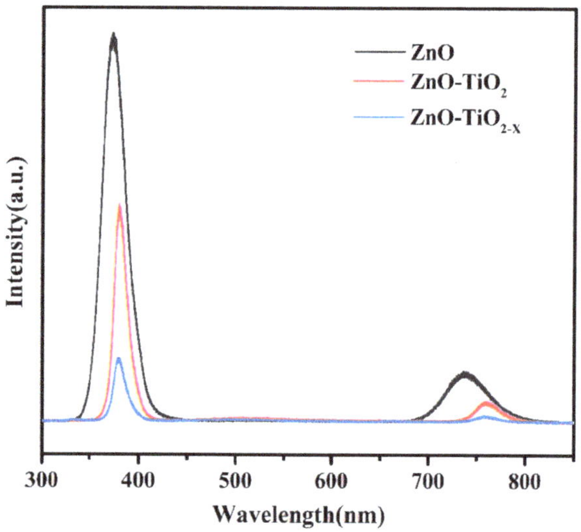

Figure 6. PL spectra of different photoanodes recorded with the excitation wavelength of 325 nm.

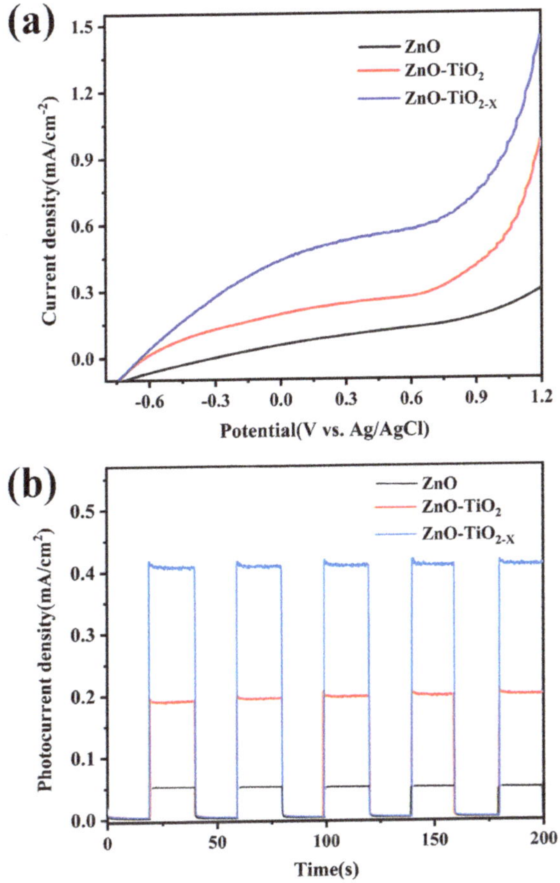

Figure 7. (**a**) Linear sweep voltammetry (LSV) characteristics and (**b**) photocurrent responses of pure ZnO, ZnO/TiO_2, ZnO/TiO_{2-X} homojunction with on/off radiation in a 1 M KOH solution under AM 1.5.

As revealed in the above experimental results of ZnO/TiO_{2-X}, the plausible photocatalytic mechanism is proposed for the improved hydrogen production activity, as shown in Figure 9. Since the valence bands (VB) positions of TiO_{2-X} are lower than that of ZnO before TiO_{2-X} is contacted with ZnO, and the conduction band (CB) positions of pure ZnO are higher than that of TiO_{2-X}. For this type-II structure in our work, under the simulated solar light irradiation, electrons accumulate in the CB of TiO_{2-X} and the holes are transferred from TiO_2 to ZnO (in VB), respectively. Moreover, black TiO_{2-X} nanostructures with oxygen vacancies enhance light absorption under full sunlight. On the other hand, oxygen vacancies caused by the reduction can be considered as electron donors of TiO_{2-X}, which can further effectively separate and transport light-excited electron-hole pairs. This proper band alignment can be applied to separate the active photogenerated electrons and holes, thus increasing the photocatalytic H_2 rate.

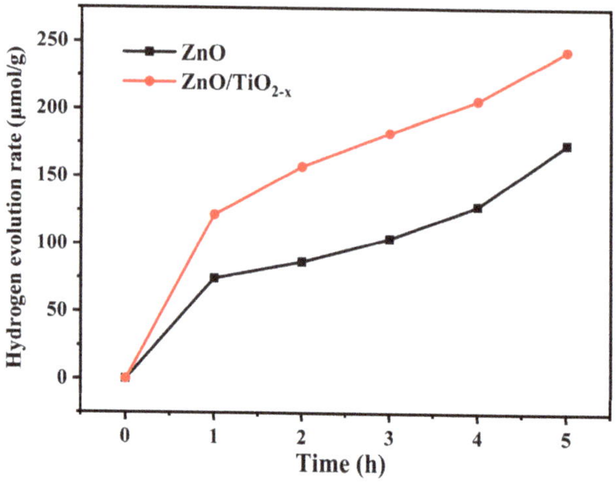

Figure 8. Temporal H_2 evolution of the pure ZnO and ZnO/TiO_{2-X} under AM 1.5.

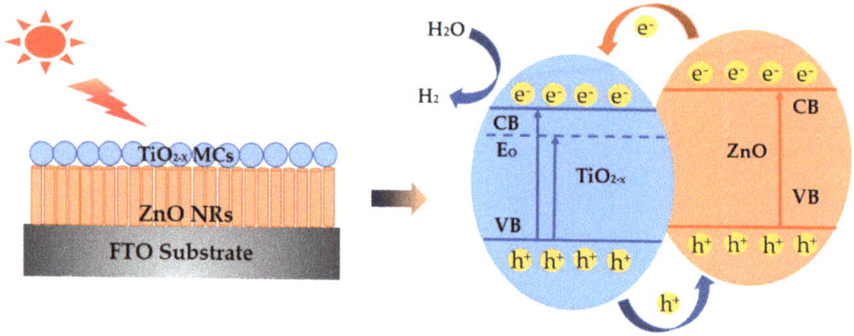

Figure 9. Mechanism model for the enhanced property of ZnO/TiO_{2-X} composite photocatalysts for the photocatalytic H_2 evolution reaction.

4. Conclusions

In summary, a novel structure on black TiO_{2-X} mesoporous spheres/ZnO nanorods heterojunction has been prepared and the photocatalytic hydrogen evolution performance has been explored. The morphology, composition, and crystallinity test results confirmed the formation of a clear hybrid structure on the FTO substrate. The properties of the ZnO NRs/TiO_{2-X} MSs hetero-junction and the photocatalytic performance have been carefully performed through various experimental methods. The photocurrent and hydrogen production performance of ZnO NRs/TiO_{2-X} MSs have been significantly enhanced compared with pure ZnO NRs. The solar-driven hydrogen evolution rate of ZnO NRs/TiO_{2-X} MSs is several times higher than that of ZnO. All these suggest that these novel ZnO NRs/TiO_{2-X} MSs are prospective, next-generation photocatalytic materials of low-cost, large area, and energy-efficiency for practical applications.

Supplementary Materials: The following are available online at http://www.mdpi.com/2079-4991/10/11/2096/s1, Figure S1: EDS elemental mapping images of the ZnO NRs/TiO_{2-X} MCs; Figure S2: Cross-section of pure ZnO nanorods; Figure S3: XPS survey spectrum of the ZnO/TiO_{2-X}. Inset: exact XPS analysis of the inner Ti 2p orbital electrons(peak area Ti^{3+}/Ti^{4+} = 0.0758); Figure S4: The plot of (ααhνν)1/2 versus the energy of light (hνν); Figure S5: Linear sweep voltammogram under dark conditions; Figure S6: Temporal H2 evolution evolution of the ZnO/TiO_2.

Author Contributions: Conceptualization, J.W.; methodology, B.Z., Y.L., and C.Z.; formal analysis, D.W. and Z.Z.; investigation, D.L.; data curation, S.J. and B.J.; writing—original draft preparation, S.G. and Z.X.; writing—review and editing, B.Z. and Q.L.; supervision, L.Z. and X.F.; funding acquisition, D.W. and Y.Z. All authors have read and agreed to the published version of the manuscript. Please turn to the CRediT taxonomy for the term explanation. Authorship must be limited to those who have contributed substantially to the work reported.

Funding: This research was funded by the National Key Research and Development Program of China, grant number 2019YFA0705201. This work was supported by the National Natural Science Foundation of China (62074018), the China Postdoctoral Science Foundation funded project (2019M652176, 2019M661680), the Developing Project of Science and Technology of Jilin Province (20200301052RQ, 20180519017JH, 20180520177JH) and the Shenzhen Fundamental Research Fund (JCYJ20180307151538972), and the Natural Science Foundation of Guangdong Province (2020A1515010868).

Conflicts of Interest: The authors declare no conflict of interest.

References

1. Tian, Z.R.; Voigt, J.A.; Liu, J.; Mckenzie, B.; Mcdermott, M.J.; Rodriguez, M.A.; Konishi, H.; Xu, H. Complex and oriented ZnO nanostructures. *Nat. Mater.* **2003**, *2*, 821–826. [CrossRef] [PubMed]
2. Hambali, N.A.; Hashim, A.M. Synthesis of zinc oxide nanostructures on graphene/glass substrate via electrochemical deposition: Effects of potassium chloride and hexamethylenetetramine as supporting reagents. *Nano-Micro Lett.* **2015**, *7*, 317–324. [CrossRef] [PubMed]
3. Bai, H.; Zan, X.; Zhang, L.; Sun, D.D. Multi-functional CNT/ZnO/TiO_2 nanocomposite membrane for concurrent filtration and photocatalytic degradation. *Sep. Purif. Technol.* **2015**, *156*, 922–930. [CrossRef]
4. Van de Krol, R.; Liang, Y.; Schoonman, J. Solar hydrogen production with nanostructured metal oxides. *J. Mater. Chem.* **2008**, *18*, 2311. [CrossRef]
5. Zhang, Y.; Huang, X.; Yeom, J. A floatable piezo-photocatalytic platform based on semi-embedded ZnO nanowire array for high-performance water decontamination. *Nano-Micro Lett.* **2019**, *11*, 11. [CrossRef]
6. Hu, D.W.; Song, C.F.; Jin, X.; Huang, Q. Polymer solution-assisted assembly of hierarchically nano-structured ZnO onto 2D neat graphene sheets with excellent photocatalytic performance. *J. Alloys Compd.* **2020**, *843*, 156030. [CrossRef]
7. Kayaci, F.; Vempati, S.; Ozgit-Akgun, C.; Donmez, I.; Biyikli, N.; Uyar, T. Transformation of polymer-ZnO core-shell nanofibers into ZnO hollow nanofibers: Intrinsic defect reorganization in ZnO and its influence on the photocatalysis. *Appl. Catal. B Environ.* **2015**, *176*, 646–653. [CrossRef]
8. Zheng, H.B.; Wu, D.; Wang, Y.L.; Liu, X.P.; Gao, P.Z.; Liu, W.; Wen, J.; Rebrov, E.V. One-step synthesis of ZIF-8/ZnO composites based on coordination defect strategy and its derivatives for photocatalysis. *J. Alloys Compd.* **2020**, *838*, 155219. [CrossRef]
9. Shimpi, N.T.; Rane, Y.N.; Shende, D.A.; Gosavi, S.R.; Ahirrao, P.B. Synthesis of rod-like ZnO nanostructure: Study of its physical properties and visible -light driven photocatalytic activity. *Optik* **2020**, *217*, 164916. [CrossRef]
10. Mohd, A.M.A.; Julkapli, N.M.; Abd, H.S.B. Review on ZnO hybrid photocatalyst: Impact on photocatalytic activities of water pollutant degradation. *Rev. Inorg. Chem.* **2016**, *36*, 77–104.
11. Vaiano, V.; Jaramillo-Paez, C.A.; Matarangolo, M.; Navio, J.A.; Hidalgo, M.D. UV and visible-light driven photocatalytic removal of caffeine using ZnO modified with different noble metals (Pt, Ag and Au). *Mater. Res. Bull.* **2019**, *112*, 251–260. [CrossRef]
12. Pankaj, R.; Anita, S.; Pardeep, S. Photocatalytic water decontamination using graphene and ZnO coupled photocatalysts: A review. *Mater. Sci. Energy Technol.* **2019**, *2*, 509–525.
13. Liu, J.L.; Wang, Y.H.; Ma, J.Z.; Peng, Y.; Wang, A.Q. A review on bidirectional analogies between the photocatalysis and antibacterial properties of ZnO. *J. Alloys Compd.* **2019**, *783*, 898–918. [CrossRef]
14. Zhang, B.; Wang, F.Z.; Zhu, C.Q.; Li, Q.; Song, J.N.; Zheng, M.J.; Ma, L.; Shen, W.Z. A facile self-assembly synthesis of hexagonal ZnO nanosheet films and their photoelectrochemical properties. *Nano-Micro Lett.* **2016**, *8*, 137–142. [CrossRef]
15. Li, Q.; An, M.; Li, D. In situ growth of ZnO nanowire film on nanocrystalline zinc electrodeposit via a low-temperature hydrothermal reaction. *Results Phys.* **2019**, *12*, 1446–1449. [CrossRef]

16. Reinosa, J.J.; Docio, C.M.A.; Ramirez, V.Z.; Fernandez, J.F. Hierarchical nano ZnO-micro TiO_2 composites: High UV protection yield lowering photodegradation in sunscreens. *Ceram. Int.* **2018**, *44*, 2827–2834. [CrossRef]
17. Hernandez, S.; Hidalgo, D.; Sacco, A.; Chiodoni, A.; Lamberti, A.; Cauda, V.; Tresso, E.; Saracco, G. Comparison of photocatalytic and transport properties of TiO_2 and ZnO nanostructures for solar-driven water splitting. *Phys. Chem.* **2015**, *17*, 7775–7786. [CrossRef]
18. Haghighatzadeh, A.; Hosseini, M.; Mazinani, B.; Shokouhimehr, M. Improved photocatalytic activity of ZnO-TiO_2 nanocomposite catalysts by modulating TiO_2 thickness. *Mater. Res. Express* **2019**, *6*, 115060. [CrossRef]
19. Li, X.F.; Lv, K.L.; Deng, K.J.; Tang, J.F.; Su, R.; Sun, J.; Chen, L.Q. Synthesis and characterization of ZnO and TiO_2 hollow spheres with enhanced photoreactivity. *Mater. Sci. Eng. B* **2009**, *158*, 40–47. [CrossRef]
20. Wang, Y.Z.; Zhu, S.P.; Chen, X.R.; Tang, Y.G.; Jiang, Y.F.; Peng, Z.G.; Wang, H.Y. One-step template-free fabrication of mesoporous ZnO/TiO_2 hollow microspheres with enhanced photocatalytic activity. *Appl. Surf. Sci.* **2014**, *307*, 263–271. [CrossRef]
21. Tan, H.; Zhao, Z.; Niu, M.; Mao, C.; Cao, D.; Cheng, D.; Feng, P.; Sun, Z.A. A facile and versatile method for preparation of colored TiO_2 with enhanced solar-driven photocatalytic activity. *Nanoscale* **2014**, *6*, 10216–10223. [CrossRef] [PubMed]
22. Chen, X.; Liu, L.; Yu, P.Y.; Mao, S.S. Increasing solar absorption for photocatalysis with black hydrogenated titanium dioxide nanocrystals. *Science* **2011**, *331*, 746–750. [CrossRef] [PubMed]
23. Chen, X.; Liu, L.; Huang, F. Black titanium dioxide (TiO_2) nanomaterials. *Chem. Soc. Rev.* **2015**, *44*, 1861–1885. [CrossRef] [PubMed]
24. Liu, X.; Zhu, G.; Wang, X.; Yuan, X.; Lin, T.; Huang, F. Progress in black titania: A new material for advanced photocatalysis. *Adv. Energy Mater.* **2016**, *6*, 1600452. [CrossRef]
25. Chen, X.; Mao, S.S. Titanium dioxide nanomaterials: Synthesis, properties, modifications, and applications. *Chem. Rev.* **2007**, *107*, 2891–2959. [CrossRef] [PubMed]
26. Schneider, J.; Matsuoka, M.; Takeuchi, M.; Zhang, J.L.; Horiuchi, Y.; Anpo, M.; Bahnemann, D.W. Understanding TiO_2 photocatalysis: Mechanisms and materials. *Chem. Rev.* **2014**, *114*, 9919–9986. [CrossRef] [PubMed]
27. Dahl, M.; Liu, Y.; Yin, Y. Composite titanium dioxide nanomaterials. *Chem. Rev.* **2014**, *114*, 9853–9889. [CrossRef]
28. Liu, G.; Kong, L.P.; Hu, Q.Y.; Zhang, S.J. Diffused morphotropic phase boundary in relaxor-$PbTiO_3$ crystals: High piezoelectricity with improved thermal stability. *Appl. Phys. Rev.* **2020**, *7*, 021405. [CrossRef]
29. Chen, D.; Huang, F.; Cheng, Y.-B.; Caruso, R.A. Mesoporous anatase TiO_2 beads with high surface areas and controllable pore sizes: A superior candidate for high-performance dye-sensitized solar cells. *Adv. Mater.* **2009**, *21*, 2206–2210. [CrossRef]
30. Zeng, Z.; Wang, D.B.; Wang, J.Z.; Jiao, S.J.; Huang, Y.W.; Zhao, S.X.; Zhang, B.K.; Ma, M.Y.; Gao, S.Y.; Feng, X.G.; et al. Self-Assembly Synthesis of the MoS_2/PtCo Alloy Counter Electrodes for High-Efficiency and Stable Low-Cost Dye-Sensitized Solar Cells. *Nanomaterials* **2020**, *10*, 1725. [CrossRef]
31. Sauvage, F.; Chen, D.; Huang, P.; Comte, F.; Heiniger, L.-P.; Cheng, Y.-B.; Caruso, R.A.; Grätzel, M. Dye-sensitized solar cells employing a single film of mesoporous TiO_2 beads achieve power conversion efficiencies over 10%. *ACS Nano* **2010**, *4*, 4420–4425. [CrossRef] [PubMed]
32. Wang, X.; Bai, L.C.; Liu, H.Y.; Yu, X.F.; Yin, Y.D.; Gao, C.B. A Unique Disintegration-Reassembly Route to Mesoporous Titania Nanocrystalline Hollow Spheres with Enhanced Photocatalytic Activity. *Adv. Funct. Mater.* **2018**, *28*, 1704208. [CrossRef]
33. Huang, Y.W.; Yu, Q.J.; Wang, J.Z.; Wang, J.A.; Yu, C.L.; Abdalla, J.T.; Zeng, Z.; Jiao, S.J.; Wang, D.B.; Gao, S.Y. Plasmon-Enhanced Self-Powered UV Photodetectors Assembled by Incorporating Ag@SiO_2 Core-Shell Nanoparticles into TiO_2 Nanocube Photoanodes. *ACS Sustain. Chem. Eng.* **2018**, *6*, 438–446. [CrossRef]
34. Chen, J.Q.; Huang, S.L.; Long, Y.J.; Wu, J.H.; Li, H.; Li, Z.; Zeng, Y.J.; Ruan, S.C. Fabrication of ZnO/Red Phosphorus Heterostructure for Effective Photocatalytic H_2 Evolution from Water Splitting. *Nanomaterials* **2018**, *8*, 835. [CrossRef] [PubMed]

35. Lettieri, S.; Gargiulo, V.; Alfe, M.; Amati, M.; Zeller, P.; Maraloiu, V.A.; Borbone, F.; Pavone, M.; Munoz-Garcia, A.B.; Maddalena, P. Simple Ethanol Refluxing Method for Production of Blue-Colored Titanium Dioxide with Oxygen Vacancies and Visible Light-Driven Photocatalytic Properties. *J. Phys. Chem. C* **2020**, *124*, 3564–3576. [CrossRef]
36. Zhang, Q.; Li, C.Y. TiO_2 Coated ZnO Nanorods by Mist Chemical Vapor Deposition for Application as Photoanodes for Dye-Sensitized Solar Cells. *Nanomaterials* **2019**, *9*, 1339. [CrossRef]
37. Doğan, İ.; van de Sanden, M.C.M. Direct characterization of nanocrystal size distribution using Raman spectroscopy. *J. Appl. Phys.* **2013**, *114*, 134310. [CrossRef]
38. Rajalakshmi, M.; Arora, A.K.; Bendre, B.S. Optical phonon confinement in zinc oxide nanoparticles. *Phys. Rev.* **2000**, *87*, 2445–2448. [CrossRef]
39. Ursaki, V.V.; Tiginyanu, I.M.; Zalamai, V.V.; Rusu, E.V.; Emelchenko, G.A.; Masalov, V.M.; Samarov, E.N. Multiphonon resonant Raman scattering in ZnO crystals and nanostructured layers. *Phys. Rev. B.* **2004**, *70*, 155204. [CrossRef]
40. Zhou, W.; Sun, F.F.; Pan, K.; Tian, G.H.; Jiang, B.J.; Ren, Z.Y.; Tian, C.G.; Fu, H.G. Well-Ordered Large-Pore Mesoporous Anatase TiO_2 with Remarkably High Thermal Stability and Improved Crystallinity: Preparation, Characterization, and Photocatalytic Performance. *Adv. Funct. Mater.* **2011**, *21*, 1922–1930. [CrossRef]
41. Vasu, K.; Sreedhara, M.B.; Ghatak, J.; Rao, C.N.R. Atomic layer deposition of p-type epitaxial thin films of undoped and N-doped anatase TiO_2. *ACS Appl. Mater. Interfaces* **2016**, *8*, 7897–7901. [CrossRef]
42. Hwang, Y.J.; Yang, S.; Lee, H. Surface analysis of N-doped TiO_2 nanorods and their enhanced photocatalytic oxidation activity. *Appl. Catal. B.* **2017**, *204*, 209–215. [CrossRef]
43. O'Brien, S.; Koh, L.H.K.; Crean, G.M. ZnO thin films prepared by a single step sol-gel process. *Thin Solid Films* **2008**, *516*, 1391–1395. [CrossRef]
44. Ullattil, S.G.; Narendranath, S.B.; Pillai, S.C.; Periyat, P. Black TiO_2 Nanomaterials: A Review of Recent Advances. *Chem. Eng. J.* **2018**, *343*, 708–736. [CrossRef]
45. Lima, S.A.M.; Sigoli, F.A.; Jafelicci, M.; Davolos, M.R. Luminescent properties and lattice correlation defects on zinc oxide. *Int. J. Inorg. Mater.* **2001**, *3*, 749–754. [CrossRef]
46. Shi, Z.-F.; Zhang, Y.-T.; Cui, X.-J.; Zhuang, S.-W.; Wu, B.; Chu, X.-W.; Dong, X.; Zhang, B.-L.; Du, G.-T. Photoluminescence performance enhancement of ZnO/MgO heterostructured nanowires and their applications in ultraviolet laser diodes. *Phys. Chem. Chem. Phys.* **2015**, *17*, 13813–13820. [CrossRef]
47. Ma, D.; Shi, J.-W.; Zou, Y.; Fan, Z.; Ji, X.; Niu, C.; Wang, L. Rational design of CdS@ZnO core-shell structure via atomic layer deposition for drastically enhanced photocatalytic H_2 evolution with excellent photostability. *Nano Energy* **2017**, *39*, 183–191. [CrossRef]
48. Amany, A.; Wang, D.B.; Wang, J.Z.; Zeng, Z.; Jiao, S.J.; Hao, C.L.; Zhao, Y.C.; Xie, Y.; Gao, S.Y.; Ni, S.M.; et al. Enhanced the UV response of AlN coated ZnO nanorods photodetector. *J. Alloys Compd.* **2018**, *776*, 111–115. [CrossRef]
49. Pastore, M.; De Angelis, F. Computational Modeling of Stark Effects in Organic Dye-Sensitized TiO_2 Heterointerfaces. *J. Phys. Chem. Lett.* **2011**, *2*, 1261–1267. [CrossRef]
50. Sun, L.H.; Dong, A.N.; Li, J.M.; Hao, D.; Tang, X.Q.; Yan, S.C.; Guo, Y.; Shan, X.Y.; Lu, X.H. Optical Stark effect of a local defect on the TiO_2 (110) surface. *Phys. Rev. B.* **2018**, *98*, 081402. [CrossRef]
51. Sun, C.Y.; Xu, Q.H.; Xie, Y.; Ling, Y.; Hou, Y. Designed synthesis of anatase–TiO_2(B) biphase nanowire/ZnO nanoparticle heterojunction for enhanced photocatalysis. *J. Mater. Chem. A* **2018**, *6*, 8289. [CrossRef]
52. Han, H.S.; Wang, W.Z.; Yao, L.Z.; Hao, C.C.; Liang, Y.J.; Fu, J.L.; Zeng, P.B. Photostable 3D heterojunction photoanode made of ZnO nanosheets coated onto TiO_2 nanowire arrays for photoelectrochemical solar hydrogen generation. *Catal. Sci. Technol.* **2019**, *9*, 1989. [CrossRef]

nanomaterials

MDPI

Article

Enhanced UV Photoresponsivity of ZnO Nanorods Decorated with Ag_2S/ZnS Nanoparticles by Successive Ionic Layer Adsorption and Reaction Method

Yimin Jin, Shujie Jiao *, Dongbo Wang *, Shiyong Gao and Jinzhong Wang

School of Materials Science and Engineering, Harbin Institute of Technology, Harbin 150001, China; jym921023@foxmail.com (Y.J.); gaoshiyong@hit.edu.cn (S.G.); jinzhong_wang@hit.edu.cn (J.W.)
* Correspondence: shujiejiao@hit.edu.cn (S.J.); wangdongbo@hit.edu.cn (D.W.)

Abstract: Recently, different kinds of energy band structures have been utilized to improve the photoelectric properties of zinc oxide (ZnO). In this work, ZnO nanorods were prepared by the hydrothermal method and then decorated with silver sulfide (Ag_2S)/zinc sulfide (ZnS) via two-step successive ionic layer adsorption and reaction method. The photoelectric properties of nanocomposites are investigated. The results show that ZnO decorated with Ag_2S/ZnS can improve the photocurrent of photodetectors from 0.34 to 0.56 A at bias of 9 V. With the immersion time increasing from 15 to 60 minutes, the photocurrent of photodetectors increases by 0.22 A. The holes in the valence band of ZnO can be transferred to the valence band of ZnS and Ag_2S, which promotes the separation and suppresses the recombination of hole-electron pairs generated in ZnO. Moreover, electrons excited by ultraviolet (UV) light in Ag_2S can also be injected into the conduction band of ZnO, which causes the photocurrent to increase more than the ZnO photodetector.

Keywords: ZnO; Ag_2S; successive ionic layer adsorption and reaction; photodetector

Citation: Jin, Y.; Jiao, S.; Wang, D.; Gao, S.; Wang, J. Enhanced UV Photoresponsivity of ZnO Nanorods Decorated with Ag_2S/ZnS Nanoparticles by Successive Ionic Layer Adsorption and Reaction Method. *Nanomaterials* **2021**, *11*, 461. https://doi.org/10.3390/nano11020461

Academic Editor: Vincent Consonni
Received: 30 December 2020
Accepted: 8 February 2021
Published: 11 February 2021

1. Introduction

Recently, increasing demand for clean energy, portable electronics, space and astronomical research, optical communications, and fire monitoring has caused UV photodetectors to be the subject of considerable attention due to their good flexibility, low-cost fabrication, and high sensitivity [1–6]. Many wide-bandgap semiconductors associated with this type of photodetectors have been explored [5,7–12]. Among wide-bandgap materials, zinc oxide has become a good candidate for UV photodetection due to its wide bandgap (3.37 eV), high exciton binding energy (60 meV), high chemical and thermal stability, low cost, and strong emission at room temperature [13–16]. To improve the properties of ZnO, incorporating noble metal nanostructures, compounding with carbon nanomaterials, and decorating with semiconductor quantum dots are effective approaches [4,17–19]. With decoration of semiconductors with different band structures, holes or electrons can be transferred from ZnO, which promotes separation and suppresses the recombination of hole-electron pairs generated in ZnO [20–23]. Ag_2S as a direct, narrow bandgap (1.1 eV) semiconductor with optical absorption similar to silicon has attracted much attention as a promising candidate for photocatalysis and photoconduction [24–26]. Studies also proved that after accompanying Ag_2S with ZnO, type-II heterojunctions are formed. Due to the different valence bands of Ag_2S and ZnO, the holes in the valence band of ZnO can be transferred to the valence band of Ag_2S. This process can promote separation and suppress the recombination of hole-electron pairs generated in ZnO, leading to the improved optoelectrical properties of ZnO [27–30]. Recently, Li and colleagues prepared Ag_2S-coupled ZnO microspheres with 1.68 $\mu A/cm^2$ at 0.2 V under visible light [31]. Chen and colleagues prepared a Ag_2S/ZnO core-shell nanoheterojunction with high photosensitivity in the wide spectral range from 400 to 1100 nm and a response time as short as 5 ms [32].

In this work, ZnO nanorods are prepared with the hydrothermal method. Then, ZnS and Ag_2S are deposited via successive ionic layer adsorption and reaction method on the surface of ZnO nanorods. ZnS is an n-type [33–35] material, and Ag_2S is a p-type [36–39] material. The nanocomposites demonstrate potential application in the fields of photodetection, photocatalysis, and solar cells [31,40–42]. In this work, the optoelectrical properties of nanocomposites are investigated.

2. Materials and Methods

The ZnO nanomaterials in this paper were prepared by the hydrothermal method on glass substrates with a conductive thin film of indium-doped tin oxides (ITO) on one side. The size of substrates was 1×1 cm. Before preparation, the substrates were cleaned with ultrasound successively in acetone, ethanol, and deionized water for 30 min. To prepare ZnO seed layers on ITO substrates by the sol-gel dip-coating method, the substrates were immersed in precursor solution for 15 min, and then, the samples were dried for 15 min. This process was repeated six times. Finally, an annealing treatment was performed in air at 150 °C for 30 min. Then ZnO nanorod arrays were formed on the substrates in a solution consisting of 0.03 M zinc acetate dehydrate ($Zn(AC)_2{\cdot}2H_2O$) and 0.03 M hexamethylenetetramine (HMT) at 90 °C for 4 h.

S^{2-} was incorporated into ZnO to form ZnS by immersing ZnO samples in aqueous solution containing 0.02 M Na_2S for 5 min, 15 min, 30 min, 45 min, and 60 min, respectively and rinsing with pure ethanol. The ZnO nanorod arrays were decorated with Ag_2S quantum dots (QDs) through the facile successive ionic layer adsorption and reaction (SILAR) method. ZnO samples were successively immersed in two different aqueous solutions, one containing 0.02 M Na_2S and the other one containing 0.02 M $AgNO_3$ aqueous solution for 30 min and different time, respectively. During immersion, the solution should be stirred. After immersion, the samples were rinsed with pure ethanol to remove excess precursors and blown dry at room temperature. Then, the Ag_2S-modified samples were fabricated as metal-semiconductor-metal (MSM) photodetectors with indium (In) electrode. Electrodes were exploited to form ohmic contact between them and the nanocomposites. One electrode was prepared on the ZnO seed layer and the other one was prepared on the top of nanorod arrays.

Surface morphologies of the nanocomposites were characterized using scanning electron microscopy (SEM, HITACH SU70, Tokyo, Japan). Finer details of the nanocomposites were characterized using transmission electron microscopy (TEM, FEI, Hillsboro, OR, USA) and high-resolution transmission electron microscopy (HRTEM, FEI, Hillsboro, OR, USA). The composition and bond band properties of the samples were measured by X-ray photoelectric spectroscopy (XPS, ESCALAB 250Xi, Thermo Fisher, Waltham, MA, USA). Ultraviolet–visible spectroscopy (Shimadzu UV1700-visible spectrophotometer) was utilized to characterize the optical properties. I–V characterization of the as-synthesized devices was measured by an electrochemical workstation (CHI660e, Chenhua instruments Ins., Shanghai, China) with a three-electrode system under UV led (λ = 365 nm). The photoresponsivity spectrum of the devices was obtained by measuring the photocurrent (calibrated with a standard Si photodiode) under the illumination of a UV-enhanced Xe lamp spectrum from 300 to 600 nm using a scanning monochromator (DSR600, Zolix, Beijing, China).

3. Results and Discussion

The morphologies of ZnO and ZnS/ZnO heterojunctions array prepared by immersing ZnO samples in Na_2S solution for 5, 15, 30, 45, and 60 min, respectively, are shown in Figure 1a–f. The hexagonal nanorods are not uniform, and the dominant diameter of the nanorods is about 260 nm. The XRD spectrum of the as-synthesized samples is shown in the inset of Figure 1a. Typical peaks belong to the wurtzite hexagonal phase of ZnO (JCPDS 36-1451), as presented in Figure 1a. When ZnO nanorods are immersed in Na_2S solution, as shown in Figure 1b–f, the solution provides sulfide ions to react with zinc ions dissolved from the ZnO nanorods in order to form ZnS. During this process, the concentration of sulfide ions can be adjusted to influence the formation of ZnS. However, if the concentration is too large, many defects can be formed on the nanocomposites, leading to decreased efficiency. The formation of a ZnS shell can be determined by XPS spectra in Figure 2. The XPS wide-survey spectrum of the sample with an immersion time of 30 min is shown in Figure 2a. The characteristic peaks in the XPS spectrum can be assigned as Zn, S, C, or O, respectively. The ratio of S/Zn is about 0.16. No other impurity peaks can be found, showing that the obtained sample is of high purity. For ZnS/ZnO in Figure 2b, the S peak located at 162 eV corresponds to S 2p from ZnS. The above XPS analysis demonstrates the process of the formation of ZnS after immersion. Therefore, ZnO nanorods were covered by the ZnS shell through immersion.

Figure 1. (**a**)–(**f**) SEM images of ZnO and ZnS/ZnO nanocomposites with different immersion time of 5, 15, 30, 45, and 60 min, respectively.

Based on the ZnS/ZnO nanocomposites, the ZnS/ZnO samples with the immersion time of 30 min in Na_2S solution were then immersed in 0.02 M $AgNO_3$ aqueous solution for 15, 30, 45, and 60 min, respectively, which allows for the Ag_2S/ZnS/ZnO nanocomposites to be obtained. The morphologies of the as-synthetized Ag_2S/ZnS/ZnO nanorod arrays are shown in Figure 3a–d. It can be seen that after immersion, the nanorods were covered with spherical nanoparticles at the short immersion time. Because of the difference between the solubility product constant (K_{sp}) of ZnS (2.93×10^{-25}) and K_{sp} of Ag_2S (6.69×10^{-50}) [43], the cation exchange process occurs where zinc ions are replaced by silver ions, leading to the formation of Ag_2S, which indicates that the spherical nanoparticles are Ag_2S. With the increasing immersion time, the nanorods demonstrate a complete shape change from hexagon to ellipse. These results suggest that Ag_2S was successfully deposited onto the surface of the nanorods.

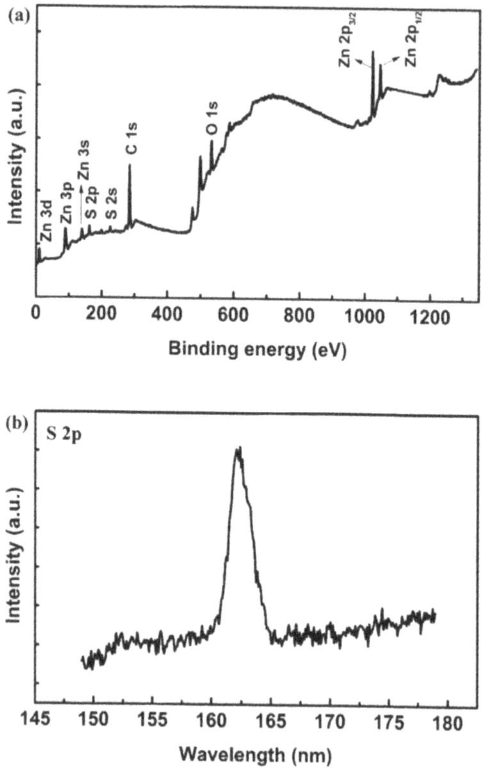

Figure 2. (**a**) XPS wide-survey spectrum; (**b**) S peak of ZnS/ZnO nanocomposites with the immersion time of 30 min.

Figure 3. (**a**)–(**d**) SEM images of Ag_2S/ZnO nanocomposites with different immersion time of 15, 30, 45, and 60 min, respectively.

In order to confirm the formation of $Ag_2S/ZnS/ZnO$ composites, TEM and HRTEM measurements were performed on the nanocomposite with the immersion time of 60 min, as illustrated in Figure 4. A large number of quantum dots evenly deposited onto the nanorod surface is shown in Figure 4a,b, which show that densely distributed QDs are formed on the surface of nanorods, and the diameter of the quantum dots is about 10–15 nm. Figure 4c shows the HRTEM image of the quantum dots deposited on the nanorod surface. It can be observed that the quantum dots have a spherical shape with a diameter of about 10 nm. The obvious lattice arrangement can be found in the quantum dots, and the d-spacing estimated to be 0.253 nm is indexed to the (−103) orientation of Ag_2S crystalline.

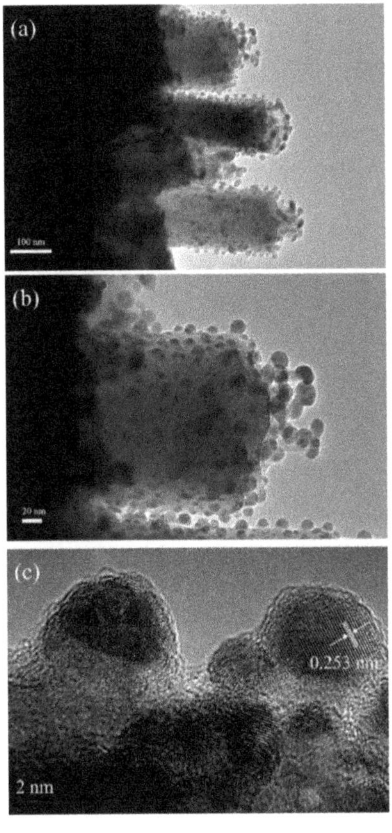

Figure 4. (**a**) and (**b**) TEM images of $Ag_2S/ZnS/ZnO$ nanocomposites with an immersion time of 60 min; (**c**) HRTEM image of nanoparticles.

To further investigate the structures of $Ag_2S/ZnS/ZnO$ nanocomposites, XPS measurements were measured. Figure 5a shows the Ag 3d region of the XPS spectra. The peak position of Ag $3d_{5/2}$ is located at about 368 eV. This value is in good agreement with the reported values for Ag_2S. Then, the peak area of Ag elements with different immersion time was calculated, as shown in Figure 5b. It is obvious that the Ag concentration increases with the increasing immersion time. Meanwhile, according to the XPS spectrum, the Ag/Zn ratio increasing from 0.08 to 0.22 indicates the same results. This result is consistent with the SEM images shown in Figure 3. Figure 5c shows the S 2p region of the XPS spectrum with the immersion time of 60 min. The black line represents the experimental data, and the red dots correspond to the fitted curve. Four labelled fitting Gaussian peaks were used to fit the experimental data. The binding energy of the S $2p_{3/2}$ peak located at 161.5 eV is in accordance with the binding energy of ZnS. The lowest energy peak of S $2p_{3/2}$ is located at 160.93 eV, which corresponds to Ag_2S [44–47]. These results illustrate that $Ag_2S/ZnS/ZnO$ nanocomposites were formed after immersion.

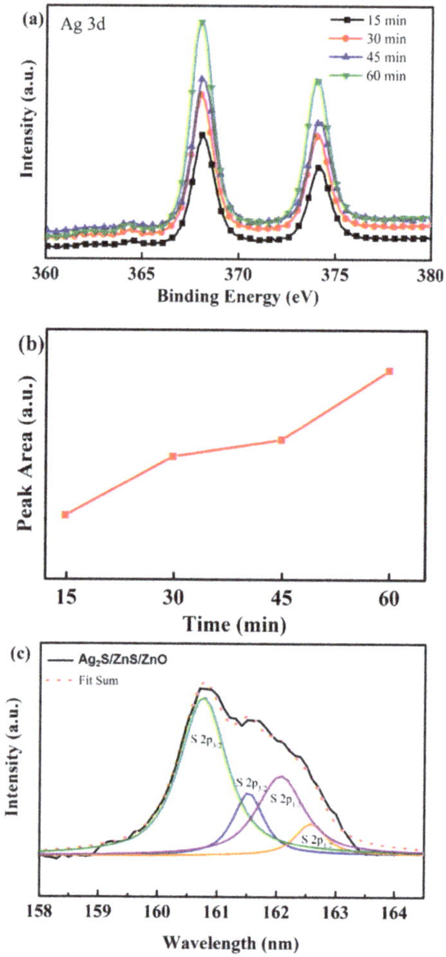

Figure 5. (**a**) XPS spectra of Ag 3d for $Ag_2S/ZnS/ZnO$ nanocomposites with various immersion time; (**b**) peak area of Ag $3d_{5/2}$ with different immersion time; (**c**) XPS spectra of S 2p for $Ag_2S/ZnS/ZnO$ nanocomposites with the immersion time of 60 min.

To investigate the optical properties of $Ag_2S/ZnS/ZnO$ nanocomposites, UV-vis absorption spectra from 350 to 600 nm were examined. For comparison, the spectrum of pure ZnO nanorods was also measured, indicated by the black line. Figure 6 shows the UV-vis absorption spectra of $Ag_2S/ZnS/ZnO$ nanocomposites with various immersion time, revealing that the absorption edge of ZnO is extended to the visible region by decoration with Ag_2S because of its narrow bandgap of ~1.1 eV. When visible light is observable on the nanocomposites, hole-electron pairs are generated in Ag_2S, leading to the absorption in visible light. With the increasing immersion time, more amounts of Ag_2S are deposited, thereby causing the increase in absorption in the visible light region.

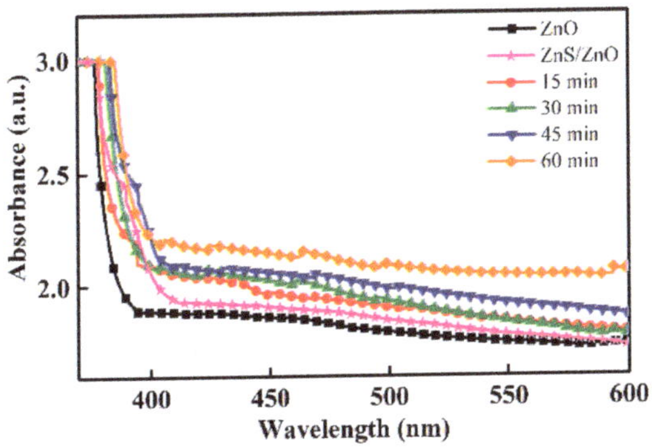

Figure 6. Absorption spectra of $Ag_2S/ZnS/ZnO$ nanocomposites with various immersion time.

A photodetector was fabricated with In electrodes to investigate the optoelectrical properties of $Ag_2S/ZnS/ZnO$ nanocomposites. The I-V characteristics of photodetectors with various immersion time under 365 nm UV LED are shown in Figure 7a–d. It can be observed that photodetectors have a photoresponse under UV illumination. The reverse current is high and of the same order of magnitude as the direct current, demonstrating that photoconductive photodetectors were fabricated. At the same voltage, the photocurrent of photodetectors with various immersion time changes from 0.34 to 0.56 A under UV illumination. Meanwhile, the dark current of the photodetectors, which can be ascribed to oxygen vacancy in ZnO, is 0.22, 0.20, 0.21, and 0.22 A, respectively. Thus, the on–off ratio of the photodetectors at a bias of 9 V is 1.53, 1.77, 1.94, and 2.47, respectively. Figure 7e shows the photocurrent of photodetectors with various immersion time under 9 V bias. It is observed that the photocurrent increases with the increase in immersion time. This phenomenon can be attributed to the replacement of ZnS by Ag_2S. The photoresponsivity of the photodetector with the immersion of 60 min is shown in Figure 7f. It can be observed that the photoresponsivity increases for the $Ag_2S/ZnS/ZnO$ nanocomposites when compared with pure ZnO photodetectors not only in UV regions but also in the visible wavelength region. In addition, the UV-to-visible rejection ratio was improved from 1.48 to 1.82.

In order to explain the mechanism of photoresponse of $Ag_2S/ZnS/ZnO$ nanocomposites, the energy band schematic diagrams of different nanocomposites are shown in Figure 8. Considering the fact that not all amounts of ZnS are replaced by Ag_2S, there are two types of energy bands in nanocomposites [48,49]. Figure 8a shows the energy band schematic diagrams of ZnS/ZnO. Under the illumination of 365 nm UV light, electrons are excited from the valence band to the conduction band of ZnO, resulting in an increase in the photocurrent. Because of the different valence bands of ZnS and ZnO, the holes in the valence band of ZnO can be transferred to the valence band of ZnS. This process can promote the separation of photogenerated hole-electron pairs and suppress their recombination in ZnO. Meanwhile, because the photon energy of UV light is smaller than the bandgap of ZnS, no electrons are excited to the conduction band in ZnS. Figure 8b shows the energy band schematic diagrams of $Ag_2S/ZnS/ZnO$ nanocomposites. It can be seen that the cascade structure represents the stepwise positions of band edges via the redistribution of ZnS and Ag_2S in order to align Fermi levels. This structure is suitable for the injection of photogenerated electrons from Ag_2S to ZnO and the transfer of holes from ZnO to Ag_2S, and it is advantageous for the separation and transmission of hole-electron pairs. Generally, this structure can further increase the photocurrent compared with the ZnS/ZnO structure because of the injection of electrons. With the increasing immersion time, the first type is gradually replaced by the second type, which further increases the photocurrent. These results are consistent with the I–V characteristics of $Ag_2S/ZnO/ZnO$ nanocomposites with different immersion time.

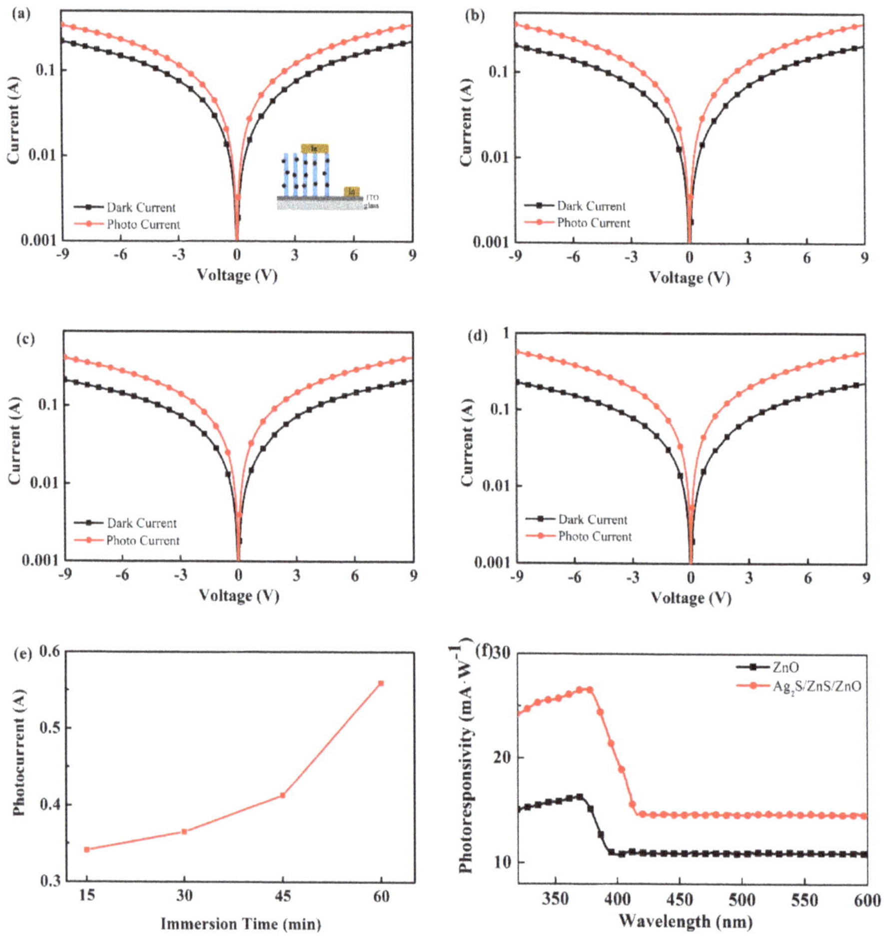

Figure 7. (**a**)–(**d**) I-V characteristics of photodetectors with various immersion time ((**a**) 15 min, (**b**) 30 min, (**c**) 45 min, and (**d**) 60 min) under UV light; the figure inset (**a**) is a schematic diagram of the photodetector; (**e**) photocurrent of photodetectors with various immersion time under 9 V bias; (**f**) photoresponsivity of the photodetector with the immersion time of 60 min.

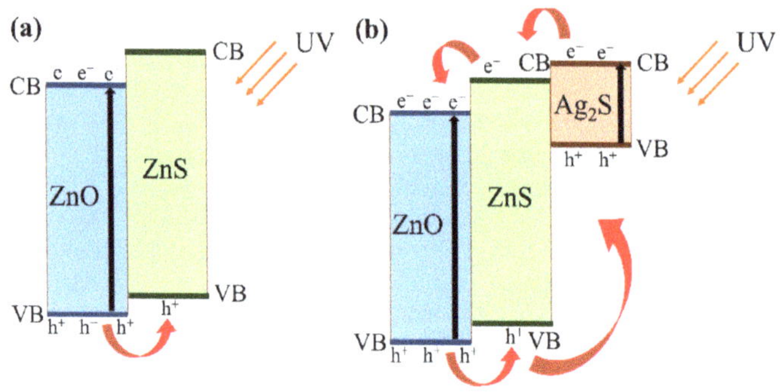

Figure 8. Energy band schematic diagrams of different nanocomposites: (**a**) ZnS/ZnO; (**b**) Ag_2S/ZnS/ZnO.

4. Conclusions

In summary, $Ag_2S/ZnS/ZnO$ nanocomposites were prepared on ITO substrates via two-step facile successive ionic layer adsorption and reaction method with different immersion time. SEM and TEM images illustrate that the ZnS and Ag_2S were evenly deposited on ZnO nanorods. The optical properties of $Ag_2S/ZnS/ZnO$ nanocomposites were investigated by UV-vis absorption spectra, which show that the absorption of $Ag_2S/ZnS/ZnO$ nanocomposites was extended to the visible light region due to the narrow bandgap of Ag_2S. Then MSM photodetectors were fabricated. The influence of ZnS and Ag_2S on the photocurrent of the photodetectors was investigated. The photocurrent increased with the increasing immersion time of AgNO3 solutions due to the increasing electrons injected from Ag_2S into ZnO. The energy band schematic diagrams were used to explain the photoresponse of the photodetectors. The transfer of holes and the injection of electrons can both enhance the photoresponse compared with pure ZnO. Compared to other similar systems, the nanocomposites improved the photocurrent under UV illumination and demonstrate potential applications in other fields [50–52].

Author Contributions: Conceptualization, S.J. and D.W.; methodology, D.W.; software, Y.J.; validation, Y.J., S.J. and J.W.; formal analysis, Y.J. and S.J.; investigation, Y.J.; resources, S.G.; data curation, S.J.; writing—original draft preparation, Y.J.; writing—review and editing, S.J.; visualization, D.W.; supervision, S.G.; project administration, J.W.; funding acquisition, S.J. All authors have read and agreed to the published version of the manuscript.

Funding: This research was supported by the National Key Research and Development Program of China (Grant No. 2019YFA0705201) and the National Natural Science Foundation of China (Grant No. 61774051 and 61574051).

Data Availability Statement: The data presented in this study are available on request from the corresponding author.

Conflicts of Interest: The authors declare no conflict of interest.

References

1. Gedamu, D.; Paulowicz, I.; Kaps, S.; Lupan, O.; Wille, S.; Haidarschin, G.; Mishra, Y.K.; Adelung, R. Rapid fabrication technique for interpenetrated ZnO nanotetrapod networks for fast UV sensors. *Adv. Mater.* **2014**, *26*, 1541–1550. [CrossRef]
2. Guo, D.; Wu, Z.; Li, P.; An, Y.; Liu, H.; Guo, X.; Yan, H.; Wang, G.; Sun, C.; Li, L.; et al. Fabrication of β-Ga_2O_3 thin films and solar-blind photodetectors by laser MBE technology. *Opt. Mater. Express* **2014**, *4*, 1067–1076. [CrossRef]
3. Li, X.; Rui, M.; Song, J.; Shen, Z.; Zeng, H. Carbon and graphene quantum dots for optoelectronic and energy devices: A review. *Adv. Funct. Mater.* **2015**, *25*, 4929–4947. [CrossRef]
4. Ning, Y.; Zhang, Z.; Teng, F.; Fang, X. Novel transparent and self-powered UV photodetector based on crossed ZnO nanofiber array homojunction. *Small* **2018**, *14*, e1703754. [CrossRef]
5. Lu, H.; Dong, H.; Jiao, S.; Nie, Y.; Wang, X.; Wang, D.; Gao, S.; Wang, J.; Su, S. The influence of annealing temperature on structure, morphology and optical properties of $(In_xGa_{1-x})_2O_3$ films. *ECS J. Solid State Sci. Technol.* **2019**, *8*, Q3171–Q3175. [CrossRef]
6. Lu, H.; Jiao, S.; Nie, Y.; Liu, S.; Gao, S.; Wang, D.; Wang, J.; Li, L.; Wang, X. Defect photoluminescence and structure properties of undoping $(In_xGa_{1-x})_2O_3$ films and their dependence on sputtering pressure. *J. Alloys Compd.* **2020**, *823*, 153903. [CrossRef]
7. Fang, X.; Bando, Y.; Gautam, U.K.; Zhai, T.; Zeng, H.; Xu, X.; Liao, M.; Golberg, D. ZnO and ZnS Nanostructures: Ultraviolet-light emitters, lasers, and sensors. *Crit. Rev. Solid State Mater. Sci.* **2009**, *34*, 190–223. [CrossRef]
8. Kong, W.Y.; Wu, G.A.; Wang, K.Y.; Zhang, T.F.; Zou, Y.F.; Wang, D.D.; Luo, L.B. Graphene-beta-Ga_2O_3 heterojunction for highly sensitive deep UV photodetector application. *Adv. Mater.* **2016**, *28*, 10725–10731. [CrossRef] [PubMed]
9. Nie, Y.; Jiao, S.; Meng, F.; Lu, H.; Wang, D.; Li, L.; Gao, S.; Wang, J.; Wang, X. Growth and properties analysis of $Al_xGa_{2-x}O_3$ thin film by radio frequency magnetron sputtering using Al/Ga_2O_3 target. *J. Alloys Compd.* **2019**, *798*, 568–575. [CrossRef]
10. Lopez-Sanchez, O.; Lembke, D.; Kayci, M.; Radenovic, A.; Kis, A. Ultrasensitive photodetectors based on monolayer MoS_2. *Nat. Nanotechnol.* **2013**, *8*, 497–501. [CrossRef]
11. Mastro, M.A.; Kuramata, A.; Calkins, J.; Kim, J.; Ren, F.; Pearton, S.J. Perspective—Opportunities and future directions for Ga_2O_3. *ECS J. Solid State Sci. Technol.* **2017**, *6*, P356–P359. [CrossRef]
12. Sang, L.; Liao, M.; Sumiya, M. A comprehensive review of semiconductor ultraviolet photodetectors: From thin film to one-dimensional nanostructures. *Sensors* **2013**, *13*, 10482–10518. [CrossRef]
13. Gogurla, N.; Sinha, A.K.; Santra, S.; Manna, S.; Ray, S.K. Multifunctional Au-ZnO plasmonic nanostructures for enhanced UV photodetector and room temperature NO sensing devices. *Sci. Rep.* **2014**, *4*, 6483. [CrossRef]

14. Guo, F.; Yang, B.; Yuan, Y.; Xiao, Z.; Dong, Q.; Bi, Y.; Huang, J. A nanocomposite ultraviolet photodetector based on interfacial trap-controlled charge injection. *Nat. Nanotechnol.* **2012**, *7*, 798–802. [CrossRef] [PubMed]
15. Hu, L.; Yan, J.; Liao, M.; Xiang, H.; Gong, X.; Zhang, L.; Fang, X. An optimized ultraviolet-A light photodetector with wide-range photoresponse based on ZnS/ZnO biaxial nanobelt. *Adv. Mater.* **2012**, *24*, 2305–2309. [CrossRef]
16. Jeong, I.S.; Kim, J.H.; Im, S. Ultraviolet-enhanced photodiode employing n-ZnO/p-Si structure. *Appl. Phys. Lett.* **2003**, *83*, 2946–2948. [CrossRef]
17. Peng, L.; Hu, L.; Fang, X. Energy harvesting for nanostructured self-powered photodetectors. *Adv. Funct. Mater.* **2014**, *24*, 2591–2610. [CrossRef]
18. Rai, S.C.; Wang, K.; Ding, Y.; Marmon, J.K.; Bhatt, M.; Zhang, Y.; Zhou, W.; Wang, Z.L. Piezo-phototronic effect enhanced UV/visible photodetector based on fully wide band gap type-II ZnO/ZnS core/shell nanowire array. *ACS Nano* **2015**, *9*, 6419–6427. [CrossRef]
19. Víctor-Román, S.; García-Bordejé, E.; Hernández-Ferrer, J.; González-Domínguez, J.M.; Ansón-Casaos, A.; Silva, A.M.T.; Maser, W.K.; Benito, A.M. Controlling the surface chemistry of graphene oxide: Key towards efficient ZnO-GO photocatalysts. *Catal. Today* **2020**, *357*, 350–360. [CrossRef]
20. Tian, W.; Zhang, C.; Zhai, T.; Li, S.L.; Wang, X.; Liu, J.; Jie, X.; Liu, D.; Liao, M.; Koide, Y.; et al. Flexible ultraviolet photodetectors with broad photoresponse based on branched ZnS-ZnO heterostructure nanofilms. *Adv. Mater.* **2014**, *26*, 3088–3093. [CrossRef]
21. Zhang, Q.; Jie, J.; Diao, S.; Shao, Z.; Zhang, Q.; Wang, L.; Deng, W.; Hu, W.; Xia, H.; Yuan, X.; et al. Solution-processed graphene quantum dot deep-UV photodetectors. *ACS Nano* **2015**, *9*, 1561–1570. [CrossRef] [PubMed]
22. Zhao, B.; Wang, F.; Chen, H.; Wang, Y.; Jiang, M.; Fang, X.; Zhao, D. Solar-blind avalanche photodetector based on single ZnO-Ga(2)O(3) core-shell microwire. *Nano Lett.* **2015**, *15*, 3988–3993. [CrossRef] [PubMed]
23. Zhou, J.; Gu, Y.; Hu, Y.; Mai, W.; Yeh, P.H.; Bao, G.; Sood, A.K.; Polla, D.L.; Wang, Z.L. Gigantic enhancement in response and reset time of ZnO UV nanosensor by utilizing Schottky contact and surface functionalization. *Appl. Phys. Lett.* **2009**, *94*, 191103. [CrossRef]
24. Hsu, M.-H.; Chang, C.-J.; Weng, H.-T. Efficient H_2 production using Ag_2S-coupled ZnO@ZnS core-shell nanorods decorated metal wire mesh as an immobilized hierarchical photocatalyst. *ACS Sustain. Chem. Eng.* **2016**, *4*, 1381–1391. [CrossRef]
25. Khanchandani, S.; Srivastava, P.K.; Kumar, S.; Ghosh, S.; Ganguli, A.K. Band gap engineering of ZnO using core/shell morphology with environmentally benign Ag_2S sensitizer for efficient light harvesting and enhanced visible-light photocatalysis. *Inorg. Chem.* **2014**, *53*, 8902–8912. [CrossRef]
26. Liu, S.; Wang, X.; Zhao, W.; Wang, K.; Sang, H.; He, Z. Synthesis, characterization and enhanced photocatalytic performance of Ag_2S-coupled ZnO/ZnS core/shell nanorods. *J. Alloys Compd.* **2013**, *568*, 84–91. [CrossRef]
27. Shuai, X.M.; Shen, W.Z. A facile chemical conversion synthesis of ZnO/ZnS core/shell nanorods and diverse metal sulfide nanotubes. *J. Phys. Chem. C* **2011**, *115*, 6415–6422. [CrossRef]
28. Subash, B.; Krishnakumar, B.; Pandiyan, V.; Swaminathan, M.; Shanthi, M. An efficient nanostructured Ag_2S–ZnO for degradation of acid black 1 dye under day light illumination. *Sep. Purif. Technol.* **2012**, *96*, 204–213. [CrossRef]
29. Wang, P.; Menzies, N.W.; Lombi, E.; Sekine, R.; Blamey, F.P.; Hernandez-Soriano, M.C.; Cheng, M.; Kappen, P.; Peijnenburg, W.J.; Tang, C.; et al. Silver sulfide nanoparticles (Ag_2S-NPs) are taken up by plants and are phytotoxic. *Nanotoxicology* **2015**, *9*, 1041–1049. [CrossRef]
30. Zhai, T.; Li, L.; Wang, X.; Fang, X.; Bando, Y.; Golberg, D. Recent developments in one-dimensional inorganic nanostructures for photodetectors. *Adv. Funct. Mater.* **2010**, *20*, 4233–4248. [CrossRef]
31. Li, D.; Wu, W.; Zhao, Y.; Qiao, R. Type-II heterojunction constructed by Ag_2S-coupled ZnO microspheres with visible light-responsive antibacterial activity. *Mater. Lett.* **2020**, *271*. [CrossRef]
32. Chen, D.; Wei, L.; Wang, D.; Chen, Y.; Tian, Y.; Yan, S.; Mei, L.; Jiao, J. Ag_2S/ZnO core-shell nanoheterojunction for a self-powered solid-state photodetector with wide spectral response. *J. Alloys Compd.* **2018**, *735*, 2491–2496. [CrossRef]
33. Bengas, R.; Lahmar, H.; Redha, K.M.; Mentar, L.; Azizi, A.; Schmerber, G.; Dinia, A. Electrochemical synthesis of n-type ZnS layers on p-Cu_2O/n-ZnO heterojunctions with different deposition temperatures. *RSC Adv.* **2019**, *9*, 29056–29069. [CrossRef]
34. Sui, M.-r.; Gu, X.-q.; Shi, M.-l.; Wang, Y.; Liu, L.-l. Improved photoelectrochemical performance by forming a ZnO/ZnS core/shell nanorod array. *Optoelectron. Lett.* **2019**, *15*, 241–244. [CrossRef]
35. Wei, C.; Gu, X.; Li, K.; Song, J.; Zhao, Y.; Qiang, Y. Enhanced photoelectrochemical activities of ZnO nanorod arrays after a modification of ZnS or $ZnIn_2S_4$. *J. Electron. Mater.* **2019**, *48*, 7345–7351. [CrossRef]
36. Bose, R.; Manna, G.; Jana, S.; Pradhan, N. Ag_2S-$AgInS_2$: P-n junction heteronanostructures with quasi type-II band alignment. *Chem. Commun.* **2014**, *50*, 3074–3077. [CrossRef] [PubMed]
37. Du, C.; Tian, J.; Liu, X. Effect of intrinsic vacancy defects on the electronic properties of monoclinic Ag2S. *Mater. Chem. Phys.* **2020**, *249*, 122961. [CrossRef]
38. Shen, X.; Yang, J.; Zheng, T.; Wang, Q.; Zhuang, H.; Zheng, R.; Shan, S.; Li, S. Plasmonic p-n heterojunction of Ag/Ag_2S/Ag_2MoO_4 with enhanced Vis-NIR photocatalytic activity for purifying wastewater. *Sep. Purif. Technol.* **2020**, *251*, 117347. [CrossRef]
39. Yang, J.; Miao, H.; Li, W.; Li, H.; Zhu, Y. Designed synthesis of a p-Ag_2S/n-PDI self-assembled supramolecular heterojunction for enhanced full-spectrum photocatalytic activity. *J. Mater. Chem. A* **2019**, *7*, 6482–6490. [CrossRef]
40. Al-Zahrani, A.A.; Zainal, Z.; Talib, Z.A.; Lim, H.N.; Holi, A.M.; Bahrudin, N.N. Enhanced photoelectrochemical performance of Bi_2S_3/Ag_2S/ZnO novel ternary heterostructure nanorods. *Arab. J. Chem.* **2020**, *13*, 9166–9178. [CrossRef]

41. Lan, J.; Gao, M.; Haw, C.; Khan, I.; Zhao, J.; Wang, Z.; Guo, S.; Huang, S.; Li, S.; Kang, J. Layer-by-layer assembly of Ag_2S quantum dots-sensitized ZnO/SnO_2 core-shell nanowire arrays for enhanced photocatalytic activity. *Phys. Lett. A* **2020**, *384*, 126708. [CrossRef]
42. Solís-Cortés, D.; Navarrete-Astorga, E.; Schrebler, R.; Peinado-Pérez, J.J.; Martín, F.; Ramos-Barrado, J.R.; Dalchiele, E.A. A solid-state integrated photo-supercapacitor based on ZnO nanorod arrays decorated with Ag_2S quantum dots as the photoanode and a PEDOT charge storage counter-electrode. *RSC Adv.* **2020**, *10*, 5712–5721. [CrossRef]
43. Babeer, A.M.; Aamir, L. Zinc oxide/silver sulfide (ZnO/Ag_2S) core-shell type composite for wide range absorption of visible spectra: Synthesis and characterization. *Nano Hybrids Compos.* **2019**, *25*, 84–89. [CrossRef]
44. Chava, R.K.; Kang, M. Ag_2S quantum dot sensitized zinc oxide photoanodes for environment friendly photovoltaic devices. *Mater. Lett.* **2017**, *199*, 188–191. [CrossRef]
45. Feng, Y.; Wang, Y.; Li, M.; Lv, S.; Li, W.; Li, Z. Novel visible light induced $Ag_2S/g\text{-}C_3N_4/ZnO$ nanoarrays heterojunction for efficient photocatalytic performance. *Appl. Surf. Sci.* **2018**, *462*, 896–903. [CrossRef]
46. Huang, J.; Yang, C.; Song, Q.; Liu, D.; Li, L. Photocatalytic performance of $Ag_2S/ZnO/ZnS$ nanocomposites with high visible light response prepared via microwave-assisted hydrothermal two-step method. *Water Sci. Technol.* **2018**, *78*, 1802–1811. [CrossRef]
47. Anjum, M.; Kumar, R.; Barakat, M.A. Visible light driven photocatalytic degradation of organic pollutants in wastewater and real sludge using $ZnO\text{-}ZnS/Ag_2O\text{-}Ag_2S$ nanocomposite. *J. Taiwan Inst. Chem. Eng.* **2017**, *77*, 227–235. [CrossRef]
48. Holi, A.M.; Zainal, Z.; Ayal, A.K.; Chang, S.-K.; Lim, H.N.; Talib, Z.A.; Yap, C.-C. Ag_2S/ZnO Nanorods composite photoelectrode prepared by hydrothermal method: Influence of growth temperature. *Optik* **2019**, *184*, 473–479. [CrossRef]
49. Han, J.; Liu, Z.; Yadian, B.; Huang, Y.; Guo, K.; Liu, Z.; Wang, B.; Li, Y.; Cui, T. Synthesis of metal sulfide sensitized zinc oxide-based core/shell/shell nanorods and their photoelectrochemical properties. *J. Power Sources* **2014**, *268*, 388–396. [CrossRef]
50. Zhou, H.; Gui, P.; Yang, L.; Ye, C.; Xue, M.; Mei, J.; Song, Z.; Wang, H. High performance, self-powered ultraviolet photodetector based on a ZnO nanoarrays/GaN structure with a CdS insert layer. *New J. Chem.* **2017**, *41*, 4901–4907. [CrossRef]
51. Huang, G.; Zhang, P.; Bai, Z. Self-powered UV-visible photodetectors based on ZnO/graphene/CdS/electrolyte heterojunctions. *J. Alloys Compd.* **2019**, *776*, 346–352. [CrossRef]
52. Holi, A.M.; Al-Zahrani, A.A.; Najm, A.S.; Chelvanathan, P.; Amin, N. PbS/CdS/ZnO nanowire arrays: Synthesis, structural, optical, electrical, and photoelectrochemical properties. *Chem. Phys. Lett.* **2020**, *750*, 137486. [CrossRef]

nanomaterials

MDPI

Article

Dimensional Roadmap for Maximizing the Piezoelectrical Response of ZnO Nanowire-Based Transducers: Impact of Growth Method

Andrés Jenaro Lopez Garcia [1], Mireille Mouis [1], Vincent Consonni [2] and Gustavo Ardila [1,*]

1 University Grenoble Alpes, Univ. Savoie Mont Blanc, CNRS, Grenoble INP, IMEP-LaHC, F-38000 Grenoble, France; andres-jenaro.lopez-garcia@grenoble-inp.fr (A.J.L.G.); mouis@minatec.grenoble-inp.fr (M.M.)

2 University Grenoble Alpes, CNRS, Grenoble INP, LMGP, F-38000 Grenoble, France; vincent.consonni@grenoble-inp.fr

* Correspondence: ardilarg@minatec.grenoble-inp.fr; Tel.: +33-456-529-532

Abstract: ZnO nanowires are excellent candidates for energy harvesters, mechanical sensors, piezotronic and piezophototronic devices. The key parameters governing the general performance of the integrated devices include the dimensions of the ZnO nanowires used, their doping level, and surface trap density. However, although the method used to grow these nanowires has a strong impact on these parameters, its influence on the performance of the devices has been neither elucidated nor optimized yet. In this paper, we implement numerical simulations based on the finite element method combining the mechanical, piezoelectric, and semiconducting characteristic of the devices to reveal the influence of the growth method of ZnO nanowires. The electrical response of vertically integrated piezoelectric nanogenerators (VING) based on ZnO nanowire arrays operating in compression mode is investigated in detail. The properties of ZnO nanowires grown by the most widely used methods are taken into account on the basis of a thorough and comprehensive analysis of the experimental data found in the literature. Our results show that the performance of VING devices should be drastically affected by growth method. Important optimization guidelines are found. In particular, the optimal nanowire radius that would lead to best device performance is deduced for each growth method.

Keywords: finite element method; piezoelectric sensor; mechanical energy harvesting; nanogenerator; surface Fermi level pinning; surface traps; chemical synthesis; doping level

Citation: Lopez Garcia, A.J.; Mouis, M.; Consonni, V.; Ardila, G. Dimensional Roadmap for Maximizing the Piezoelectrical Response of ZnO Nanowire-Based Transducers: Impact of Growth Method. *Nanomaterials* **2021**, *11*, 941. https://doi.org/10.3390/nano11040941

Academic Editor: Nikos Boukos

Received: 3 March 2021

Accepted: 2 April 2021

Published: 7 April 2021

1. Introduction

With the rapid advancement in smart wearable systems and biomimetic robot technology, piezoelectric nanogenerators (PENGs) have received significant attention, for instance in the field of self-powered sensors [1] and artificial skin [2]. PENGs can be used as energy harvesting devices or mechanical sensors that convert the mechanical energy available in ambient environment (human motion, vibration, wind, etc.) into electric energy [3]. The first PENG based on ZnO nanowire (NW) arrays was built by Wang et al. [4], achieving an energy conversion efficiency between 17 and 30%. Since then, several groups have worked on PENGs based on piezoelectric nanostructures with different configurations, in particular laterally integrated nanogenerators (LING) [5,6] and vertically integrated nanogenerators (VING) [6–10], this last one being the most commonly used configuration due to its easy manufacturing process and its high performance [11]. The VING configuration consists of an array of vertical NWs, grown on a flexible or rigid substrate and immerged into a dielectric matrix, contacted by bottom and top electrodes. Depending on the substrate, the devices can be operated under bending [12–14] or compressive forces [6–10], the compressive mode being so far the most widely studied one. In order to quantify the output potential in compressive mode, most reports have used open-circuit conditions and evaluated the voltage generated under a mechanical load. In the case of VING devices integrated on a rigid substrate, Xu et al. [6] have used ZnO NWs with 150 nm radius and 4 μm length embedded into a polymer matrix made of poly(methylmethacrylate) (PMMA). This device has produced an output potential of about 90 mV under an applied pressure of 6.25 MPa. In Ref. [8], the authors have reported a higher generated output potential of about 290 mV under an applied pressure of about 5 kPa on a device integrating ZnO NWs featuring 100 nm radius and 3 μm length. In another example, Zhu et al. [10] have shown that the fabrication of small units of VING devices and their connection in parallel produce an output voltage of about 35 V under an applied pressure of 1 MPa. The NWs in this last device had an estimated length of 10 μm and diameters between 60 nm and 300 nm. As for VING devices integrated on a flexible substrate, Deng et al. [15] have applied a pressure of about 1 MPa on a device integrating ZnO NWs of 50 nm radius and 600 nm length over a Kapton foil, achieving an output potential of about 350 mV. Lin et al. [16] have integrated ZnO NWs of 250 nm radius and 6 μm length on a polydimethylsiloxane (PDMS) substrate. Under a compressive strain of about 0.12%, they have produced an output voltage of about 8 V. It is important to mention that all the above-mentioned experiments used chemical bath deposition (CBD) for the growth of ZnO NWs, as an easy-to-implement and low temperature process compatible with industrial applications.

In the literature, there has been very few articles reporting the effect of ZnO NW dimensions on the performance of VING or related devices, despite their critical role for optimization. Rivera et al. [17] have investigated the generation of electrical energy from ZnO NW arrays as a function of their length using a charge amplifier. They have reported that the longest NWs, with 4.7 μm in length and 130 nm in diameter, produce the highest energy of about 35 nJ. In contrast, the shortest ones, with 1.3 μm in length and 100 nm in diameter, generated an energy of only 10 nJ. Kammel et al. [18] have investigated NW length influence on the output potential of VING devices. Their device consisted of a double-sided ZnO NW array covered by PDMS. They reported that the longest NWs, with an average of 2.4 μm in length and 66 nm in diameter, produce the highest values of output potential (about 4.48 V) when a pressure was exerted on the device. In contrast, the shortest NWs, with an average of 1.0 μm in length and 317 nm in diameter, generated an output potential of only 1.84 V. These two experimental reports used the CBD technique for ZnO NW growth. Riaz et al. [19] have explored voltage generation from ZnO NW arrays by scanning a conductive Atomic Force Microscopy (AFM) (Pt/Si) tip over samples that had been grown by two different deposition techniques. They have shown that the longest NWs (3–5 μm long and 100–200 nm wide), grown by carbothermal reduction using a vapor-liquid-solid (VLS) mechanism, generate an output voltage of about 30–35 mV on average while the shortest ones (1–2 μm long and 200 nm wide), grown by CBD, generate an output voltage of about 5 mV.

Basically, it is not straighforward to predict the output potential generated from NW-based piezoelectric transducers nor their general performance, because several key parameters, such as geometrical dimensions (i.e., radius and length), doping level (Nd), and surface trap density (Nit) [20] all play a significant role, and their relative effect on general performance may not be readily decoupled. One reason for that is that these key parameters strongly depend on the growth method used to form ZnO NWs. In the field of compound semiconductors, ZnO is a unique material since it can be formed in the shape of NWs by a very large number of growth methods using a self-assembled/self-induced approach (i.e., with no catalyst) [21]. It has been reported in the last two decades that ZnO NWs can be grown by physical vapor deposition techniques [22,23], chemical vapor deposition techniques [24,25], and wet chemistry [26,27]. On the one hand, this wide variety of growth methods offers a great opportunity to form ZnO NWs with controlled morphologies, tailored dimensions, and dedicated properties. On the other hand, the different media and chemicals used in these growth methods result in the formation of ZnO NWs exhibiting a broad range of doping level and surface trap density. Overall, it is well-known that ZnO NWs exhibit a high electrical conductivity and thus a high doping level, regardless of the growth method involved [28]. However, charge carrier density values spread over several decades from 10^{17} cm^{-3} to 10^{20} cm^{-3} [28–50]. The main reason is related to a large incorporation of residual impurities (i.e., Al, Ga, In, etc.) which act as shallow donors in the vapour phase deposition technique [31], as well as to the specific role of hydrogen which can form a wide variety of defects acting as shallow donors in the wet chemistry techniques [37]. Although ZnO NWs exhibit a top polar c-face and six non-polar m-plane sidewalls regardless of the growth methods used, the main characteristics of these surfaces (e.g., surface roughness) are not equivalent. Surface trap density is also, to a significant extent, affected by the growth method used [28,33,49,51,52]. It is thus very important to take the growth method into consideration in the design and optimisation of piezoelectric mechanical transducers in view of its effect on doping level and surface trap density. However, this line of research has not been explored so far.

From a theoretical point of view, the geometrical parameters of ZnO NWs have been shown to affect their overall piezoelectric performance [53,54]. These studies reveal that thinner and longer NWs would increase the generated piezoelectric potential when sollicited under compressive forces, but they do not consider their semiconducting properties, which represent a strong limitation. The influence of doping level and free carrier screening on the piezoelectric response of semiconducting (mostly ZnO) NWs has been studied theoretically by means of numerical simulations [55–57]. The presence of free charge carriers in the core of NWs basically screens the piezoelectric potential induced by the strain generated under mechanical solicitations. In these simulations, single ZnO NWs have been simulated under a compressive force. Compression creates a depletion region at the top of the NW. The piezoelectric potential can be generated only in this depleted region, making the performance practically independent of NW length [56]. These reports have also shown that an increase in doping level reduces the piezoelectric response. In Ref. [57], the authors added an external surface charge density at the top of the NW, extending the depletion region from the top and thus increasing the piezoelectric response of the NW but keeping it independent of length for practical length values. Experimental reports have confirmed the role of doping level on the VING device performances [58]. However, they have also demonstrated a dependence of the performance on NW length [17–19] which could not be explained by these theoretical models. It is then very important to correlate theoretical models and experiment data to elucidate the key parameters affecting the device performance. Understanding the effect of those parameters will allow the development of guidelines for the design of new devices along with their optimization.

It has been shown recently by numerical simulation that Fermi level pinning, resulting from the presence of surface traps at the interface between ZnO and the matrix material could explain the experimentally observed length dependence [20]. The presence of surface and interface traps has been widely acknowledged in III–V and II–VI semiconductors [59–62]. Their impact on device operation has been explained as follows [20]. In the absence of surface traps, or for low trap densities, there is no surface charge, and thus no band bending at the surface of ZnO. In this case, the application of a pressure to the VING generates a polarization field that depletes only the top of the NWs, while the core of the NW remains neutral, with free carriers screening the polarization field. This is known as the free surface Fermi level assumption, which is the usual assumption in most numerical simulations [56,57]. However, in the presence of a large density of surface traps along the NW (m-plane) sidewalls, which is expected in ZnO NWs [63], thermodynamic equilibrium results in surface band bending, with a balance between surface charges and depletion charges (Figure S1 in the Supplementary Materials). If the surface trap density is large enough, surface Fermi level can be considered as pinned close to mid-gap. Therefore, depending on NW diameter, doping level and surface trap density, it may become possible to deplete the NW until its core so that its whole volume contributes to the generated piezoelectric potential. It has been shown that provided surface traps can be considered as slow enough, they contribute to performance improvement by suppressing free carrier screening [20].

In the present work, we theoretically explore the role of free charge carriers, surface traps, and NW dimensions on the output potential of piezoelectric nanocomposites based on ZnO NWs with account for the latitude of variation offered by growth methods. The properties of NWs grown by the most widely used methods including thermal evaporation (TE), chemical vapor deposition (CVD), metal-organic CVD (MOCVD), and CBD with either O- or Zn-polarity are obtained from a thorough and comprehensive analysis of the experimental reports found in the literature. The theoretical investigation is carried out for a typical VING configuration integrating vertical ZnO NWs embedded into a polymer matrix material (PMMA here). The device is considered as operated in compression. This configuration is chosen as it corresponds to most of the experimental reports, but the conclusions drawn are more general. The theoretical study is performed with a finite element method (FEM) approach by solving the full set of coupled equations, describing the mechanical, piezoelectric and semiconducting properties of the structure. Important optimization guidelines are found, concerning in particular the optimal NW radius needed for each growth method in order to obtain the best device performance and specifically to maximize the output piezoresponse. The effect of the variation of the length is also analyzed, in particular for the NWs with O-polarity grown by CBD, which are offering a great potential in the field of piezoelectric devices.

2. Simulation Framework

In this work, we simulated the full set of differential equations that couple mechanical, piezoelectric, and semiconducting properties using the FEM approach. To this end, we used the FlexPDE® environment, which provides fully flexible description of geometry, differential equations to be solved, and boundary conditions.

2.1. Device under Study and Simulated Structure

The classical VING structure is depicted in Figure 1a. It is composed of an NW-based active layer, a polymer layer, and two metallic electrodes placed at the top and bottom of the structure. The active layer is typically made of a thin ZnO seed layer (several 10 nm thick) covered with a self-organized array of vertical ZnO NWs (several μm long). PMMA is typically used as matrix material to encapsulate the NWs and make the structure mechanically robust, as well as to isolate the NWs and avoid current leakages. The device works in capacitive mode. When an external mechanical load is applied on top of the device (a vertical compression in this case), strain is transferred to the active material. The input strain produces dipoles inside ZnO by the direct piezoelectric effect and a polarization field is then created charging the external electrodes.

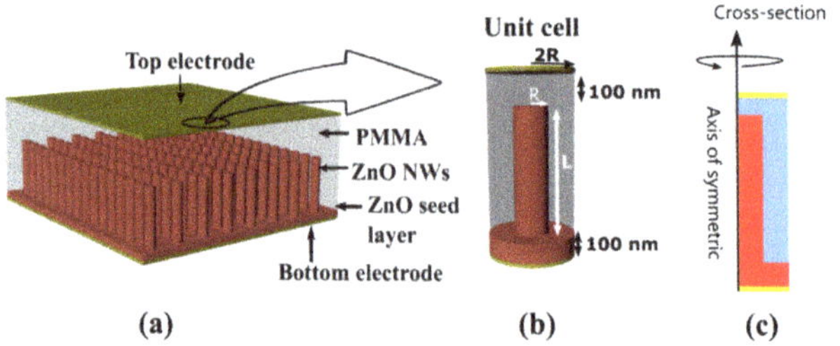

Figure 1. (**a**) Structure of the VING device; (**b**) unit cell of the VING and (**c**) cross section of the VING unit cell around the axis of symmetry.

The simulation of a VING integrating billions of NWs, arranged in a pseudo-periodic disordered array, would require an extremely high computational cost. We simplified the approach by adopting a standard strategy, already used to simulate ZnO composites under compression and bending conditions [20,54], which consists in restricting simulation space to a unit cell made up of a single ZnO NW surrounded by an insulating matrix (PMMA in this case) over a ZnO seed layer. The unit cell is sufficient to determine the generated piezoelectric potential of a whole device sollicited under compression with a high precision if the appropriate boundary conditions are applied in the model [20,54]. Furthermore, a 2D axisymmetric cylindrical model was used for the unit cell as shown in Figure 1b. The 3D cylindrical geometry which derives from the 2D axisymmetric one by rotation around the NW axis, as shown in Figure 1c, was considered as a fair enough approximation. All equations and parameters were transformed from Cartesian to cylindrical coordinates (vertical and radial coordinates, z and r, respectively).

In this work, we fixed the thickness of the ZnO seed layer to 100 nm and that of the thick insulating cap layer (PMMA) to 100 nm as well. The radius (R) of the ZnO NW was varied consistently with what is achievable with the different growth methods. The width of the unit cell was defined as two times the NW diameter (Figure 1b) based on previous work [54]. The effect of NW length (L) on device performance was also studied. The parameters related to semiconducting properties, namely doping level (N_d) and surface trap density at the interface between ZnO and PMMA (N_{it}), were also varied as a function of growth method.

2.2. System of Equations

In order to calculate the output electric potential generated by the VING device under compression, we solved the coupled system of equations for a system based on n-type semiconducting ZnO NW with piezoelectric properties, as in [20]:

$$\nabla([c][\varepsilon]) + \nabla\left([e]^T\vec{\nabla}V\right) = 0, \quad (1)$$

$$\nabla\left([\kappa]\vec{\nabla}V\right) - \nabla([e][\varepsilon]) = \rho, \quad (2)$$

where V is the electric potential, [c] is the elasticity matrix, [e] is the piezoelectric coefficient matrix, [κ] is the dielectric constant matrix, and [ε] is the strain matrix. The second terms on the left-hand side of Equations (1) and (2) represent the piezoelectric coupling terms, and ρ is the local charge density given by:

$$\rho = \begin{cases} q(n - N_d) & \text{in semiconducting regions} \\ 0 & \text{in insulating regions} \end{cases}, \quad (3)$$

where q is the electron charge and N_d is the concentration of ionized donor atoms. In previous studies [20], the free carrier concentration n was computed using Boltzmann statistics. However, depending on growth method, doping level can reach quite large values with respect to degeneracy level, which is around 10^{18} cm^{-3} in ZnO. Here, n was thus computed by considering Fermi-Dirac statistics:

$$n = N_d F_{\frac{1}{2}}\left(\frac{qV}{k_B T}\right), \quad (4)$$

where $F_{1/2}$ (x) is the Fermi-Dirac integral function of order 1/2, kB is the Boltzmann constant, q is the electric charge of one electron, and T is the temperature, considered here equal to 300 K.

The variation of the output potential resulting from the variation of external pressure on the device was obtained by solving Equations (1) and (2) for two cases: first, for the initial state (i.e., without compression) and then for the final state (i.e., under vertical compression). The output potential or "piezoresponse" in the results section was then calculated as the potential difference defined as $V_{\text{Final state}} - V_{\text{Initial state}}$.

2.3. Boundary Conditions

The mechanical boundary conditions consisted in (i) free vertical displacement and forbidden lateral displacement on the axis of symmetry, (ii) a vertical pressure of −1 MPa along the NW axis (z-axis) with free vertical and lateral displacement on top surface, (iii) free lateral displacement with forbidden vertical displacement at the bottom, (iv) free lateral and vertical displacement on the outer lateral side (Figure 2a).

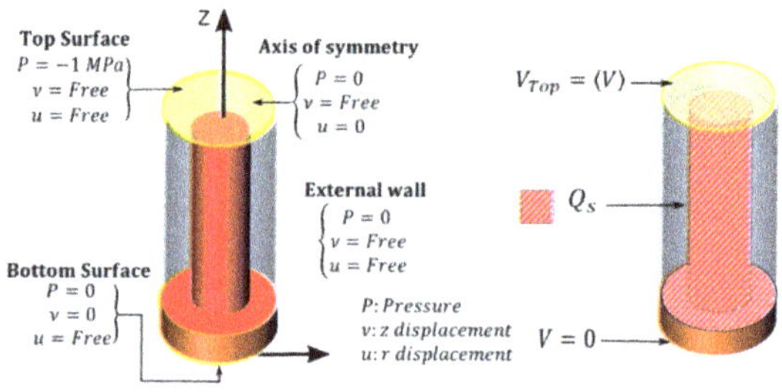

Figure 2. (**a**) Mechanical and (**b**) electrical boundary conditions of the VING unit cell.

In terms of electrical boundary conditions (Figure 2b), the bottom electrode was grounded, while Neumann conditions were applied to other boundaries. The top electrode potential V_{Top} was obtained by averaging V on the top surface. At the interface of ZnO and PMMA (i.e., the diagonal line pattern on the ZnO surface from Figure 2b), we introduced a surface charge Q_s under the assumption of a uniform trap density (N_{it}) at thermal equilibrium. In this paper, the potential used to calculate Q_s was taken from the initial state, which simulates ideally slow traps, with a charge that remains frozen during the transition from initial to final state. Q_s was expressed as a function of the local potential V_{init} as:

$$Q_s = -q^2 N_{it} (V_{init} - \varphi_{Fi}), \quad (5)$$

where φ_{Fi} is the difference between Fermi level and intrinsic level.

3. Simulation Results and Discussions

3.1. Input Experimental Data for the Simulation

A set of numerical simulations of VING devices integrating ZnO NWs was performed considering the characteristics of typical NWs grown by each method using standard conditions, namely without any intentional doping and post-deposition treatment: a typical range of radius (R), a typical length value, a typical range of doping level (N_d), and a typical surface trap density (N_{it}) value were selected in that purpose. As regards the dimensions of ZnO NWs, their typical radius was varied over a similar range of 4 to 150 nm while their length was firstly kept fixed to 5 µm in all devices. These dimensional properties of ZnO NWs are very typical and similar for each growth method. In contrast, the range of doping level in ZnO NWs strongly depends on the growth method used, as represented in Figure 3. From the large number of experimental data reported in the literature using field-effect transistor (FET) measurements [30,35,40,44–47], I–V measurements on four-terminal contacted ZnO NWs [28,29,31,36,37,50], terahertz spectroscopy [34], conductive AFM (i.e., SSRM and SCM measurements) [32,43], and electrochemical impedance spectroscopy [39,41], a range of charge carrier density values was inferred for each growth method when ZnO NWs are grown using standard conditions (i.e., typical chemical precursors, typical growth temperature and pressure). Overall, the charge carrier density of ZnO NWs typically lies in the range of 10^{17} to 10^{20} cm^{-3}. The vapour phase deposition techniques including TE, CVD, and MOCVD methods result in the formation of ZnO NWs with a lower mean charge carrier density ranging from 10^{17} to 5×10^{18} cm^{-3} at maximum [28–33,40,44–50].

In present deposition techniques, the incorporation of residual impurities (i.e., Al, Ga, In) acting as shallow donors is mainly responsible for this range of charge carrier density values [31]. Residual impurities usually occur as contaminants in the materials sources (i.e., TE) or in the growth chamber (i.e., CVD, MOCVD). The high growth temperature used in MOCVD is also favourable to the diffusion of residual impurities from the substrate, like Al from sapphire, into ZnO NWs. In contrast, wet chemistry deposition techniques, including CBD and electrodeposition, lead to the formation of ZnO NWs with a higher mean charge carrier density, ranging from 5×10^{17} cm^{-3} at minimum to 10^{20} cm^{-3} [34–39,41–43]. In the electrodeposition process, the use of zinc choride as the typical chemical precursor to enhance the morphology of ZnO NWs is favourable to the massive incorporation of chlorine acting as a shallow donor [42]. In the CBD process, the massive incorporation of hydrogen-related defects acting as shallow donors is mainly responsible for this range of charge carrier density [36]. The growth medium in water is full of hydrogen and the crystallization process resulting in the elongation of ZnO NWs through the development of their c-plane top facet basically involves a dehydration process [64]. A large number of hydrogen-related defects (e.g., interstitial hydrogen, substitutional hydrogen on the oxygen lattice site, zinc vacancy–hydrogen complexes) acting as shallow donors are thus formed systematically [37].

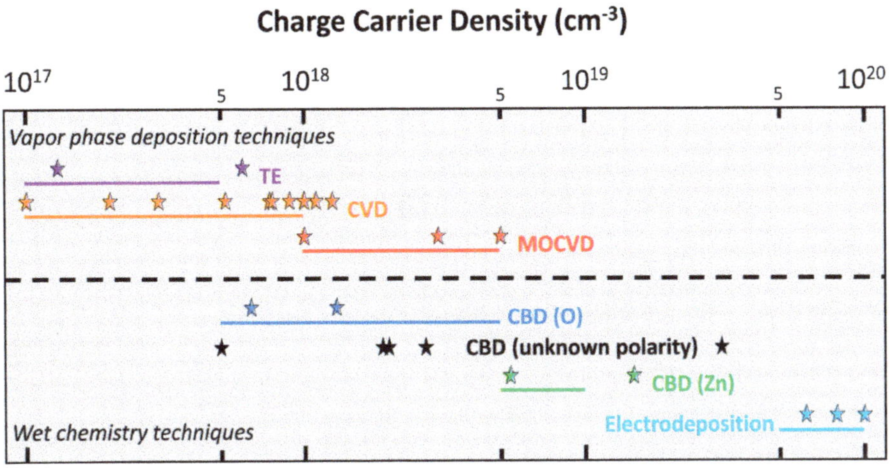

Figure 3. Schematic diagram summarizing experimental data for charge carrier density in ZnO NWs grown by TE [29,40], CVD [28,44–50], MOCVD [31–33], CBD (O) [36], CBD (unknown polarity) [34,37,43], CBD (Zn) [36], and electrodeposition [38,39,41]. A logarithmic scale is used for doping level. The coloured solid stars represent the experimental data point reported in the literature for each growth method, as an average value [28,29,33,34,38–41,44,49,50] or an interval of minimum and maximum values [31,32,36,37,43,45–48]. The coloured solid lines represent the deduced range of charge carrier density used in the numerical simulation for each growth method.

The CBD technique further offers a unique opportunity to form either O– or Zn-polar ZnO NWs [65] with a significant difference in their electrical conductivity and both of them were considered in the present theoretical investigation [36]. This is in strong contrast with physical vapour deposition techniques, for which ZnO NWs are systematically of Zn polarity when the self-assembled/self-induced approach is employed [66]. In contrast to the large number of experimental data reporting the doping level of ZnO NWs, the surface trap density and its dependence on each growth method have been much less investigated experimentally. Only a couple of investigations have been achieved, mainly by steady-state and time-resolved optical spectroscopy through the determination of the surface recombination velocity [52]. However, a consensus in the literature seems to emerge that the surface trap density of ZnO NWs grown by physical vapour deposition techniques is about 2–3 × 10^{12} cm^{-2} (i.e., about 10^{12} eV^{-1} cm^{-2}) [28,33,49,51]. In contrast, the surface trap density of ZnO NWs grown by wet chemistry is about one decade larger (1–4 × 10^{13} cm^{-2} i.e., about 10^{13} eV^{-1} cm^{-2}) [52]. The higher value when using wet chemistry is expected because their surfaces are typically rougher and thus present a larger density of defects.

3.2. *Piezoelectric Performance as a Function of the Nanowire Growth Method*

The piezoresponse was calculated for each device for a fixed NW length of 5 μm, variable NW radius and taking into account typical values of N_d and N_{it} corresponding to each NW growth method, as described in the previous section. A numerical simulation was also made to evaluate the effect of varying the length (L) for one particular growth method that is of high interest, namely the CBD growth of O-polar ZnO NWs, and with a typical radius set to 50 nm.

3.2.1. TE Method

Figure 4a shows the calculated piezoresponse of a VING transducer for different radii of ZnO NWs ranging from 20 nm up to 150 nm. A negative piezo response is obtained as expected [54] because the *c*-axis in the NW is oriented along the [0001] direction (i.e., Zn polarity) [67]. This is the case in the vast majority of the ZnO NWs grown by the numerous physical and chemical techniques we investigate here. With this growth method, Nd was evalutated in the range from 1 (blue curve) to a maximum value of 5 (red curve) × 10^{17} cm^{-3} and a value of N_{it} = 1 × 10^{12} eV^{-1} cm^{-2} was taken. These parameters correspond to ZnO NWs grown by the TE method as presented in Figure 3. Figure 4a shows a strong radius dependence of the piezoresponse, further influenced by the doping level. At the minimum value of N_d, a step in the piezoresponse was observed, with a strong increase (by about 13 times) in absolute value, as NW radius was reduced from 140 nm to 120 nm. A further reduction of the radius had little effect on the piezoresponse. A similar dependence was also found for the maximum value of N_d, but the step in piezoresponse was observed at lower radius (i.e., for a reduction from 50 nm to 40 nm). The region between the two curves in Figure 4a thus represents the range of optimization of the VING devices. VING devices integrating NWs with a radius larger than 120 nm for low doping levels (blue curve), and larger than 40 nm for high doping levels (red curve), result in poor piezoresponse. This is due to the screening effect originating from free carriers in the ZnO NW [20,54,56,68–70]. An example of this effect is shown in Figure 4b. It depicts a qualitative map of the free carrier distribution in a VING device under compression. The VING transducer integrates NWs with a radius of 140 nm and a low doping level (1 × 10^{17} cm^{-3}). A depletion region is created from the PMMA/ZnO interface and a neutral core starts from the bottom and extends towards the top of the NW. In this neutral region, the free carriers screen the piezoelectric response and the contribution of polarization electric charges is largely reduced, thus the overall voltage is reduced as well. Figure 4c shows the effect of the reduction of the NW radius down to 80 nm at the same doping level. In this case, the depletion region is large enough to fully deplete the NWs from its sides and from its top, drastically increasing the performance of the device. We can thus identify a critical value for the ZnO NW radius, below which the performance of VING transducers can be largely improved, for given doping level and trap density. This particular NW radius will be called "NW critical radius" all along the article and summarised for every growth method in Table 1. The critical radius can be evaluated analytically from charge neutrality between surface traps and surface depletion under the additional condition that depletion region reaches the center of the NW by solving Poisson equation in cylindrical coordinates (Figure S1 and Equation (S5) in the Supplementary Materials).

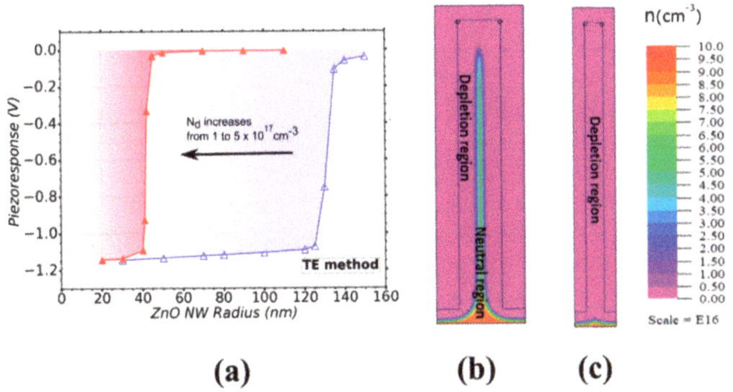

Figure 4. (**a**) Variation of the piezoresponse of a VING as a function of the ZnO NW radius for a range of doping level (N_d) from 1 (△) to 5 (▲) × 10^{17} cm^{-3} typical in the TE method. The shadowed zones represent the regions of interest for a given doping level. Free carrier distribution of a NW with N_d = 1 × 10^{17} cm^{-3} and a radius of (**b**) 140 nm and (**c**) 80 nm. A trap density N_{it} = 1 × 10^{12} eV^{-1} cm^{-2} was considered in the simulation.

Table 1. Summary of NW radius values to achieve full depletion or optimal performance for the different growth methods (NW critical radius). In the case of the CBD (O) technique, the values correspond to the radius to obtain half the optimal performance.

Growth Method	ZnO NW Radius for Full Depletion	
	Min. N_d (nm)	Max. N_d (nm)
TE	<120	<40
CVD	<120	<20
MOCVD	<22	<5
CBD (O)	<55	<15
CBD (Zn)	<18	<12
Electrodeposition	Not simulated (estimated < 4)	Not simulated (estimated < 4)

3.2.2. CVD Method

To further explore the influence of the growth method of the NWs on the VING performance, the piezoresponse was also calculated using the characteristics reported for ZnO NWs grown by the CVD method. These NWs present lower values of radius and higher doping level as compared to the NWs grown by TE although they present equivalent values of N_{it}. The doping level lies in the range from 1 × 10^{17} up to 1 × 10^{18} cm^{-3}, corresponding to the blue and red curves in Figure 5a, respectively. Figure 5a shows the weak piezoresponse obtained with NWs exhibiting radii larger than about 140 nm for low doping levels (blue curve) and 25 nm for high doping levels (red curve). A higher performance was obtained for NW radii below 120 and 20 nm for low and high doping levels, respectively.

3.2.3. MOCVD method

Figure 5b shows the piezoresponse of a VING device integrating ZnO NWs with the characteristics reported for the MOCVD method. In this case, N_d goes from 1 (blue curve) to 5 (red curve) × 10^{18} cm^{-3}. This method shares the same Nit as TE and CVD methods as a first approximation. According to the results, the piezoresponse is improved when the ZnO NWs radius is smaller than 22 nm for minimum N_d (blue curve) and smaller than 5 nm for maximum N_d (red curve). A poor performance is expected for NWs with radius larger than about 30 and 7 nm for low and high doping levels, respectively. It should be noted that the simulation of a device integrating NWs with 5 nm radius could reach the limit where continuum medium equations do not apply anymore. In this sense, it should be considered as a rough approximation. Indeed, according to studies based on first-principles calculations, ZnO nanoscale materials with radius lower than 3 nm could present significantly higher piezoelectric coefficients compared to bulk material, namely there would be a radius-dependent size effect with its piezoelectric properties [71]. However, the formation of ZnO NWs with a radius smaller than 5 nm has not been experimentally shown yet and there has been thus no experimental evidence yet of the improvement of the piezoelectric coefficients in that range of radii. A comparison of the performance obtained when using these last three growth methods involving vapour phase deposition techniques (TE, CVD, MOCVD) show that smaller radii are gradually required to obtain the optimal performances: 120 nm, 120 nm, and 22 nm respectively, for low doping levels, as well as 40 nm, 20 nm, and 5 nm, respectively, for high doping levels. This can be explained as the doping level of NWs grown by MOCVD is higher than with the other growth methods, and the doping level of the NWs grown by TE is the lowest one, while keeping the same surface trap density. Because of the higher doping levels, NWs with smaller radii are required in order to obtain the full depletion of free charge carriers and thus to reduce the screening effect.

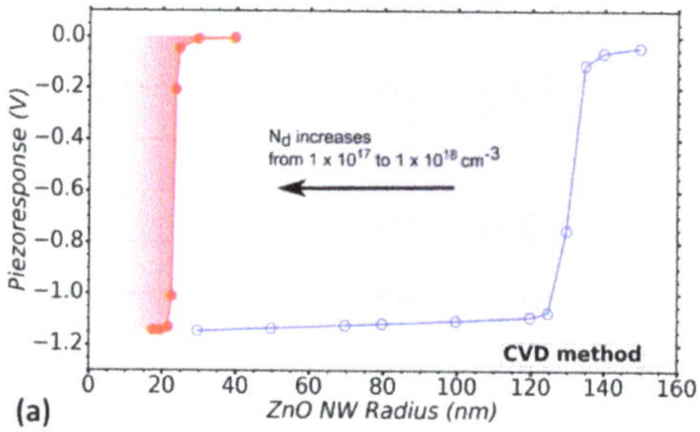

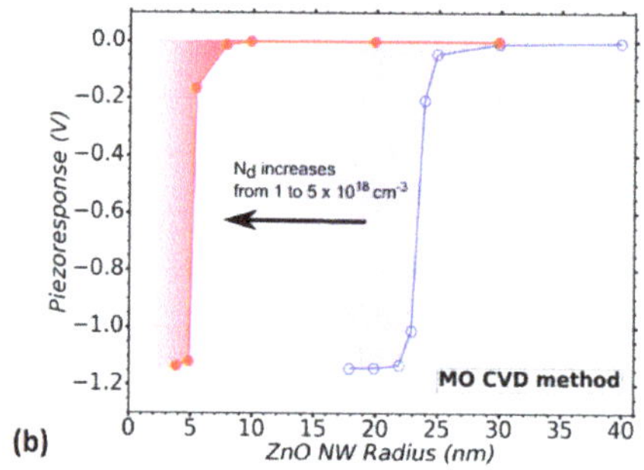

Figure 5. Variation of the piezoresponse of a VING as a function of ZnO NW radius taking into account two vapor deposition techniques for the growth: CVD and MOCVD. (**a**) A range of doping levels (N_d) from 1×10^{17} (⬡) to 1 (⬢) $\times 10^{18}$ cm^{-3} is considered, which is typical in the CVD method, (**b**) a range of doping level (N_d) from 1 (⬡) to 5 (⬢) $\times 10^{18}$ cm^{-3} is considered, which is typical in MOCVD method. A trap density $N_{it} = 1 \times 10^{12}$ eV^{-1} cm^{-2} was considered in the simulations.

3.2.4. CBD Method

The CBD technique was also considered as an important growth method for flexible devices in our theoretical study. The O- and Zn-polar NWs grown by the CBD method present a higher surface trap density compared to the previous methods (e.g., N_{it} is equal to 1×10^{13} eV^{-1} cm^{-2}. Figure 6a shows the radius dependence of the VING performance for the CBD (O) method, namely for O-polar ZnO NWs. A positive piezo response is obtained in this case because the *c*-axis in the NW is oriented along the [000–1] direction (i.e., O polarity). This also changes slightly the interaction between the piezoelectric and semiconducting effects, producing a less sharp transition from the fully screened piezo response state to the fully depleted NW state where a maximum performance is obtained. The doping level range goes from $N_d = 5 \times 10^{17}$ cm^{-3} (blue curve) to $N_d = 5 \times 10^{18}$ cm^{-3} (red curve). Owing to the higher surface trap density of the CBD method, a better performance can be obtained for larger radius compared to that of the MOCVD method. For instance, half the optimal piezo response (~0.6 V in absolute value) can be obtained for a NW grown by the CBD (O) method with a radius of ~55 nm. In contrast, a radius of ~23 nm is needed for a NW grown by MOCVD, which is larger by a factor of more than 2. Interestingly the critical radii are very similar for both methods. The best performance can be obtained for NWs with radius below 20 nm and 5 nm for low and high doping levels, respectively. For NW radius larger than 70 nm (low doping level) and 22 nm (high doping level), we expect that a poor performance is obtained.

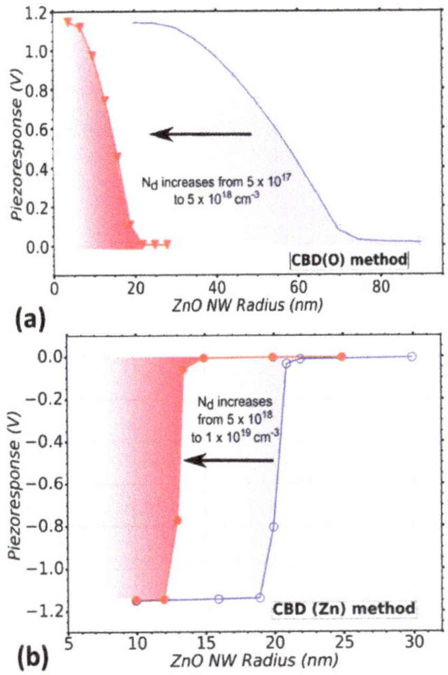

Figure 6. Variation of the piezoresponse of a VING as a function of ZnO NW radius taking into account the CBD deposition technique. (**a**) A range of doping level (N_d) from 5×10^{17} (▼) to 5 (▼) $\times 10^{18}$ cm^{-3} is considered, which is typical in CBD (O), (**b**) a range of doping level N_d from 5×10^{18} (○) to 1 (•) $\times 10^{19}$ cm^{-3} is considered, which is typical in CBD (Zn) method. A trap density $N_{it} = 1 \times 10^{13}$ eV^{-1} cm^{-2} was considered in the simulations.

Finally, the performance of VING devices integrating Zn-polar NWs grown by the CBD method presents a very low optimization window as compared to O-polar NWs grown by the same CBD technique (Figure 6b). The optimal performance can be obtained using NWs with radius below about 18 nm and 12 nm for low and high doping levels, respectively. This window of optimization of about 6 nm is around 9 times lower as compared to the optimization window for the CBD (O) method (estimated to be 52 nm). This is caused by the higher doping levels obtained in Zn-polar NWs grown by the CBD method. Up to this point, the CBD (Zn) method seems to be the most limited one in comparison with the other techniques considered in this study based on the optimization window and the very low radius that is needed to obtain optimal devices.

3.2.5. Electrodeposition: Analytical Evaluation of the Critical NW Radius

The electrodeposition method is also used to grow ZnO NWs, but it was not considered in this numerical theoretical study because of its high N_d (~10^{20} cm^{-3}) which would lead to NWs with a wide neutral core over a wide range of NW radius. Instead, an analytical approach was considered to address the case of ZnO NWs grown by electrodeposition. To assess the agreement with the numerical simulations reported in the last sub-sections, the optimal NW radius to obtain fully depleted ZnO NWs was also calculated using an analytical model proposed in [72,73] for Si and GaAs NWs. The model considered only semiconductor equations and was developed to assess the critical radius a_{crit} (Supplementary Materials), which marks the boundary between a fully depleted NW ($r < a_{crit}$) and a NW that is only depleted at its surface ($r > a_{crit}$). Figure 7 shows the values calculated within the analytical model correspond very well to the values reported in the last sub-sections for the different growth methods. In the case of the CBD (O) method, the analytical value is larger compared to the values extracted from our simulations. A theoretical value of ~70 nm and ~20 nm is calculated for low and high doping concentrations. From our simulations, we extract the values of ~20 nm and ~5 nm, respectively. This difference can be explained because the analytical models do not take into account the piezoelectric effect and the orientation of the NWs. The analytical model allows us as well to estimate the possible optimal radius of NWs grown by the electro-deposition method. The values lay well below the limit of our numerical model (estimated to be 5 nm with the piezoelectric coefficients used). This confirms that our numerical model would provide inacurate piezopotential results on NWs grown by this last method.

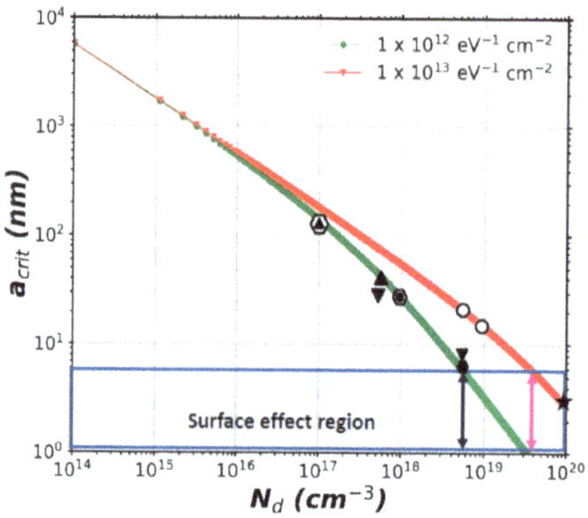

Figure 7. Critical radius a_{crit} as function of N_d for two ZnO NWs with values of N_{it} from 10^{12} (green curve) and 10^{13} eV^{-1} cm^{-2} (red curve). The black marks on the curves indicate the radius values for which full depletion was obtained as calculated using FlexPDE. The mark (▲) corresponds to the TE method, (⬤) to the CBD (Zn) method, (▼) to the CBD (O) method, (⬤) to the CVD method and (⬢) to the MOCVD method. The blue and magenta arrows indicate the limiting value of N_d for $N_{it} = 10^{12}$ eV^{-1} cm^{-2} and 10^{13} eV^{-1} cm^{-2}. Beyond this value of doping, the critical radius goes below 5nm and we consider that surface effects could modify piezoelectric coefficients. The mark (★) corresponds to the conditions of NWs grown by the electron-deposition method, which was not simulated with FlexPDE.

3.2.6. Effect of the Variation of NW Length on the VING Performance

In the simulation results from the previous sections, the piezoresponse is constant if the NW radius is smaller than a certain critical radius. This piezoresponse value is then independent upon the radius and the growth method. The value depends only on the NW length and on the ratio between the diameter of the NW and the width of the VING unit cell [54]. Here we studied the effect of the variation of the length in the particular case corresponding to the growth conditions of the CBD method forming O-polar NWs (CBD

(O)). This method presents one of the largest optimization windows and is compatible with the processing of flexible devices without any additional transfer process. For this study, the NW radius was fixed at 50 nm with $N_d = 5 \times 10^{17}$ cm^{-3}. Figure 8 shows a linear increase in the piezoresponse as the NW length increases up to 3 μm. For larger values, the piezoresponse saturates to about 0.75 V in absolute value. This means that longer NWs are not required to obtain optimal devices. This theoretical result is in accordance with the trends of experiments in energy generation [17], and output potential [18], although the output potential values are not the same, which can be due to different electrical and structural parameters. Furthermore, the inferred optimal length of O-polar NWs around 4 μm does not represent any technical challenge: it is typically achieved by the CBD method when the synthesis conditions using zinc nitrate, hexamethylenetetramine (HMTA), and chemical additives including polyethylenimine are optimized [74].

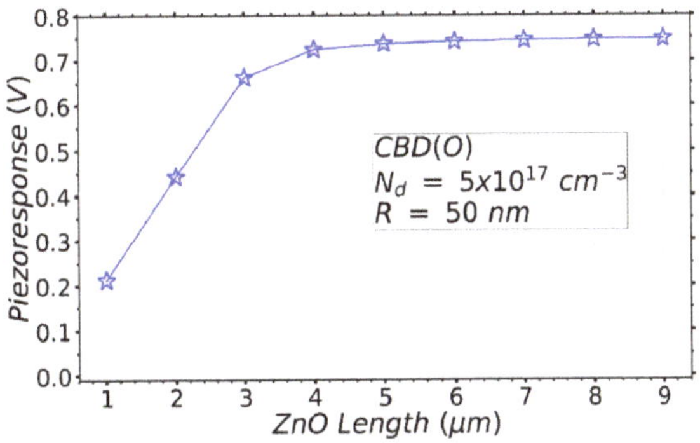

Figure 8. Variation of the piezoresponse of a VING device as a function of the NW length. The parameters of the NW used in the model correspond to ZnO NWs grown by the CBD (O) method.

3.3. Summary and Discussion about the Mechanisms at Work

As a summary of these theoretical results, Table 1 shows the NW critical radius defined as the radius value below which the ZnO NW is fully depleted and optimal piezoelectric performance can be expected. The values of critical radius were calculated for every growth method studied, both for the low and high values of N_d allowed by each method based on reported experimental data. The range of values between the critical radii obtained for the low and high values of N_d defines an optimization process window. This Table shows that the different growth methods do not offer the same potential for VING device optimization when using standard conditions. The TE and CVD methods exhibit the widest optimization process window by controlling both the radius and the doping of the NWs. In particular, they allow the use of wider NWs with radius not exceeding 120 nm for the low doping level. It should be noted here that ZnO NWs grown by the TE and CVD methods present a typical radius in that range, such that the targeted critical radius is not a technological challenge. However, for the high doping level, the critical radius drops to 40 and 20 nm for the respective TE and CVD methods, respectively. In that case, the growth of ZnO NWs with this small radius is still feasible, but deserves a particular effort to be reached. On the other hand, the MOCVD method has much smaller critical radii for the low and high doping level down to 22 and 5 nm. Although the MOCVD method is a well-known technique to get high aspect ratio ZnO NWs [66], the present small radii require working in dedicated conditions such as low VI/II ratio and relatively high growth temperature [75,76]. This can be quite challenging to be reached. In contrast, ZnO NWs grown by MOCVD have significantly been developped for optoelectronic devices [77]. The CBD method has the great advantage of emphasizing the importance of considering polarity as a critical quantity for VING devices. While ZnO NWs grown by CBD with the Zn-polarity have the narrowest optimisation window and small critical radii of 18 and 12 nm for the low and high doping levels, respectively, O-polar ZnO NWs grown with the same technique exhibit one of the largest optimization window with a significant increase in the radii to obtain increased performance, half the optimal performance is reached for NWs with radii of 55 and 15 nm for the low and high doping levels. In this technique, the typical radius of ZnO NWs lies in the range of 50–100 nm when using a zinc salt as a source of zinc ions and HMTA [78]. The further decrease in the radius of ZnO NWs is more required for Zn-polar ZnO NWs than for O-polar ZnO NWs. This is typically achieved by using chemical additives such as PEI [74] and ethylenediamine [79] to inhibit the radial growth of ZnO NWs and hence to limit their radius. However, it is evident here that the addition of chemical additives is likely not sufficient for Zn-polar ZnO NWs, which points out the strong interest in developing O-polar ZnO NWs for VING devices in the capacitive configuration. It is worth noticing that these O-polar ZnO NWs also outperforms the Zn-polar ZnO NWs to get high quality Schottky contacts with Au on their top [80], which is critical in the Schottky configuration consisting in making Schottky contacts at the bottom and at the top of the VING structure [6].

In the whole of these growth methods, the reduction of the doping level to the minimum value reported here or even further may also be considered to allow larger critical radii to be used. This can tentatively be achieved by post-deposition thermal treatments including annealing under oxygen atmosphere or plasma treatment [81]. An alternative approach would consist in compensating the doping level through the introduction of acceptors [81]. The introduction of acceptors including group-I, copper and antimony in ZnO films have been investigated to a great extent [82]. However, mastering the incorporation of dopant elements in ZnO NWs is still a big issue and correlatively strongly affects their radius during the growth phase through intricate phenomena. For instance, in the CBD method, the addition of metal salts other than zinc salt offers an opportunity to dope ZnO NWs [83], but this addition is not sufficient for the incorporation of dopant elements in their center. A further precise control of the pH conditions is required [84]. The involved physicochemical processes at work also depend on the involved dopant [84,85], which makes a general approach complicated. Eventually, the engineering of surface trap density in ZnO NWs appears as an alternative way to tune the radius to a relevant range that is compatible with the different growth methods. It has been shown that the use of chemical adsorbates on the polar and non-polar planes of ZnO has a great effect on the nature and magnitude of the band bending [86,87]. The intricate nature of all these approaches represents the major difficulty to get ZnO NWs with the optimal dedicated properties. Overall, the present findings show that the different growth methods of ZnO NWs result in a different range of optimized radii and hence exhibit a different potential for VING transducers. The TE and CBD(O) methods using standard conditions are expected to be of great interest in the piezoelectric field, the latter being compatible with the processing of flexible devices without transfer processes.

4. Conclusions

In summary, our models predict that the performance of piezoelectric transducers based on ZnO NWs in the VING configuration evaluated under compressive forces strongly depends on the growth method. Several methods have been compared theoretically in this work: TE, CVD, MOCVD, CBD (Zn), CBD (O) and electrodeposition. Each growth method produces NWs with a different range of doping level, surface trap density, and dimensions. Taking into account these parameters in our models allowed the calculation of the optimal NW radius to obtain the best performance (i.e., maximal voltage generated for a given mechanical compression). In general, at higher doping levels, a smaller NW radius is needed to reach a full depletion of its core and to increase the performance. This effect can be compensated if the surface trap density is increased, allowing the use of wider NW radius. TE and CVD methods, with the lowest doping level, allowed the use of relatively wide NWs (radius below 120 nm) to obtain optimal results. MOCVD method required relatively thinner NWs compared to the other methods to improve the performance of the transducers (radius below 22 nm at low doping level). The CBD(O) method is the only other method allowing relatively large radius (below 55 nm) at low doping level thanks to the higher surface trap density. The CBD (Zn) method requires a thinner radius for optimal devices compared to the other methods studied here (below 18 nm and 12 nm at low and high doping level, respectively). This method also presents the smallest window of optimisation in terms of optimal radius. Numerical calculations have been compared with an analytical approach to obtain the optimal radius. We found a very good correspondence between the two calculation approaches. The analytical approach was used as well in the highest doping level range corresponding to electrodeposition. The optimal NW radius for this method could be below 5 nm, suggesting a more complex model where piezoelectric properties are taken into account as a function of surface parameters. Our numerical model also predicts that an increase in the NW length increases device performance until a saturation in the output potential is reached. This means that a certain minimum length is needed to optimize the device. As an example, in the case of CBD (O) grown ZnO NWs with a radius of 50 nm and low doping level, increasing the length beyond 4 μm does not improve anymore the performance. Doping level and surface and interface traps will also affect piezotronic and piezo-phototronic devices and it would be very important to take into account the NW growth method on those applications as well. Finally, the control of surface and interface traps densities by surface engineering is key to optimize the performance of piezoelectric transducers based on piezoelectric semiconducting NWs.

Supplementary Materials: The following are available online at https://www.mdpi.com/2079-4991/11/4/941/s1, Figure S1: Energy Band diagram along half the cross-section of a n-type ZnO NW.

Author Contributions: Conceptualization, V.C. and G.A.; methodology, A.J.L.G. and M.M.; software, A.J.L.G. and M.M.; validation, A.J.L.G. and M.M.; formal analysis, A.J.L.G.; investigation, A.J.L.G.; resources, M.M. and G.A.; writing—original draft preparation, A.J.L.G.; writing—review and editing, M.M., V.C. and G.A.; visualization, A.J.L.G. and V.C.; supervision, M.M. and G.A.; project administration, V.C. and G.A.; funding acquisition, V.C. and G.A. All authors have read and agreed to the published version of the manuscript.

Funding: This work was supported by the Federation of Micro Nano Technologies (FMNT) in Grenoble, France and the French National Research Agency in the framework of the "Investissements d'avenir" program (ANR-15-IDEX-02) through the project CDP NEED and ANR SCENIC (ANR-20-CE09-0005-04). This work has received funding from project PULSE-COM of the European Union's Horizon 2020 research and innovation programme under grant agreement No 863227.

Data Availability Statement: The data are available upon request from the corresponding authors.

Conflicts of Interest: The authors declare no conflict of interest.

References

1. Wang, Z.L. Toward self-powered sensor networks. *Nano Today* **2010**, *5*, 512–514. [CrossRef]
2. Lee, T.I.; Jang, W.S.; Lee, E.; Kim, Y.S.; Wang, Z.L.; Baik, H.K.; Myoung, J.M. Ultrathin self-powered artificial skin. *Energy Environ. Sci.* **2014**, *7*, 3994–3999. [CrossRef]
3. Anton, S.R.; Sodano, H.A. A review of power harvesting using piezoelectric materials (2003–2006). *Smart Mater. Struct.* **2007**, *16*, R1–R21. [CrossRef]
4. Wang, Z.L. Piezoelectric Nanogenerators Based on Zinc Oxide Nanowire Arrays. *Science* **2006**, *312*, 242–246. [CrossRef] [PubMed]
5. Zhu, G.; Yang, R.; Wang, S.; Wang, Z.L. Flexible High-Output Nanogenerator Based on Lateral ZnO Nanowire Array. *Nano Lett.* **2010**, *10*, 3151–3155. [CrossRef] [PubMed]
6. Xu, S.; Qin, Y.; Xu, C.; Wei, Y.; Yang, R.; Wang, Z.L. Self-powered nanowire devices. *Nat. Nanotechnol.* **2010**, *5*, 366–373. [CrossRef] [PubMed]
7. Yu, A.; Li, H.; Tang, H.; Liu, T.; Jiang, P.; Wang, Z.L. Vertically integrated nanogenerator based on ZnO nanowire arrays. *Phys. Status Solidi RRL* **2011**, *5*, 162–164. [CrossRef]
8. Tao, R.; Parmar, M.; Ardila, G.; Oliveira, P.; Marques, D.; Montès, L.; Mouis, M. Performance of ZnO based piezo-generators under controlled compression. *Semicond. Sci. Technol.* **2017**, *32*, 064003. [CrossRef]
9. Hu, Y.; Lin, L.; Zhang, Y.; Wang, Z.L. Replacing a Battery by a Nanogenerator with 20 V Output. *Adv. Mater.* **2011**, *24*, 110–114. [CrossRef] [PubMed]
10. Zhu, G.; Wang, A.C.; Liu, Y.; Zhou, Y.; Wang, Z.L. Functional Electrical Stimulation by Nanogenerator with 58 V Output Voltage. *Nano Lett.* **2012**, *12*, 3086–3090. [CrossRef]

11. Wang, Z.; Pan, X.; He, Y.; Hu, Y.; Gu, H.; Wang, Y. Piezoelectric Nanowires in Energy Harvesting Applications. *Adv. Mater. Sci. Eng.* **2015**, *2015*, 165631. [CrossRef]
12. Yang, R.; Qin, Y.; Li, C.; Zhu, G.; Wang, Z.L. Converting Biomechanical Energy into Electricity by a Muscle-Movement-Driven Nanogenerator. *Nano Lett.* **2009**, *9*, 1201–1205. [CrossRef] [PubMed]
13. Yim, M.; Jeon, B.; Yoon, G. Fabrication of Tandem-type Vertically Integrated Nanogenerator by In-situ Deposition of AlN/ZnO Films. *J. Semicond. Technol. Sci.* **2019**, *19*, 233–238. [CrossRef]
14. Dahiya, A.S.; Morini, F.; Boubenia, S.; Nadaud, K.; Alquier, D.; Poulin-Vittrant, G. Organic/Inorganic Hybrid Stretchable Piezoelectric Nanogenerators for Self-Powered Wearable Electronics. *Adv. Mater. Technol.* **2018**, *3*, 1–11. [CrossRef]
15. Deng, W.; Jin, L.; Zhang, B.; Chen, Y.; Mao, L.; Zhang, H.; Yang, W. A flexible field-limited ordered ZnO nanorod-based self-powered tactile sensor array for electronic skin. *Nanoscale* **2016**, *8*, 16302–16306. [CrossRef] [PubMed]
16. Lin, L.; Hu, Y.; Xu, C.; Zhang, Y.; Zhang, R.; Wen, X.; Wang, Z.L. Transparent flexible nanogenerator as self-powered sensor for transportation monitoring. *Nano Energy* **2013**, *2*, 75–81. [CrossRef]
17. Rivera, V.F.; Auras, F.; Motto, P.; Stassi, S.; Canavese, G.; Celasco, E.; Bein, T.; Onida, B.; Cauda, V. Length-Dependent Charge Generation from Vertical Arrays of High-Aspect-Ratio ZnO Nanowires. *Chem. Eur. J.* **2013**, *19*, 14665–14674. [CrossRef]
18. Kammel, R.S.; Sabry, R.S. Effects of the aspect ratio of ZnO nanorods on the performance of piezoelectric nanogenerators. *J. Sci. Adv. Mater. Devices* **2019**, *4*, 420–424. [CrossRef]
19. Riaz, M.; Song, J.; Nur, O.; Wang, Z.L.; Willander, M. Study of the Piezoelectric Power Generation of ZnO Nanowire Arrays Grown by Different Methods. *Adv. Funct. Mater.* **2010**, *21*, 628–633. [CrossRef]
20. Tao, R.; Mouis, M.; Ardila, G. Unveiling the Influence of Surface Fermi Level Pinning on the Piezoelectric Response of Semiconducting Nanowires. *Adv. Electron. Mater.* **2018**, *4*, 1700299. [CrossRef]
21. Schmidt-Mende, L.; MacManus-Driscoll, J.L. ZnO—nanostructures, defects, and devices. *Mater. Today* **2007**, *10*, 40–48. [CrossRef]
22. Nobis, T.; Kaidashev, E.M.; Rahm, A.; Lorenz, M.; Lenzner, J.; Grundmann, M. Spatially Inhomogeneous Impurity Distribution in ZnO Micropillars. *Nano Lett.* **2004**, *4*, 797–800. [CrossRef]
23. Yao, B.D.; Chan, Y.F.; Wang, N. Formation of ZnO nanostructures by a simple way of thermal evaporation. *Appl. Phys. Lett.* **2002**, *81*, 757–759. [CrossRef]
24. Wu, J.-J.; Liu, S.-C. Low-Temperature Growth of Well-Aligned ZnO Nanorods by Chemical Vapor Deposition. *Adv. Mater.* **2002**, *14*, 215–218. [CrossRef]
25. Park, W.I.; Kim, D.H.; Jung, S.-W.; Yi, G.-C. Metalorganic vapor-phase epitaxial growth of vertically well-aligned ZnO nanorods. *Appl. Phys. Lett.* **2002**, *80*, 4232–4234. [CrossRef]
26. Peulon, S.; Lincot, D. Cathodic electrodeposition from aqueous solution of dense or open-structured zinc oxide films. *Adv. Mater.* **1996**, *8*, 166–170. [CrossRef]
27. Vayssieres, L.; Keis, K.; Lindquist, S.-E.; Hagfeldt, A. Purpose-Built Anisotropic Metal Oxide Material: 3D Highly Oriented Microrod Array of ZnO. *J. Phys. Chem. B* **2001**, *105*, 3350–3352. [CrossRef]
28. Lord, A.M.; Maffeis, T.G.; Walton, A.S.; Kepaptsoglou, D.M.; Ramasse, Q.M.; Ward, M.B.; Köble, J.; Wilks, S.P. Factors that determine and limit the resistivity of high-quality individual ZnO nanowires. *Nanotechnology* **2013**, *24*, 435706. [CrossRef] [PubMed]
29. Chiu, S.-P.; Chung, H.-F.; Lin, Y.-H.; Kai, J.-J.; Chen, F.-R.; Lin, J.-J. Four-probe electrical-transport measurements on single indium tin oxide nanowires between 1.5 and 300 K. *Nanotechnology* **2009**, *20*, 105203. [CrossRef] [PubMed]
30. Hong, W.-K.; Yoon, J.; Lee, T. Hydrogen plasma-mediated modification of the electrical transport properties of ZnO nanowire field effect transistors. *Nanotechnology* **2015**, *26*, 125202. [CrossRef] [PubMed]
31. Bugallo, A.D.L.; Donatini, F.; Sartel, C.; Sallet, V.; Pernot, J. Metallic core conduction in unintentionally doped ZnO nanowire. *Appl. Phys. Express* **2015**, *8*, 25001. [CrossRef]
32. Wang, L.; Chauveau, J.-M.; Brenier, R.; Sallet, V.; Jomard, F.; Sartel, C.; Bremond, G. Access to residual carrier concentration in ZnO nanowires by calibrated scanning spreading resistance microscopy. *Appl. Phys. Lett.* **2016**, *108*, 132103. [CrossRef]
33. den Hertog, M.; Donatini, F.; McLeod, R.; Monroy, E.; Sartel, C.; Sallet, V.; Pernot, J. In situ biasing and off-axis electron holography of a ZnO nanowire. *Nanotechnology* **2018**, *29*, 025710. [CrossRef] [PubMed]
34. Baxter, J.B.; Schmuttenmaer, C.A. Conductivity of ZnO Nanowires, Nanoparticles, and Thin Films Using Time-Resolved Terahertz Spectroscopy†. *J. Phys. Chem. B* **2006**, *110*, 25229–25239. [CrossRef] [PubMed]
35. Opoku, C.; Dahiya, A.S.; Cayrel, F.; Poulin-Vittrant, G.; Alquier, D.; Camara, N. Fabrication of field-effect transistors and functional nanogenerators using hydrothermally grown ZnO nanowires. *RSC Adv.* **2015**, *5*, 69925–69931. [CrossRef]
36. Cossuet, T.; Donatini, F.; Lord, A.M.; Appert, E.; Pernot, J.; Consonni, V. Polarity-Dependent High Electrical Conductivity of ZnO Nanorods and Its Relation to Hydrogen. *J. Phys. Chem. C* **2018**, *122*, 22767–22775. [CrossRef]
37. Villafuerte, J.; Donatini, F.; Kioseoglou, J.; Sarigiannidou, E.; Chaix-Pluchery, O.; Pernot, J.; Consonni, V. Zinc Vacancy–Hydrogen Complexes as Major Defects in ZnO Nanowires Grown by Chemical Bath Deposition. *J. Phys. Chem. C* **2020**, *124*, 16652–16662. [CrossRef]
38. Könenkamp, R.; Boedecker, K.; Lux-Steiner, M.C.; Poschenrieder, M.; Zenia, F.; Levy-Clement, C.; Wagner, S. Thin film semiconductor deposition on free-standing ZnO columns. *Appl. Phys. Lett.* **2000**, *77*, 2575–2577. [CrossRef]
39. Mora-Sero, I.; Fabregat-Santiago, F.; Denier, B.; Bisquert, J.; Tena-Zaera, R.; Elias, J.; Lévy-Clément, C. Determination of carrier density of ZnO nanowires by electrochemical techniques. *Appl. Phys. Lett.* **2006**, *89*, 203117. [CrossRef]

40. Mallampati, B.; Singh, A.; Shik, A.; Ruda, H.E.; Philipose, U. Electro-physical characterization of individual and arrays of ZnO nanowires. *J. Appl. Phys.* **2015**, *118*, 34302. [CrossRef]
41. Voss, T.; Bekeny, C.; Gutowski, J.; Tena-Zaera, R.; Elias, J.; Lévy-Clément, C.; Mora-Sero, I.; Bisquert, J. Localized versus delocalized states: Photoluminescence from electrochemically synthesized ZnO nanowires. *J. Appl. Phys.* **2009**, *106*, 054304. [CrossRef]
42. Fan, J.; Shavel, A.; Zamani, R.; Fábrega, C.; Rousset, J.; Haller, S.; Güell, F.; Carrete, A.; Andreu, T.; Arbiol, J.; et al. Control of the doping concentration, morphology and optoelectronic properties of vertically aligned chlorine-doped ZnO nan-owires. *Acta Mater.* **2011**, *59*, 6790–6800. [CrossRef]
43. Wang, L.; Guillemin, S.; Chauveau, J.-M.; Sallet, V.; Jomard, F.; Brenier, R.; Consonni, V.; Bremond, G. Characterization of carrier concentration in ZnO nanowires by scanning capacitance microscopy. *Phys. Status Solidi C* **2016**, *13*, 576–580. [CrossRef]
44. Goldberger, J.; Sirbuly, D.J.; Law, M.; Yang, P.; Goldberger, J. ZnO Nanowire Transistors. *J. Phys. Chem. B* **2005**, *109*, 9–14. [CrossRef] [PubMed]
45. Chang, P.-C.; Chien, C.-J.; Stichtenoth, D.; Ronning, C.; Lu, J.G. Finite size effect in ZnO nanowires. *Appl. Phys. Lett.* **2007**, *90*, 113101. [CrossRef]
46. Song, S.; Hong, W.-K.; Kwon, S.-S.; Lee, T. Passivation effects on ZnO nanowire field effect transistors under oxygen, ambient, and vacuum environments. *Appl. Phys. Lett.* **2008**, *92*, 263109. [CrossRef]
47. Hong, W.-K.; Sohn, J.I.; Hwang, D.-K.; Kwon, S.-S.; Jo, G.; Song, S.; Kim, S.-M.; Ko, H.-J.; Park, S.-J.; Welland, M.E.; et al. Tunable Electronic Transport Characteristics of Surface-Architecture-Controlled ZnO Nanowire Field Effect Transistors. *Nano Lett.* **2008**, *8*, 950–956. [CrossRef] [PubMed]
48. Soudi, A.; Dhakal, P.; Gu, Y. Diameter dependence of the minority carrier diffusion length in individual ZnO nanowires. *Appl. Phys. Lett.* **2010**, *96*, 253115. [CrossRef]
49. Soudi, A.; Hsu, C.-H.; Gu, Y. Diameter-Dependent Surface Photovoltage and Surface State Density in Single Semiconductor Nanowires. *Nano Lett.* **2012**, *12*, 5111–5116. [CrossRef]
50. Tsai, L.-T.; Chiu, S.-P.; Lu, J.G.; Lin, J.-J. Electrical conduction mechanisms in natively doped ZnO nanowires (II). *Nanotechnology* **2010**, *21*, 145202. [CrossRef] [PubMed]
51. Donatini, F.; Bugallo, A.D.L.; Tchoulfian, P.; Chicot, G.; Sartel, C.; Sallet, V.; Pernot, J. Comparison of Three E-Beam Techniques for Electric Field Imaging and Carrier Diffusion Length Measurement on the Same Nanowires. *Nano Lett.* **2016**, *16*, 2938–2944. [CrossRef] [PubMed]
52. Zhao, Q.; Yang, L.L.; Willander, M.; Sernelius, B.E.; Holtz, P.-O. Surface recombination in ZnO nanorods grown by chemical bath deposition. *J. Appl. Phys.* **2008**, *104*, 073526. [CrossRef]
53. Serairi, L.; Yu, D.; Leprince-Wang, Y. Numerical modeling and simulation of ZnO nanowire devices for energy harvesting. *Phys. Status Solidi C* **2016**, *13*, 683–687. [CrossRef]
54. Hinchet, R.; Lee, S.; Ardila, G.; Montès, L.; Mouis, M.; Wang, Z.L. Performance Optimization of Vertical Nanowire-based Piezoelectric Nanogenerators. *Adv. Funct. Mater.* **2013**, *24*, 971–977. [CrossRef]
55. Min Kim, S.; Kim, H.; Nam, Y.; Kim, S. Effects of external surface charges on the enhanced piezoelectric potential of ZnO and AlN nanowires and nanotubes. *AIP Adv.* **2012**, *2*, 042174. [CrossRef]
56. Romano, G.; Mantini, G.; Di Carlo, A.; D'Amico, A.; Falconi, C.; Wang, Z.L. Piezoelectric potential in vertically aligned nanowires for high output nanogenerators. *Nanotechnology* **2011**, *22*, 465401. [CrossRef] [PubMed]
57. Fathi, S.; Sheikholeslami, T.F. Investigation of External Charges Effects on Piezoelectric ZnO Nanogenerator. *J. Nano Electron. Phys.* **2016**, *8*, 2047. [CrossRef]
58. Wang, C.-H.; Liao, W.; Lin, Z.; Ku, N.; Li, Y.; Chen, Y.; Wang, Z.; Liu, C. Optimization of the Output Efficiency of GaN Nanowire Piezoelectric Nanogenerators by Tuning the Free Carrier Concentration. *Adv. Energy Mater.* **2014**, *4*, 1400392. [CrossRef]
59. Ebert, P.; Schaafhausen, S.; Lenz, A.; Sabitova, A.; Ivanova, L.; Dähne, M.; Hong, Y.L.; Gwo, S.; Eisele, H. Direct measurement of the band gap and Fermi level position at InN (1120). *Appl. Phys. Lett.* **2011**, *98*, 062103. [CrossRef]
60. Sabitova, A.; Ebert, P.; Lenz, A.; Schaafhausen, S.; Ivanova, L.; Dähne, M.; Hoffmann, A.; Dunin-Borkowski, R.E.; Förster, A.; Grandidier, B.; et al. Intrinsic bandgap of cleaved ZnO (1120) surfaces. *Appl. Phys. Lett.* **2013**, *102*, 021608. [CrossRef]
61. Ivanova, L.; Borisova, S.; Eisele, H.; Dähne, M.; Laubsch, A.; Ebert, P. Surface states and origin of the Fermi level pinning on nonpolar GaN (11-00) surfaces. *Appl. Phys. Lett.* **2008**, *93*, 192110. [CrossRef]
62. Dumont, J.; Hackens, B.; Faniel, S.; Mouthuy, P.O.; Sporken, R.; Melinte, S. ZnO(0001) surfaces probed by scanning tun-neling spectroscopy: Evidence for an inhomogeneous electronic structure. *Appl. Phys. Lett.* **2009**, *95*, 132102. [CrossRef]
63. Lord, A.M.; Maffeis, T.G.; Allen, M.W.; Morgan, D.; Davies, P.R.; Jones, D.R.; Evans, J.E.; Smith, N.A.; Wilks, S.P. Surface state modulation through wet chemical treatment as a route to controlling the electrical properties of ZnO nanowire arrays investigated with XPS. *Appl. Surf. Sci.* **2014**, *320*, 664–669. [CrossRef]
64. Lausecker, C.; Salem, B.; Baillin, X.; Consonni, V. Modeling the Elongation of Nanowires Grown by Chemical Bath Depo-sition Using a Predictive Approach. *J. Phys. Chem. C* **2019**, *123*, 29476–29483. [CrossRef]
65. Consonni, V.; Sarigiannidou, E.; Appert, E.; Bocheux, A.; Guillemin, S.; Donatini, F.; Robin, I.-C.; Kioseoglou, J.; Robaut, F. Selective Area Growth of Well-Ordered ZnO Nanowire Arrays with Controllable Polarity. *ACS Nano* **2014**, *8*, 4761–4770. [CrossRef] [PubMed]
66. Perillat-Merceroz, G.; Thierry, R.; Jouneau, P.-H.; Ferret, P.; Feuillet, G. Compared growth mechanisms of Zn-polar ZnO nanowires on O-polar ZnO and on sapphire. *Nanotechnology* **2012**, *23*, 125702. [CrossRef] [PubMed]

67. Consonni, V.; Lord, A.M. Polarity in ZnO nanowires: A critical issue for piezotronic and piezoelectric devices. *Nano Energy* **2021**, *83*, 105789. [CrossRef]
68. Mantini, G.; Gao, Y.; D'Amico, A.; Falconi, C.; Wang, Z.L. Equilibrium piezoelectric potential distribution in a deformed ZnO nanowire. *Nano Res.* **2009**, *2*, 624–629. [CrossRef]
69. Araneo, R.; Lovat, G.; Burghignoli, P.; Falconi, C. Piezo-semiconductive quasi-1D nanodevices with or without an-ti-symmetry. *Adv. Mater.* **2012**, *24*, 4719–4724. [CrossRef] [PubMed]
70. Gao, Y.; Wang, Z.L. Equilibrium Potential of Free Charge Carriers in a Bent Piezoelectric Semiconductive Nanowire. *Nano Lett.* **2009**, *9*, 1103–1110. [CrossRef] [PubMed]
71. Agrawal, R.; Espinosa, H.D. Giant Piezoelectric Size Effects in Zinc Oxide and Gallium Nitride Nanowires. A First Principles Investigation. *Nano Lett.* **2011**, *11*, 786–790. [CrossRef] [PubMed]
72. Schmidt, V.; Senz, S.; Gösele, U. Influence of the Si/SiO_2 interface on the charge carrier density of Si nanowires. *Appl. Phys. A* **2006**, *86*, 187–191. [CrossRef]
73. Chia, A.C.E.; Lapierre, R.R. Analytical model of surface depletion in GaAs nanowires. *J. Appl. Phys.* **2012**, *112*, 063705. [CrossRef]
74. Xu, C.; Shin, P.; Cao, L.; Gao, D. Preferential Growth of Long ZnO Nanowire Array and Its Application in Dye-Sensitized Solar Cells. *J. Phys. Chem. C* **2009**, *114*, 125–129. [CrossRef]
75. Bui, Q.C.; Ardila, G.; Sarigiannidou, E.; Roussel, H.; Jiménez, C.; Chaix-Pluchery, O.; Guerfi, Y.; Bassani, F.; Donatini, F.; Mescot, X.; et al. Morphology Transition of ZnO from Thin Film to Nanowires on Silicon and its Correlated Enhanced Zinc Polarity Uniformity and Piezoelectric Responses. *ACS Appl. Mater. Interfaces* **2020**, *12*, 29583–29593. [CrossRef] [PubMed]
76. Montenegro, D.; Souissi, A.; Martínez-Tomás, C.; Muñoz-Sanjose, V.; Sallet, V.; Martinez-Tomas, M.C. Morphology transitions in ZnO nanorods grown by MOCVD. *J. Cryst. Growth* **2012**, *359*, 122–128. [CrossRef]
77. Willander, M.; Nur, O.; Zhao, Q.X.; Yang, L.L.; Lorenz, M.; Cao, B.Q.; Pérez, J.Z.; Czekalla, C.; Zimmermann, G.; Grundmann, M.; et al. Zinc oxide nanorod based photonic devices: Recent progress in growth, light emitting diodes and lasers. *Nanotechnology* **2009**, *20*, 332001. [CrossRef] [PubMed]
78. Parize, R.; Garnier, J.D.; Chaix-Pluchery, O.; Verrier, C.; Appert, E.; Consonni, V. Effects of Hexamethylenetetramine on the Nucleation and Radial Growth of ZnO Nanowires by Chemical Bath Deposition. *J. Phys. Chem. C* **2016**, *120*, 5242–5250. [CrossRef]
79. Govender, K.; Boyle, D.S.; Kenway, P.B.; O'Brien, P. Understanding the factors that govern the deposition and morphology of thin films of ZnO from aqueous solution. *J. Mater. Chem.* **2004**, *14*, 2575–2591. [CrossRef]
80. Lord, A.M.; Consonni, V.; Cossuet, T.; Donatini, F.; Wilks, S.P. Schottky Contacts on Polarity-Controlled Vertical ZnO Nanorods. *ACS Appl. Mater. Interfaces* **2020**, *12*, 13217–13228. [CrossRef] [PubMed]
81. Özgür, Ü.; Alivov, Y.I.; Liu, C.; Teke, A.; Reshchikov, M.A.; Doğan, S.; Avrutin, V.; Cho, S.J.; Morkoç, H. A comprehensive review of ZnO materials and devices. *J. Appl. Phys.* **2005**, *98*, 041301. [CrossRef]
82. McCluskey, M.D.; Jokela, S.J. Defects in ZnO. *J. Appl. Phys.* **2009**, *106*, 071101. [CrossRef]
83. Joo, J.; Chow, B.Y.; Prakash, M.; Boyden, E.S.; Jacobson, J.M. Face-selective electrostatic control of hydrothermal zinc oxide nanowire synthesis. *Nat. Mater.* **2011**, *10*, 596–601. [CrossRef] [PubMed]
84. Verrier, C.; Appert, E.; Chaix-Pluchery, O.; Rapenne, L.; Rafhay, Q.; Kaminski-Cachopo, A.; Consonni, V. Effects of the pH on the Formation and Doping Mechanisms of ZnO Nanowires Using Aluminum Nitrate and Ammonia. *Inorg. Chem.* **2017**, *56*, 13111–13122. [CrossRef] [PubMed]
85. Gaffuri, P.; Appert, E.; Chaix-Pluchery, O.; Rapenne, L.; Salaün, M.; Consonni, V. The Path of Gallium from Chemical Bath into ZnO Nanowires: Mechanisms of Formation and Incorporation. *Inorg. Chem.* **2019**, *58*, 10269–10279. [CrossRef] [PubMed]
86. McNeill, A.R.; Bell, K.J.; Hyndman, A.R.; Gazoni, R.M.; Reeves, R.J.; Downard, A.J.; Allen, M.W. Synchrotron X-ray Photoelectron Spectroscopy Study of Electronic Changes at the ZnO Surface Following Aryldiazonium Ion Grafting: A Metal-to-Insulator Transition. *J. Phys. Chem. C* **2018**, *122*, 12681–12693. [CrossRef]
87. McNeill, A.R.; Hyndman, A.R.; Reeves, R.J.; Downard, A.J.; Allen, M.W. Tuning the Band Bending and Controlling the Surface Reactivity at Polar and Nonpolar Surfaces of ZnO through Phosphonic Acid Binding. *ACS Appl. Mater. Interfaces* **2016**, *8*, 31392–31402. [CrossRef] [PubMed]

nanomaterials

Article

Environment-Friendly Zinc Oxide Nanorods-Grown Cellulose Nanofiber Nanocomposite and Its Electromechanical and UV Sensing Behaviors

Lindong Zhai [1], Hyun-Chan Kim [1], Ruth M. Muthoka [1], Muhammad Latif [1], Hussein Alrobei [2], Rizwan A. Malik [3] and Jaehwan Kim [1,*]

1 CRC for Nanocellulose Future Composites, Inha University, Incheon 22212, Korea; duicaofei@naver.com (L.Z.); kim_hyunchan@naver.com (H.-C.K.); mwongelinruth@gmail.com (R.M.M.); mlatif8482@gmail.com (M.L.)
2 Department of Mechanical Engineering, Prince Sattam bin Abdul Aziz University, AlKharj 11942, Saudi Arabia; h.alrobei@psau.edu.sa
3 Department of Metallurgy and Materials Engineering, University of Engineering and Technology, Taxila 47050, Pakistan; rizwanmalik48@yahoo.com
* Correspondence: jaehwan@inha.ac.kr; Tel.: +82-32-874-7325

Citation: Zhai, L.; Kim, H.-C.; Muthoka, R.M.; Latif, M.; Alrobei, H.; Malik, R.A.; Kim, J. Environment-Friendly Zinc Oxide Nanorods-Grown Cellulose Nanofiber Nanocomposite and Its Electromechanical and UV Sensing Behaviors. *Nanomaterials* **2021**, *11*, 1419. https://doi.org/10.3390/nano11061419

Academic Editor: Vincent Consonni
Received: 25 April 2021
Accepted: 26 May 2021
Published: 27 May 2021

Abstract: This paper reports a genuine environment-friendly hybrid nanocomposite made by growing zinc oxide (ZnO) nanorods on cellulose nanofiber (CNF) film. The nanocomposite preparation, characterizations, electromechanical property, and ultraviolet (UV) sensing performance are explained. CNF was extracted from the pulp by combining the 2,2,6,6-tetramethylpiperidine-1-oxyl radical (TEMPO) oxidation and the aqueous counter collision (ACC) methods. The CNF film was fabricated using doctor blade casting, and ZnO nanorods were grown on the CNF film by seeding and by a hydrothermal method. Morphologies, optical transparency, mechanical and electromechanical properties, and UV sensing properties were examined. The nanocomposite's optical transparency was more than 80%, and the piezoelectric charge constant d_{31} was 200 times larger than the CNF film. The UV sensing performance of the prepared ZnO-CNF nanocomposites was tested in terms of ZnO concentration, UV irradiance intensity, exposure side, and electrode materials. A large aspect ratio of ZnO nanorods and a work function gap between ZnO nanorods and the electrode material are essential for improving the UV sensing performance. However, these conditions should be compromised with transparency. The use of CNF for ZnO-cellulose hybrid nanocomposite is beneficial not only for electromechanical and UV sensing properties but also for high mechanical properties, renewability, biocompatibility, flexibility, non-toxicity, and transparency.

Keywords: cellulose nanofiber; zinc oxide; nanocomposite; electromechanical property; UV sensing

1. Introduction

Organic and inorganic functional nanocomposites combine advantages of the individual materials that surpass parental material properties. They can achieve high mechanical strength, electrical conductivity, thermal conductivity, antibacterial, gas barrier, flame retardancy, electromagnetic shielding, optical transparency, energy harvesting, and actuating properties [1]. Cellulose nanofiber (CNF) is an outstanding organic material composed of nano-sized cellulose fibrils with a high aspect ratio [2–7]. The width of CNF is typically in the range of 5–20 nm, and its length is typically up to several micrometers. Cellulose molecules can form microfibrils during the biosynthesis process by forming inter- and intra-molecular hydrogen bonds [4]. CNF extracted from plants by the top-down approach has excellent properties: not only renewability, biodegradability, abundance, low price, and light weight but also high optical transparency, outstanding mechanical properties, and low thermal expansion coefficient [6]. Thus, CNF can be a building block of future materials applied for structural composites, coatings, cosmetics, 3D printing, sensors, soft actuators, flexible electronics, energy devices, and flexible displays [8–13].

Inorganic nanomaterials can improve the functional properties of organic-inorganic functional nanocomposites. The inorganic nanomaterials can be either blended or coated with/on the polymer nanocomposites [14,15]. Owing to its benefits in terms of wide bandgap (3.37 eV), high exciton binding energy (60 meV), ultraviolet (UV) response, optical transparency, highly electrical conductivity, and piezoelectricity, zinc oxide (ZnO) has been widely studied [16,17]. ZnO is broadly used in electronic devices, hybrid diodes, energy harvesters, piezoelectric devices, field-effect transistors, gas sensors, photovoltaics, and UV sensors [18–25]. It can be quickly grown on various substrates, including metal, glass, silicon, sapphire, plastics, polymers, and cellulose [20,26–32].

With increasing interest in wearable devices, the ZnO-cellulose hybrid composite is attractive because it can bridge ZnO's functional properties and renewability and flexibility of cellulose. Recently ZnO nanorods grown on paper and cellulose have been reported [33–38]. A flexible and transparent cellulose-ZnO hybrid nanocomposite was prepared by direct ZnO seeding and hydrothermal ZnO nanorod growth on a regenerated cellulose film [35,36]. A zinc oxide nanolayer was uniformly formed on a regenerated cellulose film using a solution-based hydrothermal process, which shows a drastic improvement of its electromechanical behavior [37]. However, they used regenerated cellulose, which was prepared using special solvents, such as LiCl/DMAc [39]. It is essential to use CNF film instead of regenerated cellulose in cellulose-ZnO hybrid nanocomposites because the particular solvent usage can be eliminated.

Thus, this research aimed to prepare a genuine environment-friendly ZnO-CNF nanocomposite (ZCN) by growing ZnO nanorods on a CNF film according to the hydrothermal method. By using CNF instead of regenerated cellulose, the genuine environment-friendly ZnO-cellulose nanocomposite was prepared. Since CNF possesses numerous hydroxyl groups on its surface, ZnO can be efficiently anchored on the CNF film surface. This paper illustrates the CNF isolation, CNF film preparation, and the ZnO nanorods growth on the CNF film. The ZnO nanorods were grown by adopting the seeding process, followed by the hygrothermal process [37]. The morphologies, optical transparency, mechanical, electromechanical, and UV sensing properties of the prepared ZCN were investigated.

2. Materials and Methods

2.1. Materials

Hardwood (HW) pulp was received from Hansol Paper and Pulp Co. (Jeonju, Korea). HW bleached kraft pulp in dried pad form is a combination of Aspen and Poplar, and its alpha-cellulose (α) content is 85.7%, and viscosity is 14.6 cPs. 2,2,6,6-tetramethylpiperidine-1-oxyl radical (TEMPO, 98%), sodium bromide (NaBr, 99%), sodium hypochlorite (NaClO, 15%), and hydrochloric acid (HCl, 37%) were purchased from Sigma-Aldrich St. Louis, MO, USA, and Sodium hydroxide (NaOH, 98%) was purchased from Daejung Chemical, Busan, Korea. They were used to oxidize HW pulp further to extract CNF. Zinc acetate dihydrate ($Zn(CH_3COO)_2{\cdot}2H_2O$, reagent grade 98%) was purchased from Sigma-Aldrich, and ethyl alcohol anhydrous (C_2H_5OH, purity 99.5%) was purchased from Daejung Chemical. Zinc nitrate hexahydrate ($Zn(NO_3)_2{\cdot}6H_2O$, reagent grade 98%) and hexamethylenetetramine (HMT, $(CH_2)_6N_4$, reagent grade 99%) were purchased from Sigma-Aldrich. All other chemicals used were analytical-reagent-grade (Purity > 99%) and used as received.

2.2. CNF Extraction and CNF Film Fabrication

The CNF was extracted by the TEMPO oxidation and aqueous counter collision system (ACC, ACCNAC-100, CNNT, Suwon, Korea) combined method, which has been reported previously [40]. The TEMPO oxidation acts as a pretreatment process for the ACC process. In brief, the HW pulp was cut into small pieces and swelled in deionized (DI) water for one day before disintegrating by a food mixer for 10 min. After that, TEMPO, NaBr, and DI water were added to the swollen HW pulp suspension. To start the oxidization process, the

NaClO was added into the mixture and stirred at room temperature. A pH meter (Orion Star A211, Thermo Scientific, Waltham, MA, USA) was used to monitor the mixture's pH value, maintaining it at 12 by adding NaOH. To stop the TEMPO oxidation reaction, 0.5 M HCl was added into the mixture and adjusted the pH value to 7. Finally, the oxidized cellulose pulp was washed with DI water to remove the chemical residues using a 90 μm mesh sieve. The TEMPO-oxidized pulp was further pulverized using the ACC system. The pulp was passed through a pair of diamond nozzles in a collision chamber with 200 MPa, such that a pair of aqueous solution jets collide against each other, resulting in an aqueous CNF suspension. The number of ejections passing through the nozzles was called "pass" As the number of passes increases, the CNF size decreases. TEMPO oxidation 60 min and ACC 30-pass-treated CNF was an optimum condition from the previous research [40], and the same condition was used in this research. The CNF suspension was degassed using a centrifuge machine (Supra 22K, Hanil Scientific Inc., Incheon, Korea) with 5000 rpm, 1 h.

The CNF film was prepared using a doctor blade casting, as shown in Figure 1. A polycarbonate (PC) substrate was used for the CNF film casting. The PC substrate, where its boundaries were covered with the polyimide tape, was treated with an oxygen plasma using an oxygen-plasma treatment system (FEMTO Science, CUTE, Hwaseong-si, Korea) for 20 s to slightly increase its hydrophilicity of the PC substrate. The CNF suspension was cast on the plasma-treated PC plate using a doctor blade and dried in a cleanroom. After drying, the pristine CNF film was immersed in an ethanol bath for separating the pristine CNF film from the PC substrate. After evaporating ethanol, the pristine CNF film was peeled off from the PC substrate.

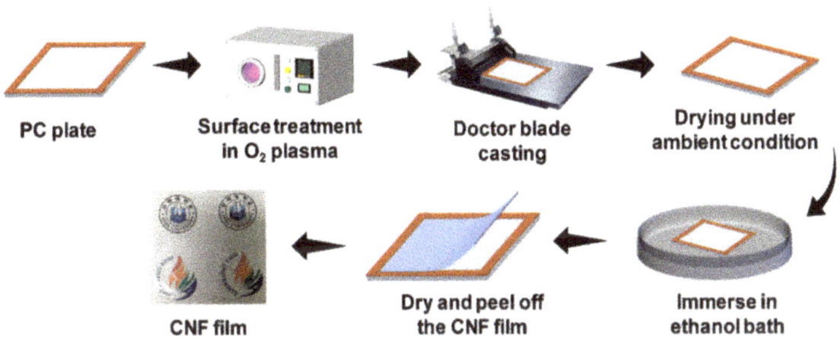

Figure 1. Schematic of CNF film fabrication process.

2.3. ZnO-CNF Nanocomposite Fabrication

The previously reported two-step process was adopted for growing ZnO nanorods on the CNF film: ZnO nanoparticles seeding and the ZnO nanorod growing [35]. In brief, the CNF film was fixed on a silicon wafer to maintain its flatness. Zinc acetate was dissolved in ethanol by stirring under 60 °C for 1 h, and the solution temperature was maintained at 60 °C to prevent ZnO crystal extraction. The solution was then spin-coated on the CNF film, followed by drying at 100 °C. This process was repeated 10 times with 3 min duration to form a dense ZnO seeding layer on top of the CNF film. After that, the ZnO-seeded CNF film was annealed at 100 °C for 30 min. Figure 2a shows the ZnO seeding process.

For growing ZnO nanorods, zinc nitrate aqueous solution and HMT aqueous solution were mixed. After the ZnO-seeded CNF film was floated on the surface of zinc nitrate aqueous solution, the solution was heated up to 90 °C for 1 h. As the nanorods grew, the CNF film color was changed from transparent to hazy then perfectly white. After growing ZnO nanorods, the nanocomposite was dried at room temperature. By controlling the chemical concentration, 25 mM and 50 mM of ZnO growing solutions were used to fabricate two ZCNs: 25 mM ZCN and 50mM ZCN. Figure 2b shows the schematic of the ZCN fabrication process.

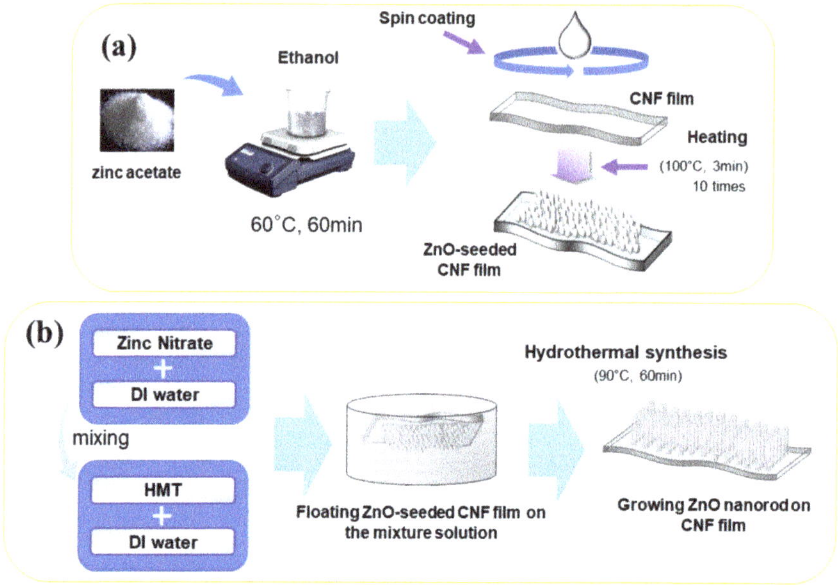

Figure 2. Schematic of ZnO-CNF nanocomposite fabrication process: (**a**) seeding process and (**b**) nanorod growing process.

2.4. Characterizations

A UV-2501PC UV-Vis spectrometer (Shimadzu, Kyoto, Japan) was used to analyze the transparency of the pristine CNF film and ZCNs. Field-emission scanning electron microscopy (FESEM) (S-4000, Hitachi, Matsuda, Japan) was used to investigate the morphologies of the pristine CNF film and ZCNs.

The mechanical and electromechanical properties were measured using a tensile test system, which consists of a load cell (UU-K010, Dacell, Nami-Myeon, Korea), a servo motor controller, a picoammeter (Keithley 6485, Tektronix, Beaverton, OR, USA), an environmental chamber, and LabVIEW software (National Instruments, Austin, TX, USA)) with a personal computer [41]. Aluminum electrodes were deposited on both sides of specimens by a thermal evaporation system (SHE-6D-350T, Samhan, Paju, Korea). Thin copper wires were connected to the electrodes and grounded to release space charges accumulated on the electrodes during the deposition. The electromechanical property was found by measuring the piezoelectric charge constant, d_{31}, in the tensile test system. The specimens were subjected to a tensile load in its length direction, and the induced current was measured across the thickness direction from the electrodes using the picoammeter. The induced current can be converted into a charge per unit electrode area, and d_{31} can be obtained as:

$$d_{31} = \left(\frac{\partial D_3}{\partial T_1}\right)_E = \frac{\text{Induced charge per unit electrod area}}{\text{Applied in} - \text{plane normal stress}} \; [\text{C/N}] \tag{1}$$

2.5. UV Sensing Test

Figure 3a shows the schematic of UV sensing from ZCN. The ZCN was cut into 1.7×1.7 cm^2, and then platinum or indium tin oxide (ITO) was coated with 1.5×1.5 cm^2 size using a sputtering system (K575X, Quorum Technologies Ltd., Lewes, UK) to form electrodes on both side of the specimens. Thin copper wires were attached to both electrodes using a conductive silver paste and then coated with a thin laminate film to protect the ZnO layer. Before proceeding with the UV sensing test, the specimen was attached on a linear stage such that the distance between the UV light source and the specimen can be adjusted (see Figure 3b). A UV light lamp (PL-S 9W/2P BLB, Philips, Eindhoven, The Netherlands) of 365 nm wavelength was used as the light source. The UV irradiance intensity was measured using a UV meter (UV-A Meter, Kuhnast, Wächtersbach, Germany), which has the highest sensitivity at 360 nm wavelength. UV irradiance intensities of 1, 1.5, 3, and 5 mW/cm^2 were applied to the specimens. For evaluating the UV sensing performance, the induced current was measured using the picoammeter. The UV irradiance was exposed to the cellulose side and ZnO side, and the exposure side effect was also investigated.

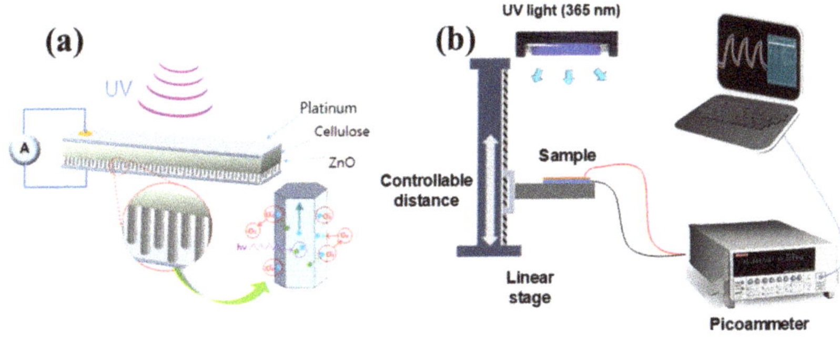

Figure 3. (**a**) Schematic of UV sensing ZCN and (**b**) UV-sensing test setup.

3. Results

3.1. Optical Transparencies and Morphologies

Figure 4 shows the transparency of the prepared ZCNs and the pristine CNF film. The pristine CNF film showed the highest transparency of 89.2% in the visible light range, and the ZCNs exhibited high transparencies of 82.9% and 80.7% for the 25 mM and 50 mM ZnO concentrations. Within the UV wavelength range up to 370 nm, the ZCNs mostly block UV light, indicating that the ZnO nanorods adsorb most UV lights.

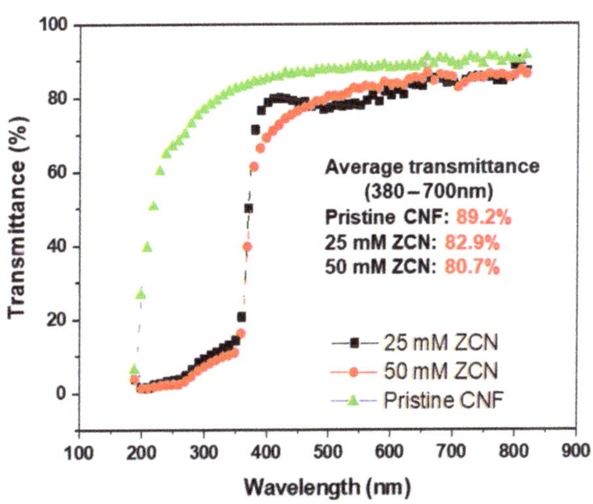

Figure 4. Optical transparencies of the pristine CNF film and ZnO-CNF nanocomposites.

Figure 5 shows the surface and cross-sectional morphologies of the pristine CNF film and ZCNs. The pristine CNF film surface is very smooth, and its cross-section exhibits a layered structure similar to the regenerated cellulose [42]. The ZnO nanorods were successfully grown vertically from the surface of CNF film, and their lengths of the 25 mM and 50 mM specimens are 500 nm and 800 nm. The vertically grown ZnO nanorods are beneficial for improving optical transparency and electromechanical properties. The ZnO nanorod morphologies are similar to the previous report [35]. The ZnO nanorod diameter was similar to the previous result, 110 nm [35]. Lower optical transparency of the 50 mM ZCN was attributed to the longer ZnO nanorods than the 25 mM ZCN.

Figure 5. Surface (**a–c**) and cross-sectional (**d–f**) FESEM images: (**a**,**d**) the pristine CNF film, (**b**,**e**) 25 mM ZCN, (**c**,**f**) 50 mM ZCN.

3.2. Mechanical and Electromechanical Properties

The mechanical and electromechanical properties of the pristine CNF and ZCNs were measured. Figure 6a shows the stress-strain curves of the specimens. The yielding occurred around 0.8–1.2% strain, and by fitting the slopes of the stress-strain curves within the elastic region, Young's moduli were found. The results are listed in Table 1. The tensile strength (St) of the pristine CNF film was 131.4 MPa, which is a bit larger than the regenerated cellulose film (120 MPa, no-stretching) [42]. The Young's modulus of the pristine CNF film was 12.8 GPa, much larger than the regenerated cellulose film (5.3 GPa, no-stretching). The CNF pristine film's mechanical properties increase might be associated with the CNF alignment in the pristine CNF film. Thus, it is beneficial to use the pristine CNF film instead of the regenerated cellulose film for the ZCN.

After ZnO nanorod growing, the tensile strength of ZCNs ranged 115–116 MPa, and Young's modulus was in the range of 9.6–9.7 GPa, a bit less than the pristine CNF film. The decreased mechanical properties might be due to the material mismatch between the CNF film and the grown ZnO nanorods. Note that the thickness of the ZnO nanorod layers was 0.5–0.8 μm, and the CNF film thickness was 20 μm. The grown ZnO nanorods cannot resist under the tensile load because they were grown perpendicular to the CNF film. Thus, although the ZnO layer thickness was smaller than the CNF film, the overall mechanical properties of ZCN were decreased.

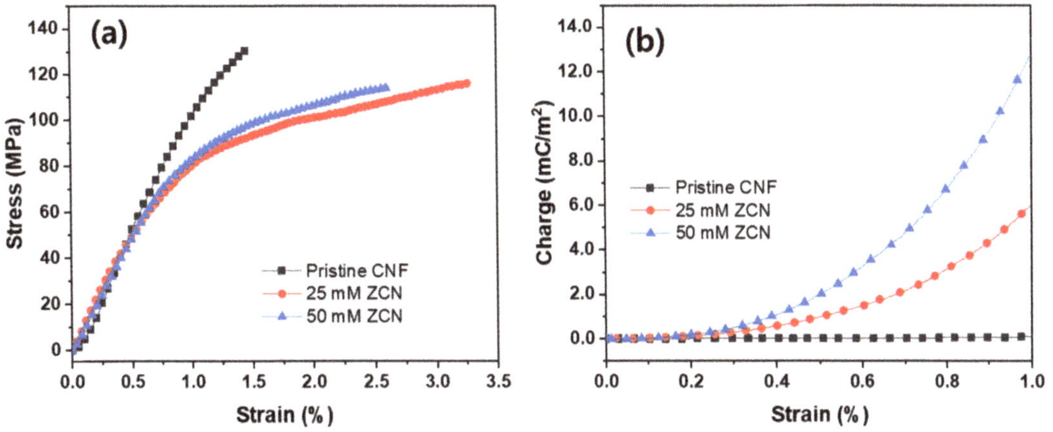

Figure 6. (**a**) Stress-strain curves and (**b**) induced charge curves of the pristine CNF film and ZCNs.

Table 1. Mechanical and electromechanical properties of the pristine CNF and ZCNs.

Sample	E (GPa)	St (MPa)	d_{31} (pC/N)
Pristine CNF film	12.8 ± 0.1	131.4 ± 3.7	0.22 ± 0.1
25 mM ZCN	9.6 ± 1.4	116.0 ± 12	26 ± 5.6
50 mM ZCN	9.7 ± 0.3	115.0 ± 13.7	48.8 ± 11.7
Regenerated cellulose EAPap [36]	5.3	120	3.4

Figure 6b shows the charge-strain curves of the pristine CNF film and ZCNs. Strain-induced charges increased with increasing the ZnO concentration within the elastic region. The piezoelectric charge constant, d_{31}, of the pristine CNF film was 0.22 pC/N. This value is lower than the regenerated-cellulose electro-active paper (EAPap) without stretching, 3.4 pC/N [42]. Note that the d_{31} of the regenerated cellulose EAPap drastically increased up to 16–17 pC/N as the stretching ratio increased to 2.0 [42,43]. The d_{31} values of the 25 mM and 50 mM ZCNs were 26 and 48.8 pC/N. The maximum d_{31} of the 50 mM ZCN is 14 times larger than the regenerated cellulose EAPap, and 200 times larger than the pristine CNF film. The remarkable electromechanical property is attributed to the dipolar orientation of ZnO nanorods and surface piezoelectricity between the CNF film and the ZnO nanorods.

3.3. UV Sensing Test

The prepared ZCNs were UV-sensing-tested. The ZnO concentration effect on the UV sensing was investigated. Figure 7 shows the induced current from the ZCNs depending on the ZnO concentration when the UV light was on and off under 5 mW/cm^2 UV irradiance. As the UV light was on, the current output from ZCNs sharply increased and saturated within 20–30 s. The 50 mM ZCN, however, gradually decreased after the saturation. After the UV light was off, the current output of ZCNs immediately decreased and returned to the nearly initial level after 120 s. Note that the current outputs are repeatable. The 50 mM ZCN exhibited almost four times higher current output than the 25 mM ZCN, which might be associated with the ZnO nanorods' length. Since bigger ZnO nanorods in the 50 mM ZCN (800 nm) than the 25 mM ZCN (500 nm) give a larger surface area, more oxygen molecules can be attracted on the ZnO nanorods, such that larger current output can be obtained from the 50 mM ZCN than the 25 mM ZCN.

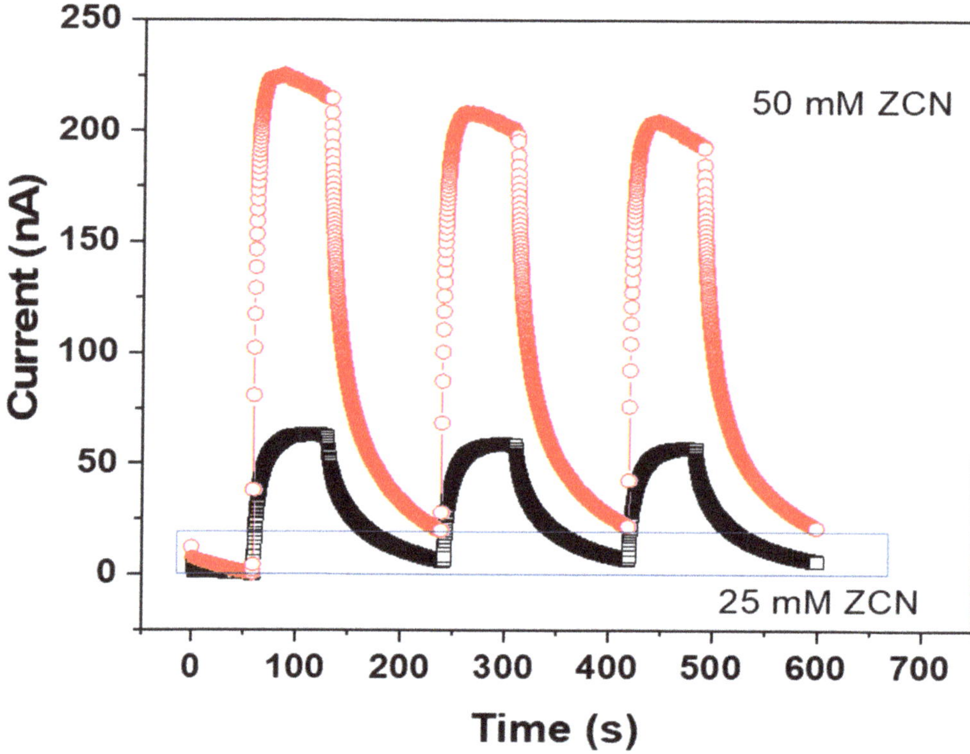

Figure 7. Induced current outputs of ZCNs with different ZnO concentrations.

Figure 8a,b show the induced current outputs of ZCNs under the UV irradiance change. By controlling the distance between the UV light source and the specimens, the UV irradiance intensity was measured using the UV-A meter, and the intensity levels were 1, 1.5, 3, and 5 mW/cm^2. Both 25 mM and 50 mM ZCNs were tested under the same condition. As the UV irradiance turned on and off with increasing the UV irradiance intensity, the current outputs followed the intensity signals and the maximum current outputs increased. Figure 8c shows the maximum current outputs of ZCNs with the UV irradiance change. The 25 mM ZCN exhibited a linear response, whereas the 50 mM ZCN represented a nonlinear response, which might be due to its large current output.

Since the ZCN has two different sides, the UV exposure side effect was investigated by irradiating UV light onto the ZnO nanorods side and cellulose side. Figure 9 represents the induced current outputs of ZCNs by exposing UV irradiance to ZnO nanorods side and cellulose side for the 25 mM ZCN (a) and 50 mM ZCN (b). The test was performed under 3 mW/cm^2 UV irradiance intensity. In the 25 mM ZCN, the current output from the cellulose side was slightly larger than the ZnO nanorods side. However, the 50 mM ZCN was different: the current output from the cellulose side was almost twice higher than the ZnO nanorods side. The CNF film is transparent, and the interface boundary between the CNF film and the ZnO seed layer is smoother than the ZnO nanorods side. The cellulose side scattered the UV irradiance less than the ZnO nanorods side, such that more current output occurred from the cellulose-side exposure [35].

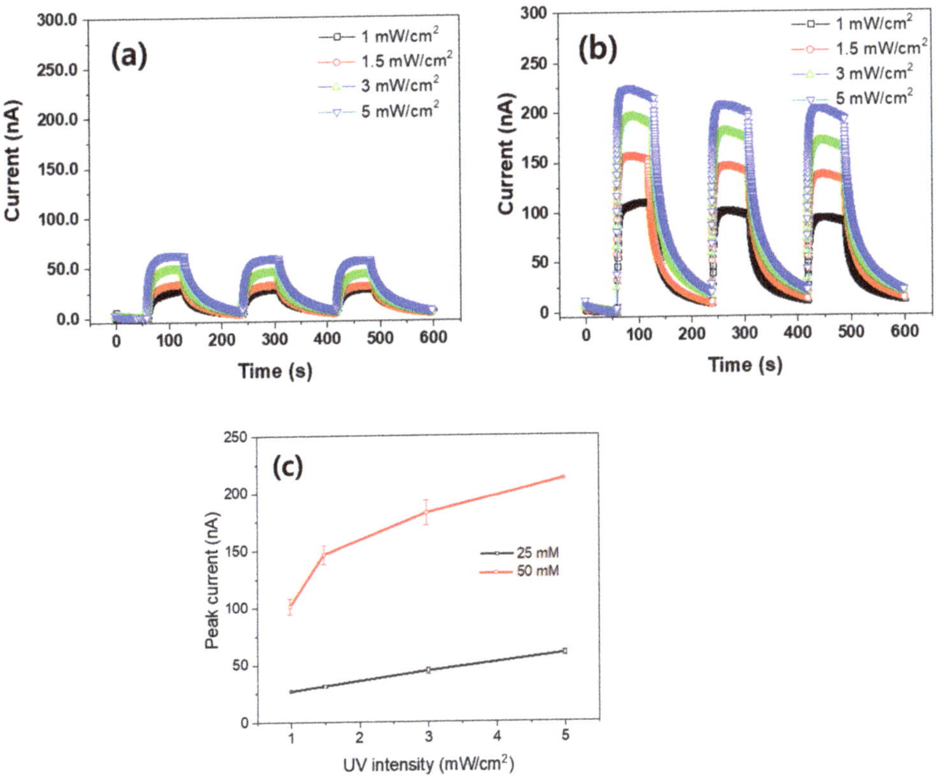

Figure 8. Induced current outputs of ZCNs: (**a**) 25 mM ZCN, (**b**) 50 mM ZCN, and (**c**) the UV irradiance intensity effect.

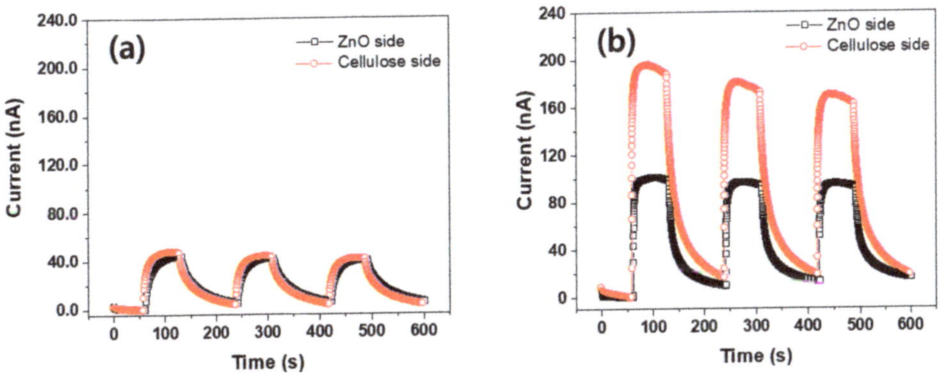

Figure 9. Current outputs of ZCNs with different UV irradiate sides: (**a**) 25 mM ZCN and (**b**) 50 mM ZCN.

So far, Pt was coated on both sides of ZCNs to form electrodes. To investigate the electrode material effect on the UV sensing performance, a transparent electrode material, ITO, was coated on both sides of the ZCNs and compared with the Pt electrode. Figure 10 shows the current output of the 25 mM ZCN (a) and the 50 mM ZCN (b) for both electrode materials. Inset photographs show the images of the Pt- and ITO-coated ZCNs. Note that ITO is known to be transparent at 365 nm in contrast to Pt. Thus the ITO-coated

ZCN is beneficial for achieving a transparent UV sensor. The Pt electrode ones showed larger current output than the ITO electrode cases in 25 mM and 50 mM ZCNs. The Pt-electrode-induced current was 1.8 times larger than the ITO electrode, and the 50 mM ZCN was 3 times larger. This phenomenon might be associated with the work function difference between the electrode and ZnO nanorods: ZnO work function is 4.3 eV, Pt is 5.7 eV, and ITO is around 4.3 eV [44]. The large work function gap between the ZnO and the electrode induced high current output under the same UV irradiance intensity. Since the work function of ITO electrode is almost the same as the ZnO, the ITO electrode did not show large current outputs. The Pt electrode on the ZnO nanorods could correspond to Schottky contact, which results in a large forward current and very low reverse bias. Additionally, the conductivity of the ITO electrode is lower than the Pt electrode. Note that the ITO electrode's current outputs dropped quickly after turning on the UV light. It might be because the oxygen atoms incorporated into the ITO electrode decrease the oxygen vacancies and then give rise to the ITO electrode's high resistivity, which resulted in fast current decay.

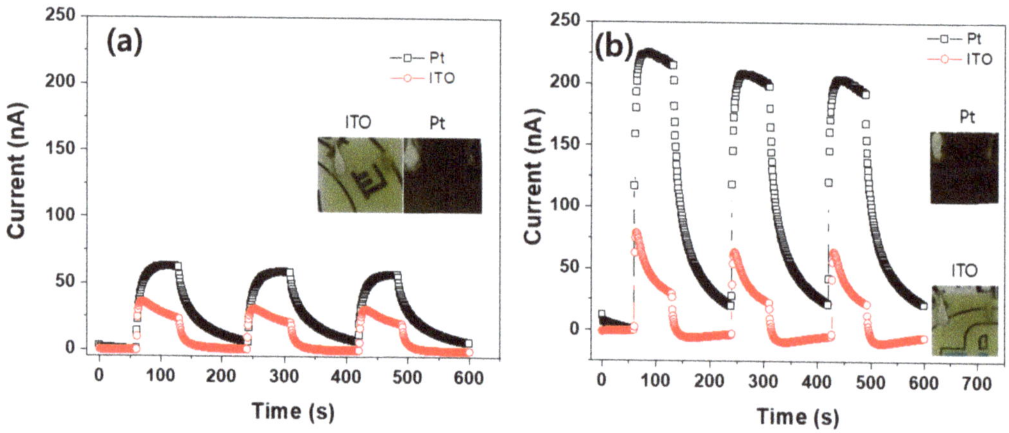

Figure 10. Platinum and ITO electrodes' effect on the ZCN: (**a**) 25 mM ZCN and (**b**) 50 mM ZCN.

The UV sensing performance of ZCN was compared with the previously reported cellulose-ZnO hybrid nanocomposite (CEZOHN) [35]. CEZOHN is a ZnO-nanorod hybrid nanocomposite grown on a regenerated cellulose film. Figure 11a shows the peak current outputs of the ZCNs compared with the CEZOHN as the UV irradiation intensity changes. The current outputs were normalized with the electrode areas, and the ZCNs were recapped from Figure 8c. The peak current outputs increased with the UV irradiation intensity, and the CEZOHN exhibited higher UV sensing performance than the ZCNs. The high performance of UV sensing was investigated with the ZnO nanorod length. The ZCNs used the same ZnO seeding and growing process as the CEZOHN. Note that the CEZOHN used a 50 mM concentration of zinc nitrate aqueous solution for growing the ZnO nanorods. Figure 11b represents the unit current output of ZCNs and CEZOHN with the length of the nanorods. The unit current output was calculated from the maximum peak current outputs divided by the UV irradiation intensity, which turned out to be nA/mW. The ZnO nanorods' length of CEZOHN was 1 μm, and the ZC was 500 nm and 800 nm. Table 2 represents the photosensitivities of the prepared ZCNs comparing with CEZOHN. The peak current increased from 59.69 ± 2.57 to 212.53 ± 9.09 nA for the 25 mM and 50 mM ZCNs, and their unit current output increased from 11.94 ± 0.51 to 42.51 ± 1.82 nA/mW/cm^2. It is clearly shown that the current output is strongly related to the ZnO nanorods' length. Thus, a large aspect ratio of ZnO nanorods is crucial for improving the UV sensing performance. A significant work function difference between ZnO nanorods and the electrode material should be made for performance improvement. However, the large aspect ratio of ZnO nanorods and considerable work function electrode material could deteriorate the ZCN's transparency. Thus, they should be compromised with the transparency of ZCN. Nevertheless, the use of CNF for ZCN is beneficial for high mechanical properties, renewability, biocompatibility, flexibility, non-toxicity, and transparency.

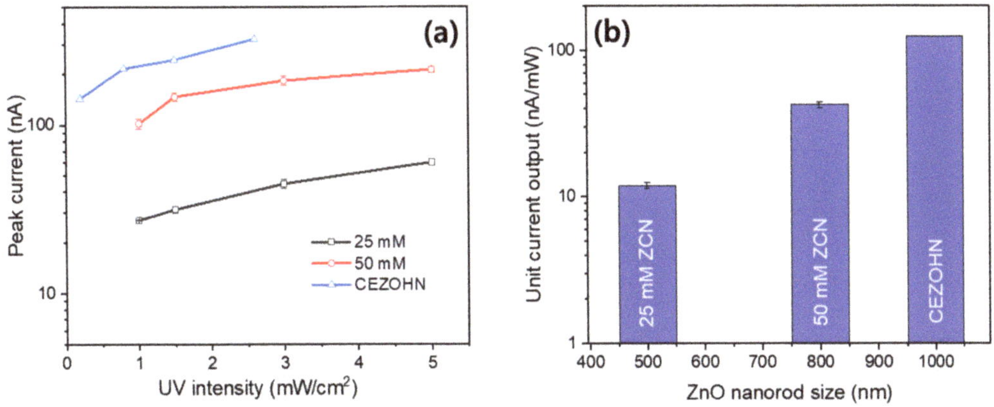

Figure 11. Current output comparison of ZCNs and CEZOHN: (**a**) peak current outputs with UV intensity changing and (**b**) the maximum unit current output with the ZnO nanorods' length.

Table 2. Photosensitivities of the ZCNs compared with CEZOHN.

Sample	ZCNs		CEZOHN
	25 mM ZCN	50 mM ZCN	
Peak current (nA/cm²)	59.69 ± 2.57	212.53 ± 9.09	142.5
Unit current output (nA/mW/cm²)	11.94 ± 0.51	42.51 ± 1.82	124.0
ZnO nanorod size (nm)	500	800	1000
Response time (s)	70	35	25

4. Conclusions

In this study, the environment-friendly ZnO-CNF nanocomposites were fabricated by growing ZnO nanorods on the CNF film using two-step methods: seeding and nanorod growing by the hydrothermal method. The CNF extraction, the film casting, the ZnO nanorod growing, and all characterization were performed. The tensile strength and Young's modulus of the CNF film (120 MPa, 12.8 GPa) were higher than the regenerated cellulose film (120 MPa, 5.3 GPa). Thus, it is beneficial to use the CNF film instead of the regenerated cellulose film to form the nanocomposites. Transparency of the ZnO-CNF nanocomposites slightly decreased from the pristine CNF film but still maintained more than 80%. The SEM images exhibited that the ZnO nanorods were grown uniformly on the CNF film surface, and the length of the nanorods was 500 nm and 800 nm for 25 mM and 50 mM ZnO concentration nanocomposites. After growing ZnO nanorods on the CNF film, the electromechanical property was increased more than 200 times compared to the pristine CNF film. As the ZnO concentration increased, the electromechanical property increased.

The UV sensing performance of the prepared ZnO-CNF nanocomposites was tested in terms of ZnO concentration, UV irradiance intensity, exposure side, and electrode materials. When increasing the ZnO concentration and UV irradiance intensity, the induced current output also increased. When the cellulose side was exposed to the UV irradiance, the induced current was higher than the ZnO nanorods-side exposure. Due to the transparency and smooth surface of the CNF film, it scattered UV irradiance less than the ZnO nanorods side. Regarding the electrode materials, the Pt electrode generated larger current output than the ITO electrode, which is associated with the work function difference between the electrode material and the ZnO nanorods.

The UV sensing performance of the ZnO-CNF nanocomposite and the previously reported regenerated cellulose-based ZnO hybrid nanocomposite was compared. The current output from the regenerated cellulose-based one was higher than the ZnO-CNF nanocomposites, attributed to the ZnO nanorods' length. When increasing the length of the nanorods, the current output also increased. The high aspect ratio of ZnO nanorods in conjunction with the work function gap between ZnO nanorods and the electrode material are essential to improve the UV sensing performance, although they should be compromised with the transparency. The use of CNF for ZnO-cellulose hybrid nanocomposites is beneficial not only for electromechanical and UV sensing properties but also for high mechanical properties, renewability, biocompatibility, flexibility, non-toxicity, and transparency.

Author Contributions: Conceptualization, L.Z. and J.K.; Data curation and experiments, R.M.M. and H.-C.K.; Writing—original draft preparation, L.Z.; Visualization, R.M.M. and M.L.; Writing—review and editing, H.A., R.A.M. and J.K.; Supervision, J.K. All authors have read and agreed to the published version of the manuscript.

Funding: This study was supported by the Deanship of Scientific Research at Prince Sattam Bin Abdulaziz University under research project no. 2020/1/17063.

Institutional Review Board Statement: Not applicable.

Informed Consent Statement: Not applicable.

Data Availability Statement: The data presented in this study are available on request from the corresponding author.

Conflicts of Interest: The authors declare no conflict of interest.

References

1. Tressler, J.F.; Alkoy, S.; Dogan, A.; Newnham, R.E. Functional composites for sensors, actuators and transducers. *Compos. Part. A Appl. Sci. Manuf.* **1999**, *30*, 477–482. [CrossRef]
2. Marchessault, R.H.; Liang, C.Y. Infrared spectra of crystalline polysaccharides. III. Mercerized cellulose. *J. Polym. Sci.* **1960**, *43*, 71–84. [CrossRef]
3. Moon, R.J.; Martini, A.; Nairn, J.; Simonsen, J.; Youngblood, J. Cellulose nanomaterials review: Structure, properties and nanocomposites. *Chem. Soc. Rev.* **2011**, *40*, 3941–3994. [CrossRef]
4. Klemm, D.; Kramer, F.; Moritz, S.; Lindström, T.; Ankerfors, M.; Gray, D.; Dorris, A. Nanocelluloses: A new family of nature-based materials. *Angew. Chem. Int. Ed.* **2011**, *50*, 5438–5466. [CrossRef]
5. Mokhena, T.C.; John, M.J. Cellulose nanomaterials: New generation materials for solving global issues. *Cellulose* **2020**, *27*, 1149–1194. [CrossRef]
6. Dufresne, A. Nanocellulose: A new ageless bionanomaterial. *Mater. Today* **2013**, *16*, 220–227. [CrossRef]
7. Ray, U.; Zhu, S.; Pang, Z.; Li, T. Mechanics design in cellulose-enabled high-performance functional materials. *Adv. Mater.* **2020**. [CrossRef]
8. Cherian, B.M.; Leão, A.L.; de Souza, S.F.; Costa, L.M.M.; de Olyveira, G.M.; Kottaisamy, M.; Nagarajan, E.R.; Thomas, S. Cellulose nanocomposites with nanofibres isolated from pineapple leaf fibers for medical applications. *Carbohydr. Polym.* **2011**, *86*, 1790–1798. [CrossRef]
9. Aulin, C.; Gällstedt, M.; Lindström, T. Oxygen and oil barrier properties of microfibrillated cellulose films and coatings. *Cellulose* **2010**, *17*, 559–574. [CrossRef]
10. Chen, J.; Huang, X.; Zhu, Y.; Jiang, P. Cellulose nanofiber supported 3D interconnected BN nanosheets for epoxy nanocomposites with ultrahigh thermal management capability. *Adv. Funct. Mater.* **2017**, *27*, 1604754. [CrossRef]
11. Kim, J.; Yun, S.; Ounaies, Z. Discovery of cellulose as a smart material. *Macromolecules* **2006**, *39*, 4202–4206. [CrossRef]
12. Kafy, A.; Sadasivuni, K.K.; Akther, A.; Min, S.-K.; Kim, J. Cellulose/graphene nanocomposite as multifunctional electronic and solvent sensor material. *Mater. Lett.* **2015**, *159*, 20–23. [CrossRef]
13. Kafy, A.; Sadasivuni, K.K.; Kim, H.-C.; Akther, A.; Kim, J. Designing flexible energy and memory storage materials using cellulose modified graphene oxide nanocomposites. *Phys. Chem. Chem. Phys.* **2015**, *17*, 5923–5931. [CrossRef] [PubMed]

14. Gomez-Romero, P. Hybrid organic-inorganic materials—In search of synergic activity. *Adv. Mater.* **2001**, *13*, 163–174. [CrossRef]
15. Sanchez, C.; Julián, B.; Belleville, P.; Popall, M. Applications of hybrid organic-inorganic nanocomposites. *J. Mater. Chem.* **2005**, *15*, 3559–3592. [CrossRef]
16. Özgür, Ü.; Alivov, Y.I.; Liu, C.; Teke, A.; Reshchikov, M.A.; Doğan, S.; Avrutin, V.; Cho, S.-J.; Morkoç, H. A comprehensive review of ZnO materials and devices. *J. Appl. Phys.* **2005**, *98*, 041301. [CrossRef]
17. Jayadevan, K.P.; Tseng, T.Y. One-dimensional ZnO nanostructures. *J. Nanosci. Nanotechnol.* **2012**, *12*, 4409–4457. [CrossRef]
18. Wang, X.; Song, J.; Liu, J.; Wang, Z.L. Direct-current nanogenerator driven by ultrasonic waves. *Science* **2007**, *316*, 102–105. [CrossRef]
19. Chu, S.; Li, D.; Chang, P.-C.; Lu, J.G. Flexible dye-sensitized solar cell based on vertical ZnO nanowire arrays. *Nanoscale Res. Lett.* **2010**, *6*, 38. [CrossRef]
20. Yao, I.C.; Tseng, T.-Y.; Lin, P. ZnO nanorods grown on polymer substrates as UV photodetectors. *Sens. Actuator A Phys.* **2012**, *178*, 26–31. [CrossRef]
21. Nanto, H.; Minami, T.; Takata, S. Zinc-oxide thin-film ammonia gas sensors with high sensitivity and excellent selectivity. *J. Appl. Phys.* **1986**, *60*, 482–484. [CrossRef]
22. Swanwick, M.E.; Pfaendler, S.M.L.; Akinwande, A.I.; Flewitt, A.J. Near-ultraviolet zinc oxide nanowire sensor using low temperature hydrothermal growth. *Nanotechnology* **2012**, *23*, 344009. [CrossRef]
23. Park, T.; Kim, N.; Kim, D.; Kim, S.-W.; Oh, Y.; Yoo, J.-K.; You, J.; Um, M.-K. An organic/inorganic nanocomposite of cellulose nanofibers and ZnO nanorods for highly sensitive, reliable, wireless, and wearable multifunctional sensor applications. *ACS Appl. Mater. Interfaces* **2019**, *11*, 48239–48248. [CrossRef]
24. Rodwihok, C.; Choopun, S.; Ruankham, P.; Gardchareon, A.; Phadungdhitidhada, S.; Wongratanaphisan, D. UV sensing properties of ZnO nanowires/nanorods. *Appl. Surf. Sci.* **2019**, *477*, 159–165. [CrossRef]
25. Park, T.; Lee, K.E.; Kim, N.; Oh, Y.; Yoo, J.K.; Um, M.-K. Aspect ratio-controlled ZnO nanorods for highly sensitive wireless ultraviolet sensor applications. *J. Mater. Chem. C* **2017**, *46*, 12256–12263. [CrossRef]
26. Kar, J.P.; Das, S.N.; Choi, J.H.; Lee, Y.A.; Lee, T.Y.; Myoung, J.M. Fabrication of UV detectors based on ZnO nanowires using silicon microchannel. *J. Cryst. Growth* **2009**, *311*, 3305–3309. [CrossRef]
27. Khan, A.; Hussain, M.; Nur, O.; Willander, M. Mechanical and piezoelectric properties of zinc oxide nanorods grown on conductive textile fabric as an alternative substrate. *J. Phys. D Appl. Phys.* **2014**, *47*, 345102. [CrossRef]
28. Zhou, Z.; Zhao, Y.; Cai, Z. Low-temperature growth of ZnO nanorods on PET fabrics with two-step hydrothermal method. *Appl. Surf. Sci.* **2010**, *256*, 4724–4728. [CrossRef]
29. Ladanov, M.; Ram, M.K.; Matthews, G.; Kumar, A. Structure and opto-electrochemical properties of ZnO nanowires grown on n-Si substrate. *Langmuir* **2011**, *27*, 9012–9017. [CrossRef] [PubMed]
30. Asib, N.A.M.; Husairi, F.S.; Eswar, K.A.; Afaah, A.N.; Mamat, M.H.; Rusop, M.; Khusaimi, Z. Developing high-sensitivity UV sensors based on ZnO nanorods grown on TiO_2 seed layer films using solution immersion method. *Sens. Actuators A Phys.* **2020**, *302*, 111827. [CrossRef]
31. Zhang, X.; Ren, S.; He, X.; Dong, L.; Bai, W.; Lei, T.; Chen, H. Preparation and characterization of cellulose nanofiber/zinc oxide composite Films. *J. Biobased Mater. Bioenergy* **2020**, *14*, 203–208. [CrossRef]
32. Fan, B.; Yao, Q.; Wang, C.; Xiong, Y.; Sun, Q.; Jin, C. Spawns structure of rod-like ZnO wrapped in cellulose nanofibers for electromagnetic wave absorption. *J. Nanomater.* **2017**, *2017*, 6329072. [CrossRef]
33. Manekkathodi, A.; Lu, M.-Y.; Wang, C.W.; Chen, L.-J. Direct growth of aligned zinc oxide nanorods on paper substrates for low-cost flexible electronics. *Adv. Mater.* **2010**, *22*, 4059–4063. [CrossRef] [PubMed]
34. Gullapalli, H.; Vemuru, V.S.M.; Kumar, A.; Botello-Mendez, A.; Vajtai, R.; Terrones, M.; Nagarajaiah, S.; Ajayan, P.M. Flexible piezoelectric ZnO-paper nanocomposite strain sensor. *Small* **2010**, *6*, 1641–1646. [CrossRef] [PubMed]
35. Mun, S.; Kim, H.C.; Ko, H.-U.; Zhai, L.; Kim, J.W.; Kim, J. Flexible cellulose and ZnO hybrid nanocomposite and its UV sensing characteristics. *Sci. Technol. Adv. Mater.* **2017**, *18*, 437–446. [CrossRef] [PubMed]
36. Ko, H.-U.; Mun, S.; Min, S.-K.; Kim, G.-W.; Kim, J. Fabrication of cellulose ZnO hybrid nanocomposite and its strain sensing behavior. *Materials* **2014**, *7*, 7000–7009. [CrossRef] [PubMed]
37. Mun, S.; Ko, H.-U.; Zhai, L.; Min, S.-K.; Kim, H.-C.; Kim, J. Enhanced electromechanical behavior of cellulose film by zinc oxide nanocoating and its vibration energy harvesting. *Acta Mater.* **2016**, *114*, 1–6. [CrossRef]
38. John, A.; Ko, H.-U.; Kim, D.-G.; Kim, J. Preparation of cellulose-ZnO hybrid films by a wet chemical method and their characterization. *Cellulose* **2011**, *18*, 675–680. [CrossRef]
39. Nayak, J.N.; Chen, Y.; Kim, J. Removal of impurities from cellulose films after their regeneration from cellulose dissolved in DMAc/LiCl solvent system. *Ind. Eng. Chem. Res.* **2008**, *47*, 1702–1706. [CrossRef]
40. Hai, L.V.; Zhai, L.; Kim, H.C.; Kim, J.W.; Choi, E.S.; Kim, J. Cellulose nanofibers isolated by TEMPO-oxidation and aqueous counter collision methods. *Carbohydr. Polym.* **2018**, *191*, 65–70. [CrossRef]
41. Yun, S.; Kim, J. Mechanical, electrical, piezoelectric and electro-active behavior of aligned multi-walled carbon nanotube/cellulose composites. *Carbon* **2011**, *49*, 518–527. [CrossRef]
42. Yang, C.; Kim, J.-H.; Kim, J.-H.; Kim, J.; Kim, H.S. Piezoelectricity of wet drawn cellulose electro-active paper. *Sens. Actuator A Phys.* **2009**, *154*, 117–122. [CrossRef]

43. Yun, S.; Kim, J.H.; Li, Y.; Kim, J. Alignment of cellulose chains of regenerated cellulose by corona poling and its piezoelectricity. *J. Appl. Phys.* **2008**, *103*, 083301. [CrossRef]
44. Schlaf, R.; Murata, H.; Kafafi, Z.H. Work function measurements on indium tin oxide films. *J. Electron. Spectros. Relat. Phenomena* **2001**, *120*, 149–154. [CrossRef]

nanomaterials

MDPI

Article

Low-Temperature Growth of ZnO Nanowires from Gravure-Printed ZnO Nanoparticle Seed Layers for Flexible Piezoelectric Devices

Andrés Jenaro Lopez Garcia [1,†], Giuliano Sico [2,†], Maria Montanino [2,†], Viktor Defoor [1], Manojit Pusty [1], Xavier Mescot [1], Fausta Loffredo [2], Fulvia Villani [2], Giuseppe Nenna [2,*] and Gustavo Ardila [1,*]

1 University Grenoble Alpes, University Savoie Mont Blanc, CNRS, Grenoble INP, IMEP-LaHC, F-38000 Grenoble, France; andres-jenaro.lopez-garcia@grenoble-inp.fr (A.J.L.G.); viktor.defoor@grenoble-inp.fr (V.D.); manojit.pusty@grenoble-inp.fr (M.P.); xavier.mescot@grenoble-inp.fr (X.M.)

2 ENEA, Italian National Agency for New Technologies, Energy and Sustainable Economic Development, Portici Research Centre, P.le E. Fermi 1, Portici, I-80055 Naples, Italy; giuliano.sico@enea.it (G.S.); maria.montanino@enea.it (M.M.); fausta.loffredo@enea.it (F.L.); fulvia.villani@enea.it (F.V.)

* Correspondence: giuseppe.nenna@enea.it (G.N.); ardilarg@minatec.grenoble-inp.fr (G.A.); Tel.: +33-456-529-532 (G.A.)

† These authors contributed equally to the work.

Abstract: Zinc oxide (ZnO) nanowires (NWs) are excellent candidates for the fabrication of energy harvesters, mechanical sensors, and piezotronic and piezophototronic devices. In order to integrate ZnO NWs into flexible devices, low-temperature fabrication methods are required that do not damage the plastic substrate. To date, the deposition of patterned ceramic thin films on flexible substrates is a difficult task to perform under vacuum-free conditions. Printing methods to deposit functional thin films offer many advantages, such as a low cost, low temperature, high throughput, and patterning at the same stage of deposition. Among printing techniques, gravure-based techniques are among the most attractive due to their ability to produce high quality results at high speeds and perform deposition over a large area. In this paper, we explore gravure printing as a cost-effective high-quality method to deposit thin ZnO seed layers on flexible polymer substrates. For the first time, we show that by following a chemical bath deposition (CBD) process, ZnO nanowires may be grown over gravure-printed ZnO nanoparticle seed layers. Piezo-response force microscopy (PFM) reveals the presence of a homogeneous distribution of Zn-polar domains in the NWs, and, by use of the data, the piezoelectric coefficient is estimated to be close to 4 pm/V. The overall results demonstrate that gravure printing is an appropriate method to deposit seed layers at a low temperature and to undertake the direct fabrication of flexible piezoelectric transducers that are based on ZnO nanowires. This work opens the possibility of manufacturing completely vacuum-free solution-based flexible piezoelectric devices.

Keywords: piezoelectric sensor; mechanical energy harvesting; nanogenerator chemical synthesis; gravure printing; flexible electronics

Citation: Garcia, A.J.L.; Sico, G.; Montanino, M.; Defoor, V.; Pusty, M.; Mescot, X.; Loffredo, F.; Villani, F.; Nenna, G.; Ardila, G. Low-Temperature Growth of ZnO Nanowires from Gravure-Printed ZnO Nanoparticle Seed Layers for Flexible Piezoelectric Devices. *Nanomaterials* **2021**, *11*, 1430. https://doi.org/10.3390/nano11061430

Academic Editor: Matt Cole

Received: 19 April 2021
Accepted: 19 May 2021
Published: 28 May 2021

1. Introduction

Flexible piezoelectric transducers make the development of innovative devices possible, especially in the energy and sensor fields [1]. Several piezoelectric materials have been studied and integrated into flexible substrates, including organic materials, such as poly(vinylidene fluoride) (PVDF) [2] and its co-polymers [3–5]; inorganic materials, like lead zirconate titanate (PZT) [6], potassium sodium niobate (KNN) [7], lead magnesium niobate-lead titanate (PMN-PT) [8], barium titanate ($BaTiO_3$) [9], zinc stannate ($ZnSnO_3$) [10], and bismuth ferrite ($BiFeO_3$) [11]; and semiconductors like zinc oxide (ZnO) in the shape of nanowires (NWs) [12,13]. Among those materials, ZnO has attracted a great deal of attention because it is biocompatible, abundant, and sustainable, making it interesting for future applications while reducing the use of critical and toxic materials [14]. In particular, ZnO NWs have been widely used in different applications related to the conversion of mechanical energy into electricity at the nanometric scale. Some advantages of these NWs, compared to the bulk counterpart, are the ability to withstand high deformation values without fracturing [15–17], high values for the Young's modulus [18,19], and high values for the piezoelectric coefficient [20], as well as a high surface-to-volume ratio that allows enhanced piezoelectric responses thanks to Fermi-level pinning [21].

ZnO can exist in different miniaturized forms, like quantum dots, nanorods, nanowires, nanotubes, nanosheets, nanobelts, nanocones, and twinned disk structures [22]. The ability to form different shapes in miniature sizes, along with a high surface charge density, polarity, and orientation enables ZnO nanostructures to display various applications in the fields of gas sensing, energy storage, energy harvesting, photocatalysis, flexible electronics, optoelectronics, and environmental waste treatment, etc. The high surface area of ZnO NWs promotes greater interfacial adsorption and diffusion, which enhances their physical and chemical properties. ZnO has a wide band gap of 3.37 eV, which is one of the highest in the II-IV semiconductor metal oxides, and also has a large excitation binding energy of 60 meV [23].

These superior properties make ZnO NWs an interesting material to be adopted in the high-tech renewable energy field for the development of energy conversion or storage devices, as ZnO NWs meet the requirements for power and energy density [24]. Also, the low cost of fabrication, negligible emission of pollutants, and the possibility of enhancing energy conversion efficiency make ZnO NW-based devices economically viable for industrial production [25,26].

The first energy harvesting device based on ZnO NWs (then referred to as a nanogenerator—NG) was demonstrated in 2006 by Wang et al. [27]. Since then, a number of NG devices with different structures that are grown on flexible or rigid substrates have been investigated. Recently, in the field of flexible electronics, a great deal of effort has been focused on flexible NG devices, which can be used while experiencing bending [28–30] or compressive [31–35] forces in applications like material-based flexible electronic skins [36] or as nanocomposite-based sensors [37].

Chemical bath deposition (CBD) is a particularly popular method to grow piezoelectric nanostructures due to its ease of implementation, support to scale production, and moderate processing temperatures [33,38,39]. With low processing temperatures, the method is suitable for the use of flexible plastic substrates. Usually, the substrate is covered by a thin seed layer to promote growth and lower the thermodynamic nucleation barrier [40]. Once the seed layer is deposited, the substrates are immersed in a supersaturated growth solution at a moderate temperature for several hours. Using this technique, ZnO NWs or nanorods (NRs) have been grown over metallic [41–43] and polymeric materials [30,44,45], depending on the type of ZnO seed layer deposition. In this work, zinc nitrate hexahydrate ($Zn(NO_3)_2.6H_2O$) and hexamethylenetetramine ($(CH_2)_6N_4$) are used as precursors for the synthesis of ZnO NWs using the CBD process. Hexamethylenetetramine (HMTA) is a popular reagent that is used as a hydroxide anion (OH^-) precursor for ZnO synthesis [46].

To date, a large variety of techniques are used to obtain ZnO seed layers, such as RF sputtering, atomic layer deposition (ALD), spray pyrolysis, chemical vapor deposition, and wet chemical synthetic routes, including spin and dip coatings [47,48]; however, most of them are confined to a laboratory scale and require an additional procedure for preparing patterned layers through expensive and complex processes such as photolithography and etching [49]. Moreover, the typical synthetic routes usually require high temperature annealing post-treatment for improving the crystallinity of the ZnO seeds, thus excluding polymers from the choice of flexible substrates [50–52].

On the other hand, there has been an increasingly pressing demand for easy and cost-effective methods to prepare ZnO patterned seed layers at low temperatures [49,53], which is essential for the use of plastic substrates.

In the last decade, conventional printing processes have been highly regarded as additive techniques for easy, fast, and low-cost thin film deposition at ambient conditions [54,55]. Using these techniques, patterning is carried out simultaneously with the deposition, thus substantially reducing energy, time, and material consumption [56]. In addition, the typical low temperatures of solution processes make printing compatible with most flexible substrates [55,57]. Among printing techniques, gravure techniques are widely employed in many fields and applications thanks to the ability to couple high throughput and high quality [58,59]. Recently, gravure printing has been proven to support the production of uniform large-scale nanoparticulate ZnO thin films on flexible plastic substrates [60]. The quality of the deposited films is so high that it successfully allows subsequent ZnO sintering through an innovative method at a very low temperature and in pressure less conditions [61]. Several printing techniques have been used to deposit ZnO seed layers for the subsequent growth of ZnO NWs, although none of the reports show piezoelectric measurements with the grown NWs [57]. Furthermore, to the best of our knowledge, there have been no reports on the growth of ZnO NWs from layers deposited by gravure printing.

Gravure printing techniques rely on the direct transfer of a low-viscosity ink (1–100 mPas) from the micro-engraved cells of a cylinder to a flexible substrate by the pressure of a counter cylinder. This process can be considered as a sequence of sub-processes (inking, doctoring, transfer, spreading, and drying, as shown in Figure 1a–d) [60]. Essentially, at the microscopic level, the fluid dynamics of the gravure printing process are governed by the balance between viscous and surface tension forces, where the latter are the driving force [62].

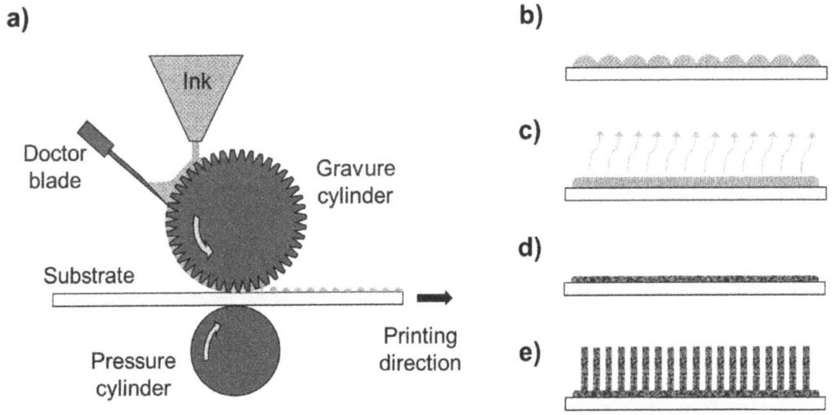

Figure 1. Fabrication process of ZnO seed layers by gravure printing and subsequent NW growth. The gravure printing process can be divided in the following steps: (**a**) inking, doctoring, and transfer; (**b**) spreading; (**c**) drying; (**d**) production of the final solid thin film. (**e**) The final growth of ZnO NWs is performed via low-temperature CBD.

In this study, gravure printing with crystalline ZnO nanoparticles (NPs) was attempted for the first time to produce a seed layer for ZnO NW growth (Figure 1e) from the perspective of flexible piezoelectric devices manufacturing. Using crystalline NPs as starting material allows the elimination of the thermal post-annealing step, making the process easier, time-effective, and compatible with polymeric substrates. Two morphologies of the seed layer were also produced (i.e., nanoparticulate and sintered) to investigate their influence on the density, the diameter, and the orientation of NW growth. The NWs grown from the gravure-printed seed layers were also compared with ZnO NWs grown from a seed layer deposited by ALD on a silicon substrate as a reference.

Piezoelectric properties, including the amplitude of the piezo-response and the phase of the grown NWs, were measured with the different samples via piezo-response force microscopy (PFM).

2. Experiments

2.1. Seed Layer Deposition

A ZnO NP colloidal suspension (Sigma-Aldrich, St. Luis, MO, USA, particle size <130 nm) was used to prepare an ink, diluting the commercial suspension with ethanol (final concentration 8 wt %), for printing seed layers of a 55 ± 6 nm size onto a commercial indium tin oxide (ITO)/poly(ethylene terephthalate) (PET) (150 nm/125 µm) substrate (2 × 1.5 cm^2) acquired from DuPont Teijin Films, Luxembourg.

Printing was performed using a lab-scale gravure printer (IGT G1-5, Amsterdam, the Netherland) equipped with a cylinder with a line density of 70 lines/cm, stylus angle of 120°, screen angle of 53°, and cell depth of 30 µm. All prints were performed in air at room temperature on corona-treated substrates. After preliminary tests, the best printing conditions were found to be a printing force of 500 N at a speed of 60 m/min. After printing, the samples were dried at 100 °C for one hour. The thickness of the dried printed layers above reported was investigated by an interferometry-based optical profilometer (Talysurf CCI HD, Taylor Hobson, Leicester, UK).

A set of ZnO NP printed samples were exposed to the vapor of 1 M acetic acid aqueous solutions for 4 h in a closed oven at 50 °C for inducing chemical bonds among the particles up to the sintering, as reported elsewhere [61].

A 40-nm-thick ZnO seed layer was also deposited on an ITO/Si (166 nm/375 µm) substrate (1.5 × 1.5 cm^2) by ALD at 250 °C as previously reported [33]. These samples were prepared as references for the growth of ZnO NWs.

2.2. Growth of Nanowires

All ZnO NWs were grown under identical conditions by CBD. Samples grown on Si/ITO substrates were firstly cleaned with acetone, ethanol, and DI water in an ultrasonic bath for 5 min and dried with an N_2 gun. For the seed layers prepared by gravure printing, the acetone and ethanol washing steps were omitted. Hexamethylenetetramine ($(CH_2)_6N_4$, Sigma-Aldrich) and zinc nitrate hexahydrate ($Zn(NO_3)_2.6H_2O$, Sigma-Aldrich, St. Luis, MO, USA) were dissolved in equimolar (50 mM) ratios in 150 mL of DI water at room temperature. The growth solution was stirred for 20 min at 1000 rpm and later put to rest for 40 min before decanting the clear solution into glass bottles. The substrates were fixed on cleaned glass slides with Kapton tape before being immersed face-down into the growth solution. The hydrothermal growth was carried out in an oven at 85 °C for 16 h. Later, the substrates were rinsed with DI water and dried with an N_2 gun before being stored for surface characterization.

2.3. Structural Property Measurement

The printed films were morphologically characterized by atomic force microscopy (AFM) and scanning electron microscopy (SEM). AFM analysis was carried out using the Veeco Dimension Digital Instruments Nanoscope IV, New York, NY, USA.

Plainview, NY 11803 apparatus in the tapping mode configuration. For the SEM investigations of the gravure films, a field emission scanning electron microscope (FEG-SEM, Leo 1530 Gemini by Zeiss, Oberkochen, Germany) was used with an operating voltage of 7–9 kV.

Raman spectroscopy was carried out by a Renishaw InVia Reflex (Renishaw, Torino, IT) spectrometer selecting the laser wavelength of 514.5 nm (laser power of 100%) and an objective of 50× for magnification. The investigated wavelength range was 300–800 cm^{-1}. For each measurement, 30 subsequent accumulations and an exposure time of 10 s were set. For each type of sample, ten measurements were taken in order to verify their homogeneity.

2.4. *Piezoelectric and Polarity Measurement*

A Bruker Dimension Icon (Santa Barbara, CA, USA) atomic force microscope instrument was used to carry out the PFM measurements on the ZnO NWs. A PtSi noncontact high-resonance frequency (NCH) tip with a high spring constant value (range values among 43–50 N/m) was used in the measurements. The PtSi-NCH tip was chosen to perform the PFM measurement as it allows a reduction of the electrostatic force involved in the piezoelectric signal [63,64]. The applied AC bias was fixed at 5 V, and the frequency was kept at 14 kHz to avoid any electrostatic contribution (typically above 50 kHz) as shown in the amplitude and phase vs. frequency plots (see Figure S1). To avoid collisions and lateral bending of the vertically grown NWs with the AFM tip, the classical contact mode PFM was not used in this study. Instead, the DataCube mode was used to measure the amplitude and phase (see the details in the Supplementary Material, including Figures S2 and S3).

3. Results and Discussion

In this work, gravure printing was investigated regarding the deposition of ZnO seed layers for the subsequent growth of NWs. On the base of a previous study [60], it was possible to obtain a very smooth, continuous, and homogeneous thin film of nanoparticulate ZnO seeds printed onto an ITO-coated PET substrate (see details in the Supplementary Material, including Table S1 and Figure S4). After printing, some of the printed samples were exposed to a vapor annealing sintering treatment at a very low temperature (50 °C) to investigate the effect of the seed layer morphology on NW growth.

Figures 2 and 3 show SEM and AFM images of the printed ZnO seed layers exposed at different annealing times of 0 and 4 h, respectively. The surface morphology changed from close-packed nanoparticulate film (as-printed) up to a dense one (sintered) over time. Indeed, thanks to a dissolution-reprecipitation mechanism during acidic vapor annealing, the chemical bonds among NPs merge particles as per a grain growth phenomenon. From these SEM images, the measured grain size in the non-sintered layers was 45 ± 12 nm. Concerning the sintered samples, although the evaluation was complex because the grain borders were not well defined, the measured values changed drastically with respect to the non-sintered layers. By the SEM analysis, the grain size ranges from 500 nm to 1000 nm with an estimated average of 547 ± 450 nm.

Figure 2. (**a**) SEM image; (**a'**) magnified SEM image; (**b**) AFM phase image; (**b'**) magnified AFM phase image of the as-printed (non-sintered) ZnO seed layer.

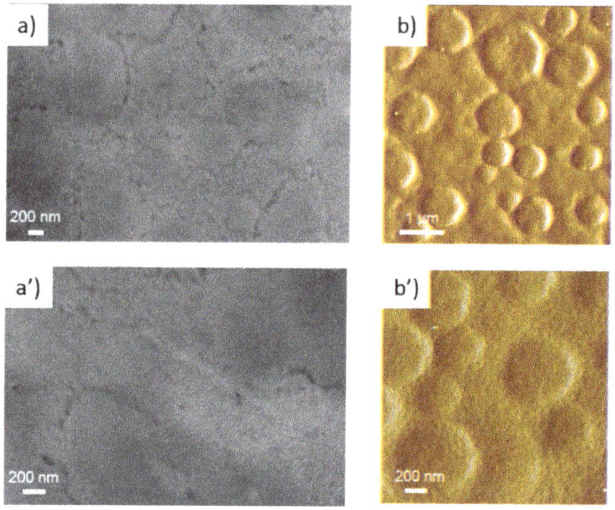

Figure 3. (**a**) SEM image; (**a'**) magnified SEM image; (**b**) AFM phase image; (**b'**) magnified AFM phase image of the sintered printed ZnO seed layer obtained after 4 h of treatment.

NWs were grown by a CBD method on both types of printed ZnO seed layers. Moreover, in order to obtain reference samples, NW growth was also performed on a Si/ITO/ZnO substrate where a ZnO seed layer was deposited by ALD. In Figure 4a–c, SEM images of the ZnO NWs on the different seed layers are shown. In Figure 4a'–c', magnified images of the ZnO NWs are shown in a similar sequence. The ZnO NWs that were grown on the ZnO (ALD)/ITO/Si and the as-printed (not sintered) ZnO seed layer/ITO/PET substrates have average diameters of nearly 199 nm and 210 nm and standard deviations of nearly 86 nm and 84 nm, respectively. Histograms of these measurements are shown in Figure S5a,b. The ZnO NWs were closely packed and showed good vertical alignment. The obtained ZnO NWs had relatively small diameters and good alignment, which is important in piezoelectric applications [21], in particular when taking into account the following: (i) low temperature techniques of seed layer deposition; (ii) deposition on flexible substrates; and (iii) the use of non-expensive chemical methods without the requirement of a vacuum (see Table S2). In Figure 4c, it may be observed that the ZnO NWs were sparsely packed and featured random growth orientations. There was a drastic change in the dimensions of the ZnO NWs, with an average diameter of 987 nm and standard deviation of 250 nm (see Figure S5c). Figure S6a,b shows cross-sectional SEM images of ZnO NWs grown on the ITO/Si and ITO/PET substrates, respectively. It was estimated that the ZnO NWs were nearly 3.5 μm in length. It can be observed that during the growth process of ZnO NWs, some of the NWs merged. The screw-like tops of the ZnO NWs in Figure 4c indicates that their growth followed a screw dislocation nucleation mechanism along the axial direction [65].

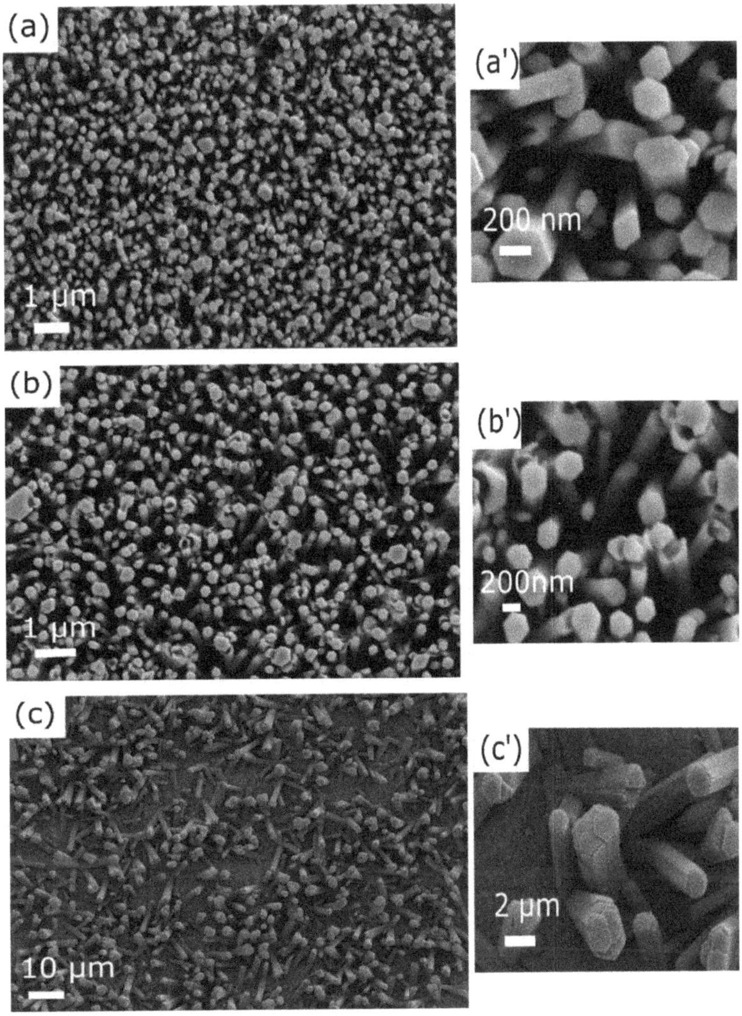

Figure 4. SEM images of the ZnO NWs grown on a (**a**) ZnO (ALD)/ITO/Si substrate; (**a′**) magnified image; (**b**) as-printed (non-sintered) ZnO seed layer/ITO/PET; (**b′**) magnified image; (**c**) sintered printed ZnO seed layer/ITO/PET; (**c′**) magnified image.

As can be seen, the growth of NWs appeared to be significantly affected by the morphology of the seed layer, as also stated in the literature [66–68]; however, the relationships between the seed layer characteristics and NW growth have not been clearly established as of yet. In particular, the seed layer thickness, orientation, and grain size have key roles on the vertical growth of ZnO NWs. As known, ZnO preferentially crystallizes in a hexagonal polar wurtzite structure [69]. The ZnO NW growth direction by CBD occurs predominantly along the polar direction of the seed layer, which provides nucleation sites [66,67].

A possible explanation for the observed results may be proposed on a thermodynamic basis, considering the processes driven by the minimization of the overall system energy. In particular, the seed layer energy governs the nucleation process and subsequently determines the density of ZnO NWs.

In general, the energy of a particle is a function of the size, shape, stress, and external environment [70]. In a NP, a significant number of atoms are localized at the surface [69]. The additional energetic term due to the high-energy under-coordinated atoms at the surface is described as the surface free energy [71,72] and becomes increasingly larger as particle size decreases. In addition, several surface defects [70,73,74] may be present, especially when the particle size is very small [75], resulting in a further increase in the particle free energy [76]. As such, due to their high energy, NPs typically show a higher reactivity compared to bulk or micrometric particles [77,78]. As a result, a highly energetic as-printed nanoparticulate seed layer can be especially chemically reactive for ZnO NW nucleation. In addition, a NP seed alignment mechanism may be present during the CBD, furthering contributing to the high NW density. As shown in Figure 2, the as-printed seed layer was composed of near-spherical NPs having a wurtzite structure, as demonstrated by Raman spectroscopy (see Figure S7 and Table S3). As with most thin polycrystalline films, nanoparticulate printed layers are not textured [66,79], namely, a random nanocrystal orientation is expected. In fact, in the gravure process, there is no particular force that is able to induce a preferential nanoparticle alignment. The final NP arrangement is determined by the strongly attractive lateral capillary forces among particles during solvent evaporation [60,80]. On the other hand, the surface energy distribution of wurtzite-type ZnO NPs is anisotropic and varies between crystallographic orientations [69,81]. Since the NPs are roughly spherical, a large variety of crystallographic planes are presumably exposed to the CBD process [82]. Among these, polar surfaces are extremely chemically reactive [83,84] and have been found to be dominant surface nucleation sites for the growth of ZnO NWs in a solution [66]. To this regard, it has been observed that a high seed layer texture along the polar axis is necessary to obtain a high NW density [67,85,86]. Furthermore, high-density and well-orientated NWs (see Figure 4b) may support a heat-triggered reorientation mechanism for random NPs during the early stages of the CBD [87]. This is possible as the NPs are stuck together by weak physical forces, where they can consequently easily rotate due to the strong electrostatic interactions between charged species (ions, particles, and embryonic dipoles) in the chemical bath, thus reducing the high-energy polar surfaces of wurtzite by nucleation. For these reasons, the as-printed nanoparticulate seed layer demonstrated here can offer a very high number of high-energy nucleation sites (potentially each seed can become a NW, unless steric hindrance interferes). This factor is responsible for the high NW density along the polar c-axis of the hexagonal wurzite crystal (see Figure 4b). Moreover, due to the small sizes of the seeds, the NW diameters were very small, which is typically a challenging result to achieve with most wet chemical methods [88]. Qualitatively, from the SEM images (Figure 4a,b), the NWs grown on the as-printed (non-sintered) layer are comparable in terms of diameter and verticality to the ones grown on the ALD deposited layers.

On the contrary, in the case of a sintered printed sample, far more nucleation sites at lower energy are exposed to the CBD process, resulting in a poor final NW density (see Figure 4c). In fact, sintering is an interface elimination process that decreases the surface area, this creating bonds between contacting particles [89,90]. The total free energy of the system is reduced by replacing the high thermodynamic chemical potential of convex free surfaces of the NPs into lower energy grain boundaries through NP merging. Furthermore, during the densification, the grain boundaries migrate for further reduction of the total free energy, thus resulting in a grain growth phenomenon [91]. Since the grain boundary energy is anisotropic, a lower energy grain boundary tends to extend the most [91–94]. As such, the surfaces of the sintered seeds are mainly dominated by electrically neutral planes [95], which are sites that are not suitable for NW nucleation. Furthermore, the grown grains become trapped by the chemical bonds between grain boundaries. As a result, only some grains with a polar surface available for growth can support nanowires. Such results appear consistent with the observations regarding the decrease in NW density as the seed layer grain size increases [96]. Nevertheless, a chemical sintering post-treatment could be investigated in future to induce potential bonding among seeds and between the seeds and substrate, as well as for tuning the nanowire characteristics.

In some recently published articles, it can be found that surfactants like trisodium citrate, ethylenediamine, poly(ethylene glycol), cetyltrimethylammonium bromide, and sodium dodecyl sulphate are used as surfactants, whose role is to change the surface free energy of the different ZnO crystal facets [97,98]; however, in this work, the role played by the surfactants was complemented by the presence of the seed layer on the substrate.

It is also necessary to discuss the role played by the HMTA in the formation of the ZnO NWs via a CBD process. Although the exact role of HMTA is still under debate, it is believed that the reagent plays multiple roles in the formation of ZnO nanocrystals. HMTA is hydrolyzed upon heating into formaldehyde and ammonia. The latter is further protonated in water to produce OH^- ions, thus increasing the solution pH. It was found that the rate of decomposition of HMTA is dependent on the concentration of protons in the solution and is independent to the precipitation of ZnO [99,100]. Consequently, it was argued that HMTA fulfills the role of a pH buffer since the rate of hydrolysis is decreases with an increased pH [101].

In more recent works, it has been shown that the crystal morphology is also strongly influenced by the chemical precursor concentration. A better crystal stoichiometry data have been obtained in the films deposited with larger amounts of HMTA in solution, thus reducing the number of oxygen vacancies and interstitial zinc in the lattice [102].

Additionally, it is believed that HMTA inhibits the growth of nonpolar m-plane sidewalls, thus favoring an anisotropic growth along the c-axis [103].

Piezoelectric Characterization

To explore the effectiveness of the gravure printing method compared to the ALD process for growing ZnO seed layers on the flexible substrates, we carried out PFM measurements on the ZnO (ALD)/ITO/Si and as-printed (not sintered) ZnO seed layer/ITO/PET samples. The PFM is a widely used tool to study the polarity and the piezo-response properties of piezoelectric thin films and NWs. In this work, the PFM was carried out using the DataCube mode to study the polarities on the top surface of ZnO NWs by recording the phase signals. The d_{33} piezoelectric coefficients were computed by analysing the piezoelectric amplitudes as presented in Figure 5a–f.

Figure 5a,b show the topography in an area of $2 \times 2\ \mu m^2$ in a top-view of ZnO NWs grown on ITO/Si and ITO/PET structures. At first view, we can see good uniformity and an excellent hexagonal shape for the ZnO NWs as per the SEM results (Figure 4). Besides, they have the same correlation regarding their diameter (between 53 and 210 nm) and roughness parameters at the top end of ZnO (e.g., root mean square (Rq) values are approximately 120 nm). Figure 5c,d show the PFM amplitudes for both samples when an AC drive voltage of 5 V was applied. The mean amplitude values (μ) were very similar (22.74 ± 8.49 pm and 20.96 ± 7.21 pm, as shown in Figure 6, for ITO/Si and ITO/PET, respectively) according to the Gaussian fitting function for piezoelectric amplitude distribution at the top surface of ZnO NWs. Based on these results, we estimated the effective piezoelectric coefficients $\left(d_{33}^{eff}\right)$ to be 4.6 and 4.1 pm/V for the ZnO NWs grown on ITO/Si and ITO/PET, respectively. It is important to highlight that both samples presented almost the same d_{33}^{eff} values and they were inside the range of the previously reported experimental results of 2–12 pm/V [104–108].

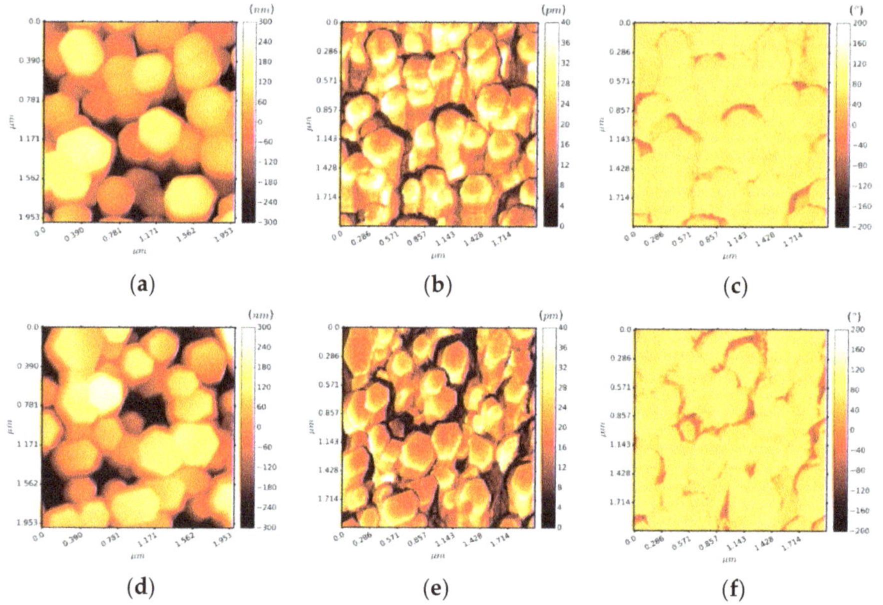

Figure 5. PFM images of the (**a**,**b**) topography, (**c**,**d**) amplitude, and (**e**,**f**) phase of the ZnO NW growth on ZnO (ALD)/ITO/Si (images lined up above) and as-printed (not sintered) ZnO seed layer/ITO/PET substrates (images lined up below).

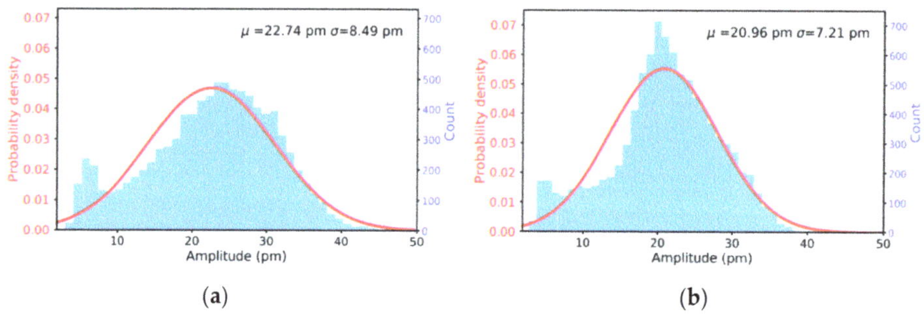

Figure 6. Piezoelectric amplitude histograms of ZnO NW growth on (**a**) ZnO (ALD)/ITO/Si and (**b**) as-printed (not sintered) ZnO seed layer/ITO/PET substrates. The red curve is the Gaussian fitting function.

On the other hand, the PFM measurements also provided results concerning the piezoelectric phase as depicted in Figure 5e,f. Both results show a single-phase value of around 75° (see Figure S3) throughout the top surface of ZnO NWs, thus demonstrating the Zn surface polarity of the crystal structure [64]. Obtaining a homogeneous surface polarity is very important at a device level, i.e., where NWs are typically integrated and placed between electrodes [33]. If the NWs had shown negative and positive phase values in their top surface, the induced negative and positive electrical charges would have cancelled out, thus reducing the overall piezoelectric response at the device level.

4. Conclusions

In summary, our experimental results show that gravure printing is a promising low-temperature ZnO seeding method, especially from the perspective of fabricating flexible piezoelectric devices. Study of the printed seed layer morphology revealed that the nanoparticulate allowed a better growth of vertical ZnO NWs by CBD with respect to the sintered one because the high-energy NPs provided a higher number of nucleation sites. To explain the observed results, a possible reorientation mechanism of the ZnO nanoparticle seeds was also proposed. PFM measurements of the NWs revealed (i) a Zn-phase polarity with a very good homogeneity, which is essential to obtaining globally optimal piezoelectric performance when integrated in devices, and (ii) an estimated piezoelectric coefficient close to 4 pm/V, which is comparable to the NWs grown on Si/ITO substrate.

The mobility and flexibility of the whole structure, provided by the use of flexible ZnO NWs, is expected to have a good impact on the device's mechanical performance and durability.

This work shows the way towards the fabrication of flexible piezoelectric transducers based on ZnO nanowires. Fabrication is completely carried out by solution processes at a low temperature.

Supplementary Materials: The following are available online at https://www.mdpi.com/2079-4991/11/6/1430/s1, Figure S1: sweep of drive frequency value of AC bias signal for PFM amplitude and phase values in different ranges of frequency; Figure S2: single frame of (a) amplitude and (b) phase measurement and average value of 20 frames of (c) amplitude and (d) phase; Figure S3: piezoelectric phase distribution value of ZnO NWs grown on (a) ZnO (ALD)/ITO/Si and (b) as-printed (non-sintered) ZnO seed layer/ITO/PET substrates. The red curve is the fitting curve using Gaussian mixture models, Figure S4: height distribution of the top surface of (a) ITO over PET, (b) ITO over Si, and (c) ZnO seed layer (deposited by ALD) over ITO/Si to extract the roughness parameters by AFM; Figure S5: diameter distribution of ZnO NWs grown on (a) ZnO (ALD)/ITO/Si and (b) as-printed (non-sintered) ZnO seed layer/ITO/PET substrates. The red curve is the Gaussian fitting function; Figure S6: SEM image of the cross-section of the ZnO NWs grown on (a) ITO/Si substrate and (b) ITO/PET substrate; Figure S7: Raman spectra of ZnO seed layer films deposited onto aluminium by gravure printing measured before and after 4 h of sintering; Table S1: arithmetical mean deviation values (Ra) and root squared values (Rq) measured for the different substrates and manufactured ZnO seed layers; Table S2: comparison table of ZnO NWs grown by chemical bath deposition (CBD) from recent articles; Table S3: assignments for the Raman bands marked in Figure S2 to specific vibrations for ZnO structures.

Author Contributions: Conceptualization, G.A.; G.N.; G.S.; methodology, A.J.L.G.; V.D.; G.S.; M.M.; software, A.J.L.G.; validation, A.J.L.G.; V.D.; G.A.; G.S.; M.M.; formal analysis, A.J.L.G.; investigation, A.J.L.G.; V.D.; M.P.; G.S.; M.M.; F.L.; F.V.; resources, X.M.; G.S.; M.M.; F.L.; F.V.; data curation, A.J.L.G.; writing—original draft preparation, A.J.L.G.; G.A.; V.D.; M.P.; G.S.; M.M.; F.L.; F.V.; G.N.; writing—review and editing, G.A.; M.P.; X.M.; G.S.; M.M.; F.L.; F.V.; G.N.; visualization, A.J.L.G.; V.D.; supervision, G.A.; project administration, G.A.; funding acquisition, G.A.; G.N. All authors have read and agreed to the published version of the manuscript.

Funding: This work has received funding from project PULSE-COM of the European Union's Horizon 2020 research and innovation program under grant agreement No. 863227.

Data Availability Statement: The data are available upon request from the corresponding authors.

Acknowledgments: This work has been performed with the help of the "Plateforme Technologique Amont" from Grenoble.

Conflicts of Interest: The authors declare no conflict of interest.

References

1. Aliqué, M.; Simão, C.D.; Murillo, G.; Moya, A. Fully-Printed Piezoelectric Devices for Flexible Electronics Applications. *Adv. Mater. Technol.* **2021**, *6*, 2001020. [CrossRef]
2. Martins, P.; Lopes, A.C.; Lanceros-Mendez, S. Electroactive phases of poly(vinylidene fluoride): Determination, processing and applications. *Prog. Polym. Sci.* **2014**, *39*, 683–706. [CrossRef]
3. Bhavanasi, V.; Kumar, V.; Parida, K.; Wang, J.; Lee, P.S. Enhanced Piezoelectric Energy Harvesting Performance of Flexible PVDF-TrFE Bilayer Films with Graphene Oxide. *ACS Appl. Mater. Interfaces* **2016**, *8*, 521–529. [CrossRef]
4. Wu, Y.; Qu, J.; Daoud, W.A.; Wang, L.; Qi, T. Flexible composite-nanofiber based piezo-triboelectric nanogenerators for wearable electronics. *J. Mater. Chem. A* **2019**, *7*, 13347–13355. [CrossRef]

5. Costa, P.; Nunes-Pereira, J.; Pereira, N.; Castro, N.; Gonçalves, S.; Lanceros-Mendez, S. Recent Progress on Piezoelectric, Pyroelectric, and Magnetoelectric Polymer-Based Energy-Harvesting Devices. *Energy Technol.* **2019**, *7*, 1–19. [CrossRef]
6. Gupta, S.; Bhunia, R.; Fatma, B.; Maurya, D.; Singh, D.; Gupta, R.; Priya, S.; Gupta, R.K.; Garg, A. Multifunctional and Flexible Polymeric Nanocomposite Films with Improved Ferroelectric and Piezoelectric Properties for Energy Generation Devices. *ACS Appl. Mater. Interfaces* **2019**, *2*, 6364–6374. [CrossRef]
7. Kang, H.B.; Han, C.S.; Pyun, J.C.; Ryu, W.H.; Kang, C.Y.; Cho, Y.S. (Na,K)NbO_3 nanoparticle-embedded piezoelectric nanofiber composites for flexible nanogenerators. *Compos. Sci. Technol.* **2015**, *111*, 1–8. [CrossRef]
8. Hwang, G.T.; Park, H.; Lee, J.H.; Oh, S.; Park, K.I.; Byun, M.; Park, H.; Ahn, G.; Jeong, C.K.; No, K.; et al. Self-powered cardiac pacemaker enabled by flexible single crystalline PMN-PT piezoelectric energy harvester. *Adv. Mater.* **2014**, *26*, 4880–4887. [CrossRef]
9. Dudem, B.; Kim, D.H.; Bharat, L.K.; Yu, J.S. Highly-flexible piezoelectric nanogenerators with silver nanowires and barium titanate embedded composite films for mechanical energy harvesting. *Appl. Energy* **2018**, *230*, 865–874. [CrossRef]
10. Rovisco, A.; dos Santos, A.; Cramer, T.; Martins, J.; Branquinho, R.; Águas, H.; Fraboni, B.; Fortunato, E.; Martins, R.; Igreja, R. Piezoelectricity Enhancement of Nanogenerators Based on PDMS and $ZnSnO_3$ Nanowires through Microstructuration. *ACS Appl. Mater. Interfaces* **2020**, *12*, 18421–18430. [CrossRef]
11. Queraltó, A.; Frohnhoven, R.; Mathur, S.; Gómez, A. Intrinsic piezoelectric characterization of $BiFeO_3$ nanofibers and its implications for energy harvesting. *Appl. Surf. Sci.* **2020**, *509*, 144760–144767. [CrossRef]
12. Yang, D.; Qiu, Y.; Jiang, Q.; Guo, Z.; Song, W.; Xu, J.; Zong, Y.; Feng, Q.; Sun, X. Patterned growth of ZnO nanowires on flexible substrates for enhanced performance of flexible piezoelectric nanogenerators. *Appl. Phys. Lett.* **2017**, *110*, 63901–63905. [CrossRef]
13. Panth, M.; Cook, B.; Alamri, M.; Ewing, D.; Wilson, A.; Wu, J.Z. Flexible zinc oxide nanowire array/graphene nanohybrid for high-sensitivity strain detection. *ACS Omega* **2020**, *5*, 27359–27367. [CrossRef] [PubMed]
14. Wei, H.; Wang, H.; Xia, Y.; Cui, D.; Shi, Y.; Dong, M.; Liu, C.; Ding, T.; Zhang, J.; Ma, Y. An overview of lead-free piezoelectric materials and devices. *J. Mater. Chem. C* **2018**, *6*, 12446–12467. [CrossRef]
15. Hughes, W.L.; Wang, Z.L. Controlled synthesis and manipulation of ZnO nanorings and nanobows. *Appl. Phys. Lett.* **2005**, *86*, 43106. [CrossRef]
16. Espinosa, H.D.; Bernal, R.A.; Minary-Jolandan, M. A review of mechanical and electromechanical properties of piezoelectric nanowires. *Adv. Mater.* **2012**, *24*, 4656–4675. [CrossRef] [PubMed]
17. Jing, G.; Zhang, X.; Yu, D. Effect of surface morphology on the mechanical properties of ZnO nanowires. *Appl. Phys. A* **2010**, *100*, 473–478. [CrossRef]
18. Chen, C.Q.; Shi, Y.; Zhang, Y.S.; Zhu, J.; Yan, Y.J. Size dependence of Young's modulus in ZnO nanowires. *Phys. Rev. Lett.* **2006**, *96*, 75505–75508. [CrossRef]
19. Soomro, M.Y.; Hussain, I.; Bano, N.; Broitman, E.; Nur, O.; Willander, M. Nanoscale elastic modulus of single horizontal ZnO nanorod using nanoindentation experiment. *Nanoscale Res. Lett.* **2012**, *7*, 1–5. [CrossRef]
20. Agrawal, R.; Espinosa, H.D. Giant piezoelectric size effects in zinc oxide and gallium nitride nanowires. A first principles investigation. *Nano Lett.* **2011**, *11*, 786–790. [CrossRef] [PubMed]
21. Tao, R.; Mouis, M.; Ardila, G. Unveiling the influence of surface fermi level pinning on the piezoelectric response of semiconducting nanowires. *Adv. Electron. Mater.* **2018**, *4*, 1700299. [CrossRef]
22. Hussain, S.; Liu, T.; Aslam, N.; Kashif, M.; Cao, S.; Rashad, M.; Zhang, Y.; Zeng, W.; Javed, M.S. Polymer-assisted co-axial multi-layered circular ZnO nanodisks. *Mater. Lett.* **2015**, *152*, 260–263. [CrossRef]
23. Hussain, S.; Liu, T.; Aslam, N.; Zhao, S.; Li, T.; Hou, D.; Zeng, W. Assembly of bulbous ZnO nanorods to bulbous nanoflowers and their high selectivity towards formaldehyde. *J. Mater. Sci. Mater. Electron.* **2016**, *27*, 4966–4971. [CrossRef]
24. Hussain, S.; Yang, X.; Aslam, M.K.; Shaheen, A.; Javed, M.S.; Aslam, N.; Aslam, B.; Liu, G.; Qiao, G. Robust TiN nanoparticles polysulfide anchor for Li–S storage and diffusion pathways using first principle calculations. *Chem. Eng. J.* **2020**, *391*, 123595. [CrossRef]
25. Hussain, S.; Khan, A.J.; Arshad, M.; Javed, M.S.; Ahmad, A.; Shah, S.S.A.; Khan, M.R.; Akram, S.; Zulfiqar; Ali, S.; et al. Charge storage in binder-free 2D-hexagonal $CoMoO_4$ nanosheets as a redox active material for pseudocapacitors. *Ceram. Int.* **2021**, *47*, 8659–8667. [CrossRef]
26. Hussain, S.; Hassan, M.; Javed, M.S.; Shaheen, A.; Ahmad Shah, S.S.; Nazir, M.T.; Najam, T.; Khan, A.J.; Zhang, X.; Liu, G. Distinctive flower-like $CoNi_2S_4$ nanoneedle arrays (CNS–NAs) for superior supercapacitor electrode performances. *Ceram. Int.* **2020**, *46*, 25942–25948. [CrossRef]
27. Wang, Z.L.; Song, J. Piezoelectric Nanogenerators Based on Zinc Oxide Nanowire. *Science* **2006**, *312*, 242–246. [CrossRef] [PubMed]
28. Yang, R.; Qin, Y.; Li, C.; Zhu, G.; Wang, Z.L. Converting biomechanical energy into electricity by a muscle-movement-driven nanogenerator. *Nano Lett.* **2009**, *9*, 1201–1205. [CrossRef]
29. Yim, M.; Jeon, B.; Yoon, G. Fabrication of tandem-type vertically integrated nanogenerator by in-situ deposition of AlN/ZnO films. *J. Semicond. Technol. Sci.* **2019**, *19*, 233–238. [CrossRef]
30. Dahiya, A.S.; Morini, F.; Boubenia, S.; Nadaud, K.; Alquier, D.; Poulin-Vittrant, G. Organic/Inorganic Hybrid Stretchable Piezoelectric Nanogenerators for Self-Powered Wearable Electronics. *Adv. Mater. Technol.* **2018**, *3*, 1–11. [CrossRef]

31. Lee, S.; Bae, S.H.; Lin, L.; Yang, Y.; Park, C.; Kim, S.W.; Cha, S.N.; Kim, H.; Park, Y.J.; Wang, Z.L. Super-flexible nanogenerator for energy harvesting from gentle wind and as an active deformation sensor. *Adv. Funct. Mater.* **2013**, *23*, 2445–2449. [CrossRef]
32. Yu, A.; Li, H.; Tang, H.; Liu, T.; Jiang, P.; Wang, Z.L. Vertically integrated nanogenerator based on ZnO nanowire arrays. *Phys. Status Solidi Rapid Res. Lett.* **2011**, *5*, 162–164. [CrossRef]
33. Tao, R.; Parmar, M.; Ardila, G.; Oliveira, P.; Marques, D.; Montès, L.; Mouis, M. Performance of ZnO based piezo-generators under controlled compression. *Semicond. Sci. Technol.* **2017**, *32*, 064003. [CrossRef]
34. Hu, Y.; Lin, L.; Zhang, Y.; Wang, Z.L. Replacing a battery by a nanogenerator with 20 v output. *Adv. Mater.* **2012**, *24*, 110–114. [CrossRef]
35. Zhu, G.; Wang, A.C.; Liu, Y.; Zhou, Y.; Wang, Z.L. Functional electrical stimulation by nanogenerator with 58 V output voltage. *Nano Lett.* **2012**, *12*, 3086–3090. [CrossRef] [PubMed]
36. Yuan, H.; Lei, T.; Qin, Y.; Yang, R. Flexible electronic skins based on piezoelectric nanogenerators and piezotronics. *Nano Energy* **2019**, *59*, 84–90. [CrossRef]
37. He, H.; Fu, Y.; Zang, W.; Wang, Q.; Xing, L.; Zhang, Y.; Xue, X. A flexible self-powered T-ZnO/PVDF/fabric electronic-skin with multi-functions of tactile-perception, atmosphere-detection and self-clean. *Nano Energy* **2017**, *31*, 37–48. [CrossRef]
38. Baxter, J.B.; Schmuttenmaer, C.A. Conductivity of ZnO nanowires, nanoparticles, and thin films using time-resolved terahertz spectroscopy. *J. Phys. Chem. B* **2006**, *110*, 25229–25239. [CrossRef]
39. Villafuerte, J.; Donatini, F.; Kioseoglou, J.; Sarigiannidou, E.; Chaix-Pluchery, O.; Pernot, J.; Consonni, V. Zinc Vacancy–Hydrogen Complexes as Major Defects in ZnO Nanowires Grown by Chemical Bath Deposition. *J. Phys. Chem. C* **2020**, *124*, 16652–16662. [CrossRef]
40. Sugunan, A.; Warad, H.C.; Boman, M.; Dutta, J. Zinc oxide nanowires in chemical bath on seeded substrates: Role of hexamine. *J. Sol-Gel Sci. Technol.* **2006**, *39*, 49–56. [CrossRef]
41. Chelu, M.; Stroescu, H.; Anastasescu, M.; Calderon-Moreno, J.M.; Preda, S.; Stoica, M.; Fogarassy, Z.; Petrik, P.; Gheorghe, M.; Parvulescu, C. High-quality PMMA/ZnO NWs piezoelectric coating on rigid and flexible metallic substrates. *Appl. Surf. Sci.* **2020**, *529*, 147135–147144. [CrossRef]
42. Yang, W.; Wang, Y.; Hou, Z.; Li, C. A facile hot-pressing process for fabricating flexible top electrodes of piezoelectric ZnO nanowire nanogenerators. *Nanotechnology* **2019**, *30*, 505402–505407. [CrossRef] [PubMed]
43. Deng, W.; Jin, L.; Zhang, B.; Chen, Y.; Mao, L.; Zhang, H.; Yang, W. A flexible field-limited ordered ZnO nanorod-based self-powered tactile sensor array for electronic skin. *Nanoscale* **2016**, *8*, 16302–16306. [CrossRef] [PubMed]
44. Yilmaz, P.; Greenwood, P.; Meroni, S.; Troughton, J.; Novák, P.; Li, X.; Watson, T.; Briscoe, J. Self-adhesive electrode applied to ZnO nanorod-based piezoelectric nanogenerators. *Smart Mater. Struct.* **2019**, *28*, 105040. [CrossRef]
45. Zhang, Y.; Liu, C.; Liu, J.; Xiong, J.; Liu, J.; Zhang, K.; Liu, Y.; Peng, M.; Yu, A.; Zhang, A. Lattice strain induced remarkable enhancement in piezoelectric performance of ZnO-based flexible nanogenerators. *ACS Appl. Mater. Interfaces* **2016**, *8*, 1381–1387. [CrossRef] [PubMed]
46. Wojnarowicz, J.; Chudoba, T.; Lojkowski, W. A Review of Microwave Synthesis of Zinc Oxide Nanomaterials: Reactants, Process Parameters and Morphoslogies. *Nanomaterials* **2020**, *10*, 1086. [CrossRef]
47. Noack, V.; Eychmüller, A. Annealing of nanometer-sized zinc oxide particles. *Chem. Mater.* **2002**, *14*, 1411–1417. [CrossRef]
48. Mosalagae, K.; Murape, D.M.; Lepodise, L.M. Effects of growth conditions on properties of CBD synthesized ZnO nanorods grown on ultrasonic spray pyrolysis deposited ZnO seed layers. *Heliyon* **2020**, *6*, e04458–e04467. [CrossRef]
49. Chang, S.-Y.; Yang, N.-H.; Huang, Y.-C.; Lin, S.-J.; Kattamis, T.Z.; Liu, C.-Y. Spontaneous growth of one-dimensional nanostructures from films in ambient atmosphere at room temperature: ZnO and TiO_2. *J. Mater. Chem.* **2011**, *21*, 4264–4271. [CrossRef]
50. Peiris, T.A.N.; Alessa, H.; Sagu, J.S.; Bhatti, I.A.; Isherwood, P.; Wijayantha, K.G.U. Effect of ZnO seed layer thickness on hierarchical ZnO nanorod growth on flexible substrates for application in dye-sensitised solar cells. *J. Nanoparticle Res.* **2013**, *15*, 2115–2124. [CrossRef]
51. Burke-Govey, C.P.; Plank, N.O.V. Review of hydrothermal ZnO nanowires: Toward FET applications. *J. Vac. Sci. Technol. B Nanotechnol. Microelectron. Mater. Process. Meas. Phenom.* **2013**, *31*, F101–F106. [CrossRef]
52. Wei, P.; Zhu, D.; Huang, S.; Zhou, W.; Luo, F. Effects of the annealing temperature and atmosphere on the microstructures and dielectric properties of ZnO/Al_2O_3 composite coatings. *Appl. Surf. Sci.* **2013**, *285*, 577–582. [CrossRef]
53. Ranjan, P.; Suematsu, H.; Chakravarthy, S.R.; Jayaganthan, R.; Sarathi, R. Thermodynamic analysis of ZnO nanoparticle formation by wire explosion process and characterization. *Ceram. Int.* **2017**, *43*, 6709–6720. [CrossRef]
54. Wu, W. Inorganic nanomaterials for printed electronics: A review. *Nanoscale* **2017**, *9*, 7342–7372. [CrossRef]
55. Abbel, R.; Galagan, Y.; Groen, P. Roll-to-Roll Fabrication of Solution Processed Electronics. *Adv. Eng. Mater.* **2018**, *20*, 1701190–1701219. [CrossRef]
56. Khan, S.; Lorenzelli, L.; Dahiya, R.S. Technologies for printing sensors and electronics over large flexible substrates: A review. *IEEE Sens. J.* **2014**, *15*, 3164–3185. [CrossRef]
57. Arrabito, G.; Aleeva, Y.; Pezzilli, R.; Ferrara, V.; Medaglia, P.G.; Pignataro, B.; Prestopino, G. Printing ZnO Inks: From Principles to Devices. *Crystals* **2020**, *10*, 449. [CrossRef]
58. Søndergaard, R.R.; Hösel, M.; Krebs, F.C. Roll-to-Roll fabrication of large area functional organic materials. *J. Polym. Sci. Part. B Polym. Phys.* **2013**, *51*, 16–34. [CrossRef]

59. Krebs, F.C. Fabrication and processing of polymer solar cells: A review of printing and coating techniques. *Sol. Energy Mater. Sol. Cells* **2009**, *93*, 394–412. [CrossRef]
60. Sico, G.; Montanino, M.; Prontera, C.T.; Del Mauro, A.D.G.; Minarini, C. Gravure printing for thin film ceramics manufacturing from nanoparticles. *Ceram. Int.* **2018**, *44*, 19526–19534. [CrossRef]
61. Sico, G.; Montanino, M.; Ventre, M.; Mollo, V.; Prontera, C.T.; Minarini, C.; Magnani, G. Pressureless sintering of ZnO thin film on plastic substrate via vapor annealing process at near-room temperature. *Scr. Mater.* **2019**, *164*, 48–51. [CrossRef]
62. Choi, Y.; Kim, G.H.; Jeong, W.H.; Kim, H.J.; Chin, B.D.; Yu, J.-W. Characteristics of gravure printed InGaZnO thin films as an active channel layer in thin film transistors. *Thin Solid Film.* **2010**, *518*, 6249–6252. [CrossRef]
63. Kim, S.; Seol, D.; Lu, X.; Alexe, M.; Kim, Y. Electrostatic-free piezoresponse force microscopy. *Sci. Rep.* **2017**, *7*, 1–8. [CrossRef] [PubMed]
64. Bui, Q.C.; Ardila, G.; Sarigiannidou, E.; Roussel, H.; Jiménez, C.; Chaix-Pluchery, O.; Guerfi, Y.; Bassani, F.; Donatini, F.; Mescot, X.; et al. Morphology Transition of ZnO from Thin Film to Nanowires on Silicon and its Correlated Enhanced Zinc Polarity Uniformity and Piezoelectric Responses. *ACS Appl. Mater. Interfaces* **2020**, *12*, 29583–29593. [CrossRef]
65. Kumar, Y.; Rana, A.K.; Bhojane, P.; Pusty, M.; Bagwe, V.; Sen, S.; Shirage, P.M. Controlling of ZnO nanostructures by solute concentration and its effect on growth, structural and optical properties. *Mater. Res. Express* **2015**, *2*, 105017–105027. [CrossRef]
66. Guillemin, S.; Consonni, V.; Appert, E.; Puyoo, E.; Rapenne, L.; Roussel, H. Critical nucleation effects on the structural relationship between ZnO seed layer and nanowires. *J. Phys. Chem. C* **2012**, *116*, 25106–25111. [CrossRef]
67. Guillemin, S.; Rapenne, L.; Roussel, H.; Sarigiannidou, E.; Brémond, G.; Consonni, V. Formation mechanisms of ZnO nanowires: The crucial role of crystal orientation and polarity. *J. Phys. Chem. C* **2013**, *117*, 20738–20745. [CrossRef]
68. Abdulrahman, A.F.; Ahmed, S.M.; Ahmed, N.M.; Almessiere, M.A. Enhancement of ZnO Nanorods Properties Using Modified Chemical Bath Deposition Method: Effect of Precursor Concentration. *Crystals* **2020**, *10*, 386. [CrossRef]
69. Zuniga-Perez, J.; Consonni, V.; Lymperakis, L.; Kong, X.; Trampert, A.; Fernandez-Garrido, S.; Brandt, O.; Renevier, H.; Keller, S.; Hestroffer, K. Polarity in GaN and ZnO: Theory, measurement, growth, and devices. *Appl. Phys. Rev.* **2016**, *3*, 41303–41402. [CrossRef]
70. Marks, L.D.; Peng, L. Nanoparticle shape, thermodynamics and kinetics. *J. Phys. Condens. Matter* **2016**, *28*, 53001–53048. [CrossRef]
71. Levitas, V.I.; Samani, K. Size and mechanics effects in surface-induced melting of nanoparticles. *Nat. Commun.* **2011**, *2*, 1–6. [CrossRef]
72. Vollath, D.; Fischer, F.D.; Holec, D. Surface energy of nanoparticles–influence of particle size and structure. *Beilstein J. Nanotechnol.* **2018**, *9*, 2265–2276. [CrossRef]
73. Biener, J.; Wittstock, A.; Baumann, T.F.; Weissmüller, J.; Bäumer, M.; Hamza, A.V. Surface chemistry in nanoscale materials. *Materials* **2009**, *2*, 2404–2428. [CrossRef]
74. Choudhary, K.; Biacchi, A.J.; Ghosh, S.; Hale, L.; Walker, A.R.H.; Tavazza, F. High-throughput assessment of vacancy formation and surface energies of materials using classical force-fields. *J. Phys. Condens. Matter* **2018**, *30*, 395901. [CrossRef] [PubMed]
75. Meulenkamp, E.A. Synthesis and growth of ZnO nanoparticles. *J. Phys. Chem. B* **1998**, *102*, 5566–5572. [CrossRef]
76. Przybyszewska, M.; Krzywania, A.; Zaborski, M.; Szynkowska, M.I. Surface properties of zinc oxide nanoparticles studied by inverse gas chromatography. *J. Chromatogr. A* **2009**, *1216*, 5284–5291. [CrossRef] [PubMed]
77. Lu, H.M.; Jiang, Q. Size-dependent surface energies of nanocrystals. *J. Phys. Chem. B* **2004**, *108*, 5617–5619. [CrossRef]
78. He, H.; Cao, J.; Duan, N. Defects and their behaviors in mineral dissolution under water environment: A review. *Sci. Total Environ.* **2019**, *651*, 2208–2217. [CrossRef] [PubMed]
79. Ching, K.-L.; Li, G.; Ho, Y.-L.; Kwok, H.-S. The role of polarity and surface energy in the growth mechanism of ZnO from nanorods to nanotubes. *CrystEngComm* **2016**, *18*, 779–786. [CrossRef]
80. Kralchevsky, P.A.; Nagayama, K. Capillary forces between colloidal particles. *Langmuir* **1994**, *10*, 23–36. [CrossRef]
81. Mudunkotuwa, I.A.; Grassian, V.H. The devil is in the details (or the surface): Impact of surface structure and surface energetics on understanding the behavior of nanomaterials in the environment. *J. Environ. Monit.* **2011**, *13*, 1135–1144. [CrossRef] [PubMed]
82. Zhang, P.; Xu, F.; Navrotsky, A.; Lee, J.S.; Kim, S.; Liu, J. Surface enthalpies of nanophase ZnO with different morphologies. *Chem. Mater.* **2007**, *19*, 5687–5693. [CrossRef]
83. Tang, C.; Spencer, M.J.S.; Barnard, A.S. Activity of ZnO polar surfaces: An insight from surface energies. *Phys. Chem. Chem. Phys.* **2014**, *16*, 22139–22144. [CrossRef] [PubMed]
84. Connolly, B.M.; Greer, H.F.; Zhou, W. Formation Mechanisms of ZnO Spherulites and Derivatives. *Cryst. Growth Des.* **2018**, *19*, 249–257. [CrossRef]
85. Demes, T.; Ternon, C.; Riassetto, D.; Stambouli, V.; Langlet, M. Comprehensive study of hydrothermally grown ZnO nanowires. *J. Mater. Sci.* **2016**, *51*, 10652–10661. [CrossRef]
86. Cossuet, T.; Roussel, H.; Chauveau, J.-M.; Chaix-Pluchery, O.; Thomassin, J.-L.; Appert, E.; Consonni, V. Well-ordered ZnO nanowires with controllable inclination on semipolar ZnO surfaces by chemical bath deposition. *Nanotechnology* **2018**, *29*, 475601–475612. [CrossRef] [PubMed]
87. Theissmann, R.; Fendrich, M.; Zinetullin, R.; Guenther, G.; Schierning, G.; Wolf, D.E. Crystallographic reorientation and nanoparticle coalescence. *Phys. Rev. B* **2008**, *78*, 205413–205421. [CrossRef]
88. Ali, A.; Ambreen, S.; Javed, R.; Tabassum, S.; Ul Haq, I.; Zia, M. ZnO nanostructure fabrication in different solvents transforms physio-chemical, biological and photodegradable properties. *Mater. Sci. Eng. C* **2017**, *74*, 137–145. [CrossRef]

89. Castro, R.H.R.; Gouvêa, D. Sintering and nanostability: The thermodynamic perspective. *J. Am. Ceram. Soc.* **2016**, *99*, 1105–1121. [CrossRef]
90. German, R.M. Sintering trajectories: Description on how density, surface area, and grain size change. *JOM* **2016**, *68*, 878–884. [CrossRef]
91. Rohrer, G.S. Grain boundary energy anisotropy: A review. *J. Mater. Sci.* **2011**, *46*, 5881–5895. [CrossRef]
92. Lee, J.; Wiederhorn, S.M. Effects of polarity on grain-boundary migration in ZnO. *J. Am. Ceram. Soc.* **2004**, *87*, 1319–1323. [CrossRef]
93. German, R.M. Coarsening in sintering: Grain shape distribution, grain size distribution, and grain growth kinetics in solid-pore systems. *Crit. Rev. Solid State Mater. Sci.* **2010**, *35*, 263–305. [CrossRef]
94. Sui, M.; Pandey, P.; Li, M.-Y.; Zhang, Q.; Kunwar, S.; Lee, J. Tuning the configuration of Au nanostructures: From vermiform-like, rod-like, triangular, hexagonal, to polyhedral nanostructures on c-plane GaN. *J. Mater. Sci.* **2017**, *52*, 391–407. [CrossRef]
95. Kuang, Q.; Wang, X.; Jiang, Z.; Xie, Z.; Zheng, L. High-energy-surface engineered metal oxide micro- and nanocrystallites and their applications. *Acc. Chem. Res.* **2014**, *47*, 308–318. [CrossRef]
96. Syrrokostas, G.; Govatsi, K.; Yannopoulos, S.N. High-Quality, Reproducible ZnO Nanowire Arrays Obtained by a Multiparameter Optimization of Chemical Bath Deposition Growth. *Cryst. Growth Des.* **2016**, *16*, 2140–2150. [CrossRef]
97. Hussain, S.; Liu, T.; Kashif, M.; Miao, B.; He, J.; Zeng, W.; Zhang, Y.; Hashim, U.; Pan, F. Surfactant dependent growth of twinned ZnO nanodisks. *Mater. Lett.* **2014**, *118*, 165–168. [CrossRef]
98. Qiu, J.; Weng, B.; Zhao, L.; Chang, C.; Shi, Z.; Li, X.; Kim, H.K.; Hwang, Y.H. Synthesis and characterization of flower-like bundles of ZnO nanosheets by a surfactant-free hydrothermal process. *J. Nanomater.* **2014**, *2014*. [CrossRef]
99. Ashfold, M.N.R.; Doherty, R.P.; Ndifor-Angwafor, N.G.; Riley, D.J.; Sun, Y. The kinetics of the hydrothermal growth of ZnO nanostructures. *Thin Solid Film.* **2007**, *515*, 8679–8683. [CrossRef]
100. Strom, J.G., Jr.; Jun, H.W. Kinetics of hydrolysis of methenamine. *J. Pharm. Sci.* **1980**, *69*, 1261–1263. [CrossRef] [PubMed]
101. Govender, K.; Boyle, D.S.; Kenway, P.B.; O'Brien, P. Understanding the factors that govern the deposition and morphology of thin films of ZnO from aqueous solution. *J. Mater. Chem.* **2004**, *14*, 2575–2591. [CrossRef]
102. Ruiz-Rojas, C.A.; Aguilar-Frutis, M.; Ramos-Brito, F.; Garduno-Wilches, I.A.; Narro-Ríos, J.; Lartundo-Rojas, L.; Alarcón-Flores, G. Synthesis, characterization, and temperature-dependent electronic properties of ZnO nanorods using CBD techniques. *J. Mater. Sci. Mater. Electron.* **2021**, 1–14. [CrossRef]
103. Parize, R.; Garnier, J.; Chaix-Pluchery, O.; Verrier, C.; Appert, E.; Consonni, V. Effects of Hexamethylenetetramine on the Nucleation and Radial Growth of ZnO Nanowires by Chemical Bath Deposition. *J. Phys. Chem. C* **2016**, *120*, 5242–5250. [CrossRef]
104. Tamvakos, D.; Lepadatu, S.; Antohe, V.-A.; Tamvakos, A.; Weaver, P.M.; Piraux, L.; Cain, M.G.; Pullini, D. Piezoelectric properties of template-free electrochemically grown ZnO nanorod arrays. *Appl. Surf. Sci.* **2015**, *356*, 1214–1220. [CrossRef]
105. Broitman, E.; Soomro, M.Y.; Lu, J.; Willander, M.; Hultman, L. Nanoscale piezoelectric response of ZnO nanowires measured using a nanoindentation technique. *Phys. Chem. Chem. Phys.* **2013**, *15*, 11113–11118. [CrossRef] [PubMed]
106. Fan, H.J.; Lee, W.; Hauschild, R.; Alexe, M.; Le Rhun, G.; Scholz, R.; Dadgar, A.; Nielsch, K.; Kalt, H.; Krost, A. Template-assisted large-scale ordered arrays of ZnO pillars for optical and piezoelectric applications. *Small* **2006**, *2*, 561–568. [CrossRef] [PubMed]
107. Scrymgeour, D.A.; Sounart, T.L.; Simmons, N.C.; Hsu, J.W.P. Polarity and piezoelectric response of solution grown zinc oxide nanocrystals on silver. *J. Appl. Phys.* **2007**, *101*, 14316–14321. [CrossRef]
108. Cavallini, D.; Fortunato, M.; De Bellis, G.; Sarto, M.S. PFM Characterization of Piezoelectric PVDF/ZnONanorod thin films. In Proceedings of the 2018 IEEE 18th International Conference on Nanotechnology (IEEE-NANO), Cork, Ireland, 23–26 July 2018; IEEE: Piscataway, NJ, USA, 2018; pp. 1–3.

nanomaterials

MDPI

Article

Fabrication of Piezoelectric ZnO Nanowires Energy Harvester on Flexible Substrate Coated with Various Seed Layer Structures

Taoufik Slimani Tlemcani *, Camille Justeau, Kevin Nadaud, Daniel Alquier and Guylaine Poulin-Vittrant

GREMAN UMR 7347, Université de Tours, CNRS, INSA Centre Val de Loire, 37071 Tours, France; camille.justeau@univ-tours.fr (C.J.); kevin.nadaud@univ-tours.fr (K.N.); daniel.alquier@univ-tours.fr (D.A.); guylaine.poulin-vittrant@univ-tours.fr (G.P.-V.)

* Correspondence: taoufik.slimani@univ-tours.fr; Tel.: +33-02-4742-8133

Abstract: Flexible piezoelectric nanogenerators (PENGs) are very attractive for mechanical energy harvesting due to their high potential for realizing self-powered sensors and low-power electronics. In this paper, a PENG that is based on zinc oxide (ZnO) nanowires (NWs) is fabricated on flexible and transparent Polydimethylsiloxane (PDMS) substrate. The ZnO NWs were deposited on two different seed layer structures, i.e., gold (Au)/ZnO and tin-doped indium-oxide (ITO)/ZnO, using hydrothermal synthesis. Along with the structural and morphological analyses of ZnO NWs, the electrical characterization was also investigated for ZnO NWs-based flexible PENGs. In order to evaluate the suitability of the PENG device structure, the electrical output performance was studied. By applying a periodic mechanical force of 3 N, the ZnO NWs-based flexible PENG generated a maximum root mean square (RMS) voltage and average power of 2.7 V and 64 nW, respectively. Moreover, the comparison between the fabricated device performances shows that a higher electrical output can be obtained when ITO/ZnO seed layer structure is adopted. The proposed ZnO NWs-based PENG structure can provide a flexible and cost-effective device for supplying portable electronics.

Keywords: ZnO nanowires; energy harvester; seed layer; flexible piezoelectric nanogenerator; hydrothermal synthesis

Citation: Slimani Tlemcani, T.; Justeau, C.; Nadaud, K.; Alquier, D.; Poulin-Vittrant, G. Fabrication of Piezoelectric ZnO Nanowires Energy Harvester on Flexible Substrate Coated with Various Seed Layer Structures. *Nanomaterials* **2021**, *11*, 1433. https://doi.org/10.3390/nano11061433

Academic Editors: Vincent Consonni and Filippo Giubileo

Received: 6 May 2021
Accepted: 26 May 2021
Published: 28 May 2021

1. Introduction

There has been a significant recent emphasis on mechanical energy harvesting for scavenging wasted or freely available ambient energy to generate electricity. In this context, piezoelectric nanogenerators (PENGs) are required to play a noteworthy role for long lifetime and reliable power supply [1,2].

During the last decade, many research studies have been implemented about the structural design, suitable material choices, understanding of the working mechanism, modeling, and performance optimization of the PENGs. Until recently, PENGs have been intensively developed and applied to harvest various forms of mechanical energies, starting with human motions to environmental scavenging [3,4]. Further, the adoption of flexible and stretchable PENGs has high research interest since it opens new opportunities for the development of flexible and wearable electronics [5,6]. Moreover, low-cost flexible substrates can be utilized for reducing production costs, which can make further progress on the mechanical energy harvesting market in terms of shortening the energy payback time. Besides low-cost flexible substrates, the use of low-cost materials is needed to obtain low production costs. Furthermore, ZnO, as a piezoelectric material, is beneficial for low cost as well as biosafety implementation, compared to lead-based PENGs [7].

ZnO is a promising material for a huge range of energy-saving applications, owing to its attractive semiconducting and piezoelectric properties [7,8]. Many reports have been already published on the physical properties of ZnO including nanorods (NRs) and nanowires (NWs) [9–14]. Ascribed to their remarkable properties, ZnO NWs have found widespread use in the development of PENGs.

In this work, we focus on the fabrication and characterization of ZnO NWs-based flexible PENGs with different seed layer structures. This report closely follows our previous work on rigid PENGs in which we reported the integration of seed layer to fabricate ZnO NWs-based rigid PENGs [11]. Herein, the growth of ZnO NWs using a low-temperature hydrothermal synthesis is reported. Furthermore, Au/ZnO and ITO/ZnO seed layer structures have been employed for the growth of ZnO NWs. Hence, in the present investigation, a proper and sincere attempt is made to understand the development of PENGs from the synthesis of ZnO NWs on flexible substrate right through to the design and fabrication of flexible PENG devices. The preparation of flexible substrates and the fabrication process of flexible PENGs are explained in detail in the Experimental Section.

2. Experimental

2.1. Flexible Substrate Preparation

Previous work from our group has already reported on the fabrication of PENGs by Dahiya et al. [15]. Originally, the PENGs were fabricated on a PDMS layer coated on a silicon substrate for easy handling during the fabrication steps. The PENGs were then released by peeling off the final devices from the Si substrate. However, Boubenia et al. [16] reported a decrease in the PENGs electrical performances after the peeling-off step of the device from the silicon substrate. The peeling-off step appears to introduce cracks and constraints in the metal layer electrode that could influence its durability and electrical performances; thus, this could lead to a decrease in the device lifetime.

Therefore, in this paper, we present a new process for PDMS substrate fabrication that helps to keep the PENGs' integrity before electrical measurements. The present proposed PENG process is different, compared with the previous device protocols [15–17]. Here, the complete PENG device was not lifted off from the rigid substrate, but only the PDMS layer was peeled off before starting the fabrication steps of the device. To avoid peeling off the entire PENG from the substrate, the PDMS film was only attached by its edges to a rigid frame. To do so, as described in Figure 1a, a rigid PET frame was placed on a plane PET surface. Then, a degassed mixture of elastomer and curing agent (10:1) forming liquid PDMS was spin-coated on the PET setup at 500 rpm for 15 s and then heated on a hot plate at 100 °C for 45 min. While curing, the PDMS was attached to the first PET frame; afterward, both elements were peeled off from the plane PET surface, freeing the PDMS surface (Figure 1b).

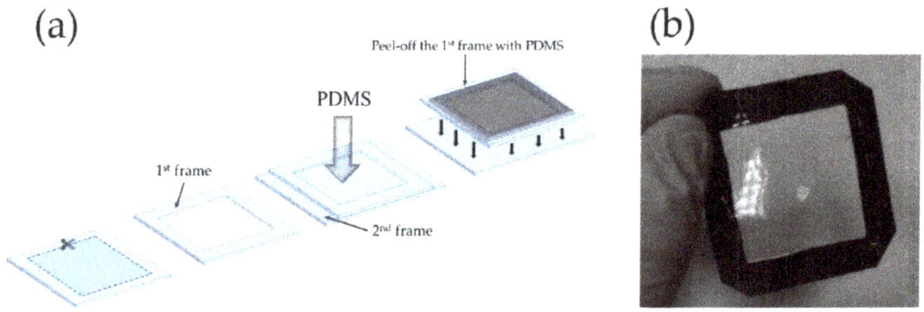

Figure 1. (**a**) Schematic illustration of PDMS substrate fabrication steps and (**b**) photograph of the PDMS substrate attached to the frame.

2.2. Bottom Electrode-Seed Layers Preparation (ITO/ZnO, Ti/Au/ZnO) and Growth of the ZnO Nanowires

In this study, two different bottom electrodes, i.e., Ti/Au and ITO, were deposited on the top of the PDMS substrate. The first electrode, Ti/Au (50 nm/100 nm), was DC sputtered at room temperature as follows: power of 500 W and pressure of 5 mTorr under argon atmosphere.

The second type of electrode consisted of coating the PDMS substrate with an ITO layer. In order to avoid cracks on its surface and ensure good conductivity, the ITO deposition protocol was based on the report of Lien et al. [18] and Casper et al. [19]. Therefore, a 400 nm thick layer of ITO was radio frequency (RF) sputtered at room temperature in two steps, as described in Table 1. The first thin ITO layer was sputtered at lower power to avoid any deformation of the PDMS substrate, which could impact the electrical integrity of ITO afterward. Once the PDMS substrate was prepared to receive the rest of the ITO layer, the second deposition step was performed at 65 W.

Table 1. ITO sputtering conditions.

Step	Pressure	Gas Flow	Power	Deposition Time	Thickness
1	5 mTorr	O_2:Ar (1:1)	50 W	5 min	≈20 nm
2	5 mTorr	Ar	65 W	58 min	380 nm

At last, on both electrode surfaces, a thin ZnO seed layer (100 nm) was deposited using RF sputtering. Specifically, the sputter power was 65 W, and the chamber pressure was 5 mTorr under Ar ambient. Therefore, to control the crystallinity and the morphology of ZnO NWs, the ZnO seed layer is required [20–22]. For the hydrothermal growth of ZnO NWs, the complete procedure was previously described in detail by Tlemcani et al. [10].

2.3. Fabrication of PENG Device

Figure 2 illustrates the fabrication process steps of the PENG device. ZnO NWs were synthesized on the PDMS substrate using the hydrothermal method, as shown in Figure 2a. Then, a parylene-C polymer matrix was deposited on the grown ZnO NWs by vapor phase pyrolysis technique following the Gorham steps [23], as seen in Figure 2b. In order to insulate the ZnO NWs from each other and create a capacitive coupling between the NWs and the top electrode, the parylene-C layer needed to be infiltrated into the sample [11,24,25].

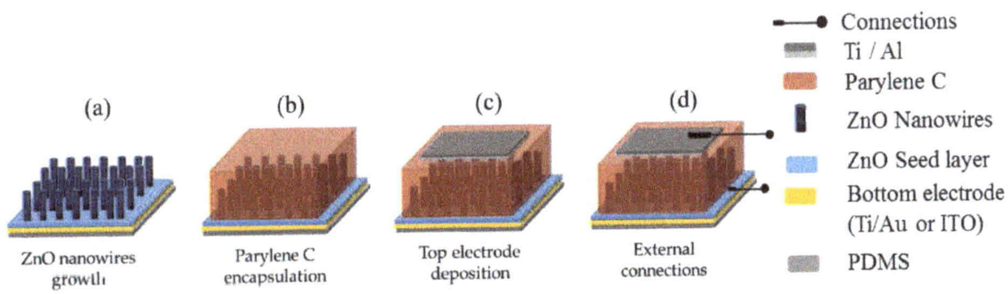

Figure 2. Fabrication steps of flexible PENG: (**a**) hydrothermal growth of ZnO NWs, (**b**) parylene-C deposition, (**c**) top electrode deposition, and (**d**) wires connection.

Then, the surface of the parylene-C was masked to deposit a 1.2 cm^2 Ti/Al (100 nm/400 nm) layer by evaporation (Figure 2c).

An example of cross-sectional scanning electron microscopy (SEM) images of the PENG can be seen in Figure 3. It clearly shows the good infiltration of parylene-C between the ZnO NWs.

Finally, the silver conductive adhesive paste was used to connect copper wires to the top and the bottom electrodes of the harvester (Figure 2d), and the whole PENG was encapsulated in PDMS to protect the devices from external factors and to improve their durability during the electrical analyses. Once the entire PENG fabrication was completed, the PET frame was removed to allow flexibility. The visual aspect of the PENGs is shown in Figure 4.

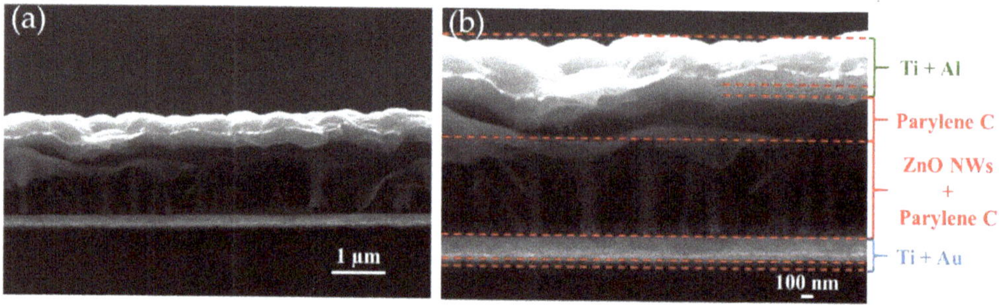

Figure 3. Cross-sectional SEM images of PENG at (**a**) 15,000× and (**b**) 30,000× magnifications.

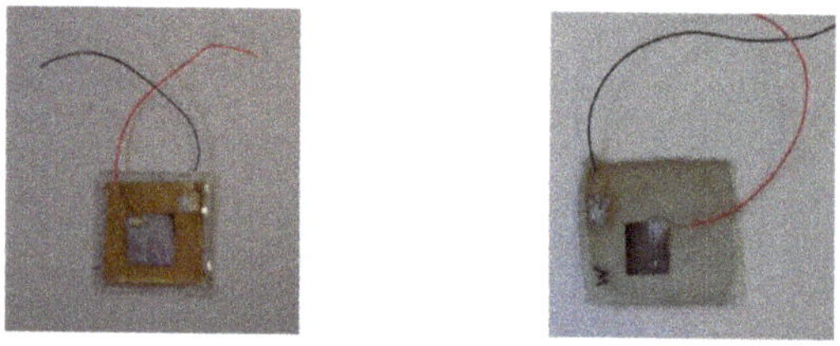

(a) **(b)**

Figure 4. Photograph of PENGs devices made from ZnO NWs deposited on (**a**) Au/ZnO and (**b**) ITO/ZnO seed layer structures.

2.4. Characterization Techniques

The crystal structure of ZnO NWs was examined using X-ray diffraction (XRD, Brucker AXS D8 discover) operated with CuKα radiation. The surface morphology was observed using scanning electron microscopy (SEM, JEOL JSM-7900F). The extraction of the morphological parameters was conducted using Image J software on SEM images [10]. The electrical measurements were performed using the test bench designed and described in detail by our team [17], as shown in Figure 5.

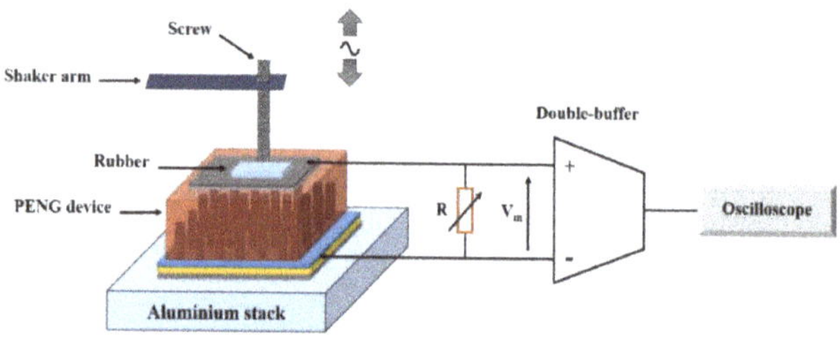

Figure 5. Test bench for PENG performance measurement.

The PENG was being solicited using the aluminum mechanical shaker, which impacted its active surface area (1 cm^2). The device was also connected to a double-buffer circuit comprising a differential amplifier, which measured the output performances without making any parasitic bias and allowed us to use a resistive load higher than the input impedance of the oscilloscope [26].

3. Results and Discussion

Figure 6 shows the XRD spectra of ZnO NWs deposited on Au/ZnO and ITO/ZnO seed layer structures, exhibiting the main diffraction peak of hexagonal würtzite ZnO structure (JCPDS Card No. 05-0664), without detection of any impurity phase. Therefore, ZnO NWs grown by hydrothermal technique led to a pure-phase material. The entire samples exhibit a strong peak at 34.4° corresponding to the lattice plane (0002) of würtzite ZnO structure, confirming that the NWs grew predominantly along the [0001] direction.

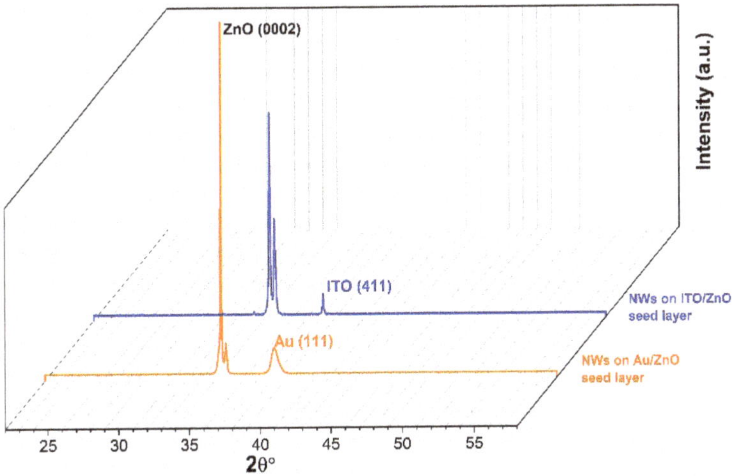

Figure 6. XRD 3D spectra of the ZnO NWs deposited on Au/ZnO and ITO/ZnO seed layer structures.

Additionally, it can be observed that the XRD pattern of ZnO NWs deposited on Au/ZnO presents the highest intensity of the (0002) peak, which is attributed to the improvement in the crystallinity of the NWs. Additionally, XRD patterns show a (0002) peak splitting for both samples. This can be related to the large lattice mismatch between seed layers and NWs, as detailed in our previous report [10].

Top and cross-sectional views of SEM micrographs of ZnO NWs deposited on the two seed layers are shown in Figure 7. It is obviously observed that the alignment and morphology of the NWs are strongly influenced by the seed layer types. Moreover, the SEM micrographs show that the ZnO NWs deposited on Au/ZnO seed layer structure have a much better alignment (Figure 7a) with a perpendicular orientation to the substrate surface, whereas the ZnO NWs deposited on ITO/ZnO seed layer structure are randomly oriented.

This result is consistent with the XRD patterns, shown in Figure 6, in which it was seen that the NWs deposited on Au/ZnO seed layer structure showed better crystallinity than those on the ITO/ZnO one. If the seed layer is a crucial structure for the preparation of high-quality ZnO NWs, it is also known that catalysts such as Cu, Ag, and Au integrated on the ZnO seed layer enhance the growth quality of ZnO NWs [27,28]. All these results are in good accordance with our previous studies [10,11] and validate the fact that the seed layer structure has an important influence on the structural and morphological characteristics of the resulting ZnO NWs. Additionally, Table 2 shows the morphological characteristics of ZnO NWs deposited on different seed layer structures. It can be noticed that the length of the NWs with ITO/ZnO seed layer structure is larger than those with Au/ZnO one. As a general observation, the morphological values of the NWs, particularly the density and diameter related to the two types of seed layer structures, are quite similar to each other.

The NWs grown on different seed layer structures have been used to realize PENGs, as described in Section 2.2. Those PENGs have been characterized by applying an alternative mechanical excitation with a magnitude of 3 N at a frequency of 5 Hz. The resistive load is connected at the output of the PENGs to emulate the current consumption of an external circuit, and this resistive load has been varied from 100 kΩ to 100 MΩ. Furthermore, the voltage at the resistive load is measured using an oscilloscope through the double buffer circuit [26]. Figure 8 shows the measured voltage at the output of the PENGs for two values of the resistive load, 100 kΩ (Figure 8a) and 100 MΩ (Figure 8b). It is clear that the resistive load value strongly affects the output voltage. We note that the curves are quite similar for both samples while, in each case, the voltage magnitude is in the mV range for 100 kΩ, and in the V range for 100 MΩ.

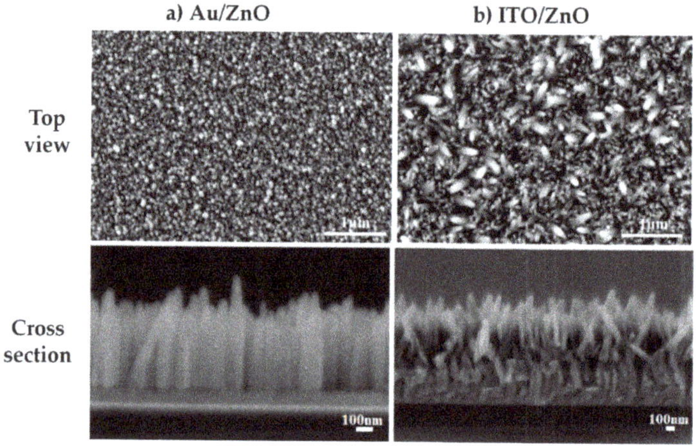

Figure 7. Top and cross-sectional views of SEM micrographs of ZnO NWs deposited on (**a**) Au/ZnO and (**b**) ITO/ZnO seed layer structures.

Table 2. Morphological characteristics of ZnO NWs deposited on Au/ZnO and ITO/ZnO seed layer structures.

Seed Layer Structure	NWs Length (μm)	NWs Diameter (μm)	Density (NWs/μm^2)	NWs Aspect Ratio
Au/ZnO	0.59 ± 0.15	0.07 ± 0.01	39.2 ± 0.3	9
ITO/ZnO	0.70 ± 0.17	0.07 ± 0.02	41.4 ± 0.2	10

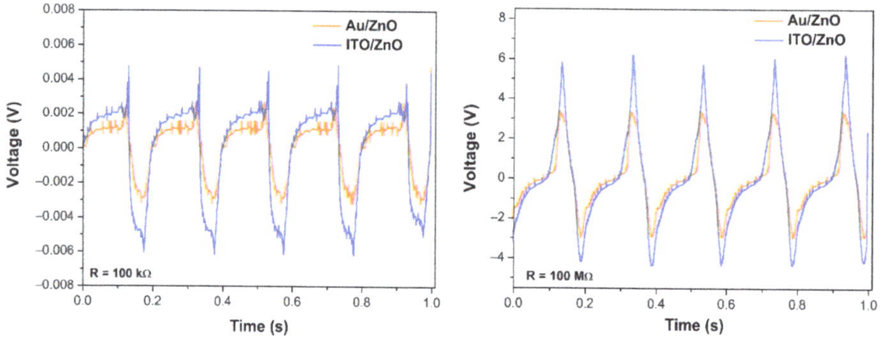

(a) (b)

Figure 8. Voltage amplitude of the fabricated PENGs with different seed layer structures at a load resistance of (**a**) 100 kΩ and (**b**) 100 MΩ, when applying an alternative mechanical excitation with a magnitude of 3 N at a frequency of 5 Hz.

Using the output voltage curves, the average electrical power was computed as a function of the load resistance, as reported in Figure 9 [26]. The root mean square (RMS) output voltage and the average power increase when the resistive load increases and the curves do not present a maximum on this range of resistive load, as generally observed, indicating that the optimal load can be certainly higher than 100 MΩ. In this case, the PENG can be used in two ways, either to charge a capacitor [29] or as an energy-autonomous wearable sensor [30].

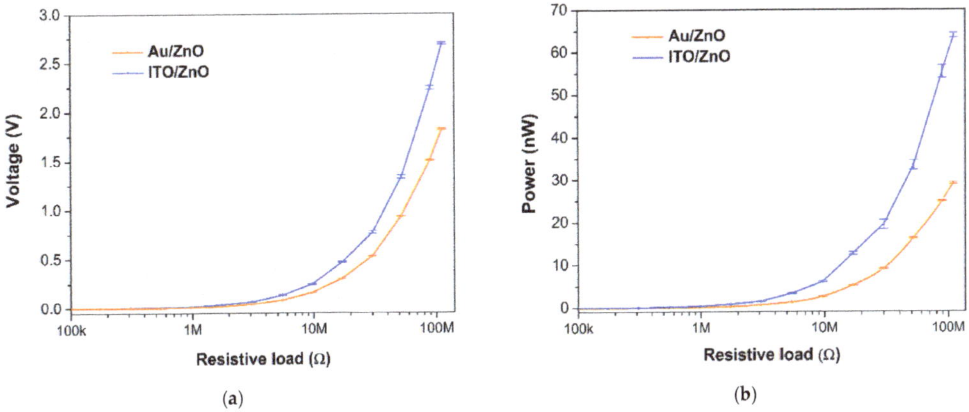

Figure 9. RMS output voltage (**a**) and average power (**b**) generated by PENGs with different seed layer structures. Measured RMS voltage and average power for a load resistance ranging from 100 kΩ to 100 MΩ, when applying an alternative mechanical excitation with a magnitude of 3 N at a frequency of 5 Hz.

The sample with the ITO/ZnO seed layer structure shows a maximum average power and RMS voltage of 64 nW and 2.7 V, respectively, which are higher than the values of the sample with Au/ZnO seed layer. The relatively low performance of the PENG device with Au layer (metal electrode) may be attributed to the interface adhesion deterioration, leading to possible delamination, wrinkling, or even cracking of the stacked film layers caused by the applied mechanical stress. It is important to note that the adherence of metal electrodes to PDMS substrate is usually poor [31]. Furthermore, among materials generally used in electronics devices, a thin metal layer is one of the weakest parts against mechanical stress and has a potential problem of cracking when stressed because of the degradation of its morphological and electrical properties during repeated mechanical excitation [32–35]. This could result in bad electrical charge transfer and increase the electrical losses, preventing the metal layer from being a functional bottom electrode. Moreover, the difference in the PENG performances might be explained by the different morphological characteristics of ZnO NWs. Actually, when comparing the aspect ratio, determined by the ratio of length over diameter (Table 2), the NWs deposited on Au/ZnO exhibited the lowest value with 9, compared to a ratio of 10 for NWs deposited on the ITO/ZnO seed layer. This reveals that the better uniformity obtained for NWs deposited on Au/ZnO has also promoted a decrease of their aspect ratio, which is needed to be high enough to afford large deformation of the NWs and to achieve efficient electromechanical energy conversion [36]. The electrical performances such as maximum peak instantaneous voltage (V_{peak}), maximum RMS voltage $(V_{RMS})_{max}$, maximum RMS short circuit current (I_{sc}), and maximum average power $(P_{av})_{max}$ of both samples, deduced from Figure 9a,b, are summarized in Table 3. In this table, the electrical energy ($W = P_{av}/f$) generated by each PENG is also indicated, where f is the frequency value of the applied mechanical force (here f = 5 Hz).

Table 3. PENG device performances for various seed layer structures: V_{peak}, $V_{RMS\ (max)}$, $I_{sc\ (RMS)}$, $P_{av\ (max)}$, energy during one period of mechanical input force (W), and corresponding optimal load (R_{opt}), when applying an alternative mechanical excitation with a magnitude of 3 N at a frequency of 5 Hz.

Seed Layer Structure	V_{peak} (V)	$V_{RMS\ (max)}$ (V)	$I_{sc\ (RMS)}$ (nA)	$P_{av\ (max)}$ (nW)	W (nJ)	R_{opt} (MΩ)
Au/ZnO	3.9	1.8	15	29	5.8	>100
ITO/ZnO	6.8	2.7	28	64	12.8	>100

To obtain a further insight into the performances of ZnO NWs-based PENGs, Table 4 shows a comparison of the obtained PENG device performances with our previously reported values and some of the interesting studies available in the literature. These reports have been devoted to the fabrication of PENG devices on a rigid silicon substrate via Au/ZnO seed layer (V_{peak} of 0.27 V) [11], flexible PDMS substrate via Au/ZnO seed layer (V_{peak} of 9.1 V [17], V_{peak} of 2.03 V [37]), and flexible PDMS substrate using ITO/ZnO seed layer (V_{peak} of 8 V [38]). It is evident from Table 4 that the value of the applied stress (mechanical loading) was not always mentioned, which is essential to compare the characteristics among the PENG devices. A notable performance has been achieved by Dahiya et al. for flexible devices. However, it can be readily observed that the mechanical stress (13 N) was four times higher, compared with the present study. Additionally, the proposed flexible PENG employs a transparent ITO electrode layer. Further comparisons with the state-of-the-art performances for ZnO NWs-based PENG devices are not easy due to the lack of a standardized testing protocol as well as insufficient information available regarding the experimental protocol in the literature.

Table 4. Comparison of the characteristics of ZnO NWs-based PENG devices fabricated on different substrates and seed layers.

Substrate	Seed Layer Structure	V_{peak} (V)	Mechanical Loading: Force (N)	Device Dimensions (cm²)	Ref.
Silicon	Au/ZnO	0.27	3	1.2	[11]
PDMS	Au/ZnO	9.1	13	1.2	[17]
PDMS	Au/ZnO	2.03	N/A	1	[37]
PDMS	ITO/ZnO	8	N/A	1.5	[38]
PDMS	ITO/ZnO	6.8	3	1.2	This work

Consequently, the difference between the performances of flexible and rigid devices might be related to several factors. An explanation is that the transition from a rigid to a flexible substrate can probably modify the way in which the NWs are constrained, and the distribution of the forces can be more three-dimensional than only along one direction (vertical) [39].

Overall, the presently proposed ZnO NWs-based flexible PENGs have shown quite interesting performances. Further thorough investigations when varying the thickness of both the seed layer structure and the PDMS substrate, and/or adoption of aluminum-doped ZnO (AZO) instead of ZnO seed layer could further increase the device performance levels.

4. Conclusions

In summary, flexible PENGs based on ZnO NWs have been fabricated employing Au/ZnO and ITO/ZnO seed layer structures. This approach involves a new protocol for the preparation of PDMS substrate with an original design of functional flexible PENG devices. These devices are based on the deposition of ZnO NWs on PDMS substrate using different seed layer structures exhibiting high crystallinity, good flexibility, and promising electrical properties. The fabricated devices with ITO/ZnO seed layer structure showed twofold higher average power than those with the Au/ZnO one. This suggests that it is possible to employ transparent and cost-effective ITO electrodes to fabricate energy-efficient flexible PENG devices. However, though this study presents a new way to fabricate flexible PENGs, exhibiting a promising prospect for transparent flexible devices, further progresses and challenges are still needed to improve the substrate flexibility, fabrication process, and electrical performances.

Author Contributions: Conceptualization, methodology, and writing—original draft preparation, T.S.T. and C.J.; formal analysis and data curation, T.S.T.; writing—review and editing, T.S.T., K.N., D.A., and G.P.-V.; supervision, K.N., D.A., and G.P.-V. All authors have read and agreed to the published version of the manuscript.

Funding: This research was funded by the EnSO project through the electronic components and systems for European leadership joint undertaking in collaboration with the European Union's H2020 framework program (H2020/2014-2020) and national authorities (grant agreement number 692482).

Institutional Review Board Statement: Not applicable.

Informed Consent Statement: Not applicable.

Data Availability Statement: The data presented in this study are available from the corresponding author upon request.

Conflicts of Interest: The authors declare no conflict of interest.

References

1. Sripadmanabhan Indira, S.; Aravind Vaithilingam, C.; Oruganti, K.S.P.; Mohd, F.; Rahman, S. Nanogenerators as a Sustain-able Power Source: State of Art, Applications, and Challenges. *Nanomaterials* **2019**, *9*, 773. [CrossRef] [PubMed]
2. Hu, Y.; Wang, Z.L. Recent progress in piezoelectric nanogenerators as a sustainable power source in self-powered systems and active sensors. *Nano Energy* **2015**, *14*, 3–14. [CrossRef]
3. Dagdeviren, C.; Yang, B.D.; Su, Y.; Tran, P.L.; Joe, P.; Anderson, E.; Xia, J.; Doraiswamy, V.; Dehdashti, B.; Feng, X.; et al. Conformal piezoelectric energy harvesting and storage from motions of the heart, lung, and diaphragm. *Proc. Natl. Acad. Sci. USA* **2014**, *111*, 1927–1932. [CrossRef]
4. Ye, S.; Cheng, C.; Chen, X.; Chen, X.; Shao, J.; Zhang, J.; Hu, H.; Tian, H.; Li, X.; Ma, L.; et al. High-performance piezoelec-tric nanogenerator based on microstructured P(VDF-TrFE)/BNNTs composite for energy harvesting and radiation protection in space. *Nano Energy* **2019**, *60*, 701–714. [CrossRef]
5. Liu, Y.; Wang, L.; Zhao, L.; Yu, X.; Zi, Y. Recent progress on flexible nanogenerators toward self-powered systems. *InfoMat* **2020**, *2*, 318–340. [CrossRef]
6. Fan, F.R.; Tang, W.; Wang, Z.L. Flexible Nanogenerators for Energy Harvesting and Self-Powered Electronics. *Adv. Mater.* **2016**, *28*, 4283–4305. [CrossRef] [PubMed]
7. Wang, Z.L.; Song, J.H. Piezoelectric Nanogenerators Based on Zinc Oxide Nanowire Arrays. *Science* **2006**, *312*, 42–46. [CrossRef]
8. Borysiewicz, M.A. ZnO as a Functional Material, a Review. *Crystals* **2019**, *9*, 505. [CrossRef]
9. Fakhri, P.; Amini, B.; Bagherzadeh, R.; Kashfi, M.; Latifi, M.; Yavari, N.; Kani, S.A.; Kong, L. Flexible hybrid structure piezo-electric nanogenerator based on ZnO nanorod/PVDF nanofibers with improved output. *RSC Adv.* **2019**, *9*, 17–23. [CrossRef]
10. Tlemcani, T.S.; Justeau, C.; Nadaud, K.; Poulin-Vittrant, G.; Alquier, D. Deposition Time and Annealing Effects of ZnO Seed Layer on Enhancing Vertical Alignment of Piezoelectric ZnO Nanowires. *Chemosensors* **2019**, *7*, 7. [CrossRef]
11. Justeau, C.; Tlemcani, T.S.; Poulin-Vittrant, G.; Nadaud, K.; Alquier, D. A Comparative Study on the Effects of Au, ZnO and AZO Seed Layers on the Performance of ZnO Nanowire-Based Piezoelectric Nanogenerators. *Materials* **2019**, *12*, 2511. [CrossRef]
12. Boubenia, S.; Dahiya, A.S.; Poulin-Vittrant, G.; Morini, F.; Nadaud, K.; Alquier, D. A facile hydrothermal approach for the density tunable growth of ZnO nanowires and their electrical characterizations. *Sci. Rep.* **2017**, *7*, 1–10. [CrossRef] [PubMed]
13. Poulin-Vittrant, G.; Oshman, C.; Opoku, C.; Dahiya, A.S.; Camara, N.; Alquier, D.; Tran Huu Hue, L.P.; Lethiecq, M. Fabri-cation and Characterization of ZnO Nanowire-based Piezoelectric Nanogenerators for Low Frequency Mechanical Energy Harvesting. *Phys. Procedia* **2015**, *70*, 909–913. [CrossRef]
14. Ou, C.; Sanchez-Jimenez, P.E.; Datta, A.; Boughey, F.L.; Whiter, R.A.; Sahonta, S.L.; Kar-Narayan, S. Template-assisted hy-drothermal growth of aligned zinc oxide nanowires for piezoelectric energy harvesting applications. *Appl. Mater. Interfaces* **2016**, *8*, 13678–13683. [CrossRef] [PubMed]
15. Dahiya, A.S.; Morini, F.; Boubenia, S.; Justeau, C.; Nadaud, K.; Rajeev, K.P.; Alquier, D.; Poulin-Vittrant, G. Zinc oxide nanowire-parylene nanocomposite based stretchable piezoelectric nanogenerators for self-powered wearable electronics. *J. Phys. Conf. Ser.* **2018**, *1052*, 012028. [CrossRef]
16. Boubenia, S. Générateurs Piézoélectriques à Base de Nanofils Piézo-Semiconducteurs: Modélisation, Fabrication et Caractérisation. Ph.D. Thesis, University of Tours, Tours, France, 2019.

17. Dahiya, A.S.; Morini, F.; Boubenia, S.; Nadaud, K.; Alquier, D.; Poulin-Vittrant, G. Organic/Inorganic Hybrid Stretchable Piezoelectric Nanogenerators for Self-Powered Wearable Electronics. *Adv. Mater. Technol.* **2018**, *3*, 1700249. [CrossRef]
18. Lien, S.-Y.; Nautiyal, A.; Lee, S.J. Optoelectronic Properties of Indium–Tin Oxide Films Deposited on Flexible and Transparent Poly(dimethylsiloxane) Substrate. *Jpn. J. Appl. Phys.* **2013**, *52*, 115801. [CrossRef]
19. Casper, M.D.; Gözen, A.Ö.; Dickey, M.D.; Genzer, J.; Maria, J.-P. Surface wrinkling by chemical modification of poly(dimethylsiloxane)-based networks during sputtering. *Soft Matter* **2013**, *9*, 7797–7803. [CrossRef]
20. Ghayour, H.; Rezaie, H.R.; Mirdamadi, S.; Nourbakhsh, A.A. The effect of seed layer thickness on alignment and morphol-ogy of ZnO nanorods. *Vaccum* **2011**, *86*, 101–105. [CrossRef]
21. Guillemin, S.; Consonni, V.; Appert, E.; Puyoo, E.; Rapenne, L.; Roussel, H. Critical Nucleation Effects on the Structural Rela-tionship Between ZnO Seed Layer and Nanowires. *J. Phys. Chem. C* **2012**, *116*, 25106–25111. [CrossRef]
22. Elzein, B.; Yao, Y.; Barham, A.S.; Dogheche, E.; Jabbour, G.E. Toward the Growth of Self-Catalyzed ZnO Nanowires Perpen-dicular to the Surface of Silicon and Glass Substrates, by Pulsed Laser Deposition. *Materials* **2020**, *13*, 4427. [CrossRef]
23. Gorham, W.F. A New, General Synthetic Method for the Preparation of Linear Poly-p-xylylenes. *J. Polym. Sci. Part A-1 Polym. Chem.* **1966**, *4*, 3027–3039. [CrossRef]
24. Yang, D.; Qiu, Y.; Jiang, Q.; Guo, Z.; Song, W.; Xu, J.; Zong, Y.; Feng, Q.; Sun, X. Patterned growth of ZnO nanowires on flexible substrates for enhanced performance of flexible piezoelectric nanogenerators. *Appl. Phys. Lett.* **2017**, *110*, 063901. [CrossRef]
25. Lee, K.Y.; Bae, J.; Kim, S.; Lee, J.-H.; Yoon, G.C.; Gupta, M.K.; Kim, S.; Kim, H.; Park, J.; Kim, S.-W. Depletion width engineering via surface modification for high performance semiconducting piezoelectric nanogenerators. *Nano Energy* **2014**, *8*, 165–173. [CrossRef]
26. Nadaud, K.; Morini, F.; Dahiya, A.S.; Justeau, C.; Boubenia, S.; Rajeev, K.P.; Alquier, D.; Poulin-Vittrant, G. Double buffer circuit for the characterization of piezoelectric nanogenerators based on ZnO nanowires. *Appl. Phys. Lett.* **2018**, *112*, 063901. [CrossRef]
27. Giri, P.; Dhara, S.; Chakraborty, R. Effect of ZnO seed layer on the catalytic growth of vertically aligned ZnO nanorod arrays. *Mater. Chem. Phys.* **2010**, *122*, 18–22. [CrossRef]
28. Li, S.Y.; Lee, C.Y.; Tseng, T.Y. Copper-catalyzed ZnO nanowires on silicon (1 0 0) grown by vapor–liquid–solid process. *J. Cryst. Growth* **2003**, *247*, 357–362. [CrossRef]
29. Van den Heever, T.S.; Perold, W.J. Comparing three different energy harvesting circuits for a ZnO nanowire based nano-generator. *Smart Mater. Struct.* **2013**, *22*, 105029. [CrossRef]
30. Dahiya, A.S.; Thireau, J.; Boudaden, J.; Lal, S.; Gulzar, U.; Zhang, Y.; Gil, T.; Azemard, N.; Ramm, P.; Kiessling, T.; et al. Review—Energy Autonomous Wearable Sensors for Smart Healthcare: A Review. *J. Electrochem. Soc.* **2020**, *167*, 037516. [CrossRef]
31. Schmid, H.; Wolf, H.; Allenspach, R.; Riel, H.; Karg, S.; Michel, B.; Delamarche, E. Preparation of Metallic Films on Elasto-meric Stamps and Their Application for Contact Processing and Contact Printing. *Adv. Funct. Mater.* **2003**, *13*, 145–153. [CrossRef]
32. Kim, T.-W.; Lee, J.-S.; Kim, Y.-C.; Joo, Y.-C.; Kim, B.-J. Bending Strain and Bending Fatigue Lifetime of Flexible Metal Electrodes on Polymer Substrates. *Materials* **2019**, *12*, 2490. [CrossRef] [PubMed]
33. Huck, W.T.S.; Bowden, N.; Onck, P.; Pardoen, T.; Hutchinson, A.J.W.; Whitesides, G.M. Ordering of Spontaneously Formed Buckles on Planar Surfaces. *Langmuir* **2000**, *16*, 3497–3501. [CrossRef]
34. Sekitani, T. Someya, Stretchable, Large-area Organic Electronics. *Adv. Mater.* **2010**, *22*, 2228–2246. [CrossRef] [PubMed]
35. Gupta, R. Fabrication of stretchable compliant electrodes on PDMS with Au nanoparticles. *Bull. Mater. Sci.* **2018**, *41*, 114. [CrossRef]
36. Gao, Y.; Wang, Z.L. Electrostatic Potential in a Bent Piezoelectric Nanowire. The Fundamental Theory of Nanogenerator and Nanopiezotronics. *Nano Lett.* **2007**, *7*, 2499–2505. [CrossRef]
37. Zhu, G.; Yang, R.; Wang, S.; Wang, Z.L. Flexible High-Output Nanogenerator Based on Lateral ZnO Nanowire Array. *Nano Lett.* **2010**, *10*, 3151–3155. [CrossRef]
38. Lin, L.; Hu, Y.; Xu, C.; Zhang, Y.; Zhang, R.; Wen, X.; Wang, Z.L. Transparent flexible nanogenerator as self-powered sensor for transportation monitoring. *Nano Energy* **2013**, *2*, 75–81. [CrossRef]
39. Hinchet, R.; Lee, S.; Ardila, G.; Montès, L.; Mouis, M.; Wang, Z.L. Performance Optimization of Vertical Nanowire-based Piezoelectric Nanogenerators. *Adv. Funct. Mater.* **2013**, *24*, 971–977. [CrossRef]

nanomaterials

MDPI

Article

Controlled Fabrication of Quality ZnO NWs/CNTs and ZnO NWs/Gr Heterostructures via Direct Two-Step CVD Method

Nicholas Schaper [1], Dheyaa Alameri [1,2], Yoosuk Kim [1], Brian Thomas [1,3], Keith McCormack [1], Mathew Chan [1], Ralu Divan [4], David J. Gosztola [4], Yuzi Liu [4] and Irma Kuljanishvili [1,*]

1 Department of Physics, Saint Louis University, St. Louis, MO 63103, USA; nicholas.schaper@slu.edu (N.S.); dheyaa.alameri@slu.edu (D.A.); yoosuk.kim@slu.edu (Y.K.); brian.thomas@slu.edu (B.T.); keith.mccormack@slu.edu (K.M.); mathewdave.chan@slu.edu (M.C.)
2 Department of Physics, College of Science, University of Misan, Maysan 62001, Iraq
3 Parks College of Engineering, Aviation and Technology, Saint Louis University, St. Louis, MO 63103, USA
4 Center for Nanoscale Materials, Argonne National Laboratory, 9700 S. Cass Avenue, Lemont, IL 60439, USA; divan@anl.gov (R.D.); Gosztola@anl.gov (D.J.G.); yuziliu@anl.gov (Y.L.)
* Correspondence: irma.kuljanishvili@slu.edu

Abstract: A novel and advanced approach of growing zinc oxide nanowires (ZnO NWs) directly on single-walled carbon nanotubes (SWCNTs) and graphene (Gr) surfaces has been demonstrated through the successful formation of 1D–1D and 1D–2D heterostructure interfaces. The direct two-step chemical vapor deposition (CVD) method was utilized to ensure high-quality materials' synthesis and scalable production of different architectures. Iron-based universal compound molecular ink was used as a catalyst in both processes (a) to form a monolayer of horizontally defined networks of SWCNTs interfaced with vertically oriented ZnO NWs and (b) to grow densely packed ZnO NWs directly on a graphene surface. We show here that our universal compound molecular ink is efficient and selective in the direct synthesis of ZnO NWs/CNTs and ZnO NWs/Gr heterostructures. Heterostructures were also selectively patterned through different fabrication techniques and grown in predefined locations, demonstrating an ability to control materials' placement and morphology. Several characterization tools were employed to interrogate the prepared heterostructures. ZnO NWs were shown to grow uniformly over the network of SWCNTs, and much denser packed vertically oriented ZnO NWs were produced on graphene thin films. Such heterostructures can be used widely in many potential applications, such as photocatalysts, supercapacitors, solar cells, piezoelectric or thermal actuators, as well as chemical or biological sensors.

Keywords: carbon nanotubes; zinc oxide nanowires; graphene; heterostructure interfaces; chemical vapor deposition; direct-write patterning

Citation: Schaper, N.; Alameri, D.; Kim, Y.; Thomas, B.; McCormack, K.; Chan, M.; Divan, R.; Gosztola, D.J.; Liu, Y.; Kuljanishvili, I. Controlled Fabrication of Quality ZnO NWs/CNTs and ZnO NWs/Gr Heterostructures via Direct Two-Step CVD Method. *Nanomaterials* **2021**, *11*, 1836. https://doi.org/10.3390/nano11071836

Academic Editor: Vincent Consonni

Received: 1 June 2021
Accepted: 9 July 2021
Published: 15 July 2021

1. Introduction

Perpetually evolving technology requires novel materials and architecture designs to keep up with the growing, industry-driven demand towards miniaturization [1]. Nanoscale materials have been suggested as attractive alternatives to traditional materials because of their unique properties, which are consequential to their low dimensionality, that can be engineered to satisfy the needs of the next generation of applications. Specifically, 1D nanostructures have been intensely studied because they could be manufactured with targeted morphological characteristics and unique physical and electronic properties, mechanical strength, optical sensitivity, etc. [2,3]. Nanomaterials play an essential role in developing life-saving therapeutics due to selective antibiotic [4,5] and anti-cancer [4,6–11] properties, in addition to their potential for promising therapies for challenging pathological conditions [12–14].

For instance, 1D CNTs have drawn much attention for their high aspect ratio, large surface area, and selective electrical properties making them ideal for scaffolds, functionalization with biomolecules, and electrode connections in devices [15–19]. SWCNTs' electrical properties can be selectively developed as either semiconducting or metallic, suggesting an ability to tailor them to targeted applications. Since all CNTs produced in our study were single-walled carbon nanotubes, we used an abbreviation CNTs to describe SWCNTs from this point forward unless otherwise specified. ZnO NWs are the example of another 1D nanomaterial that has been proposed for applications in optoelectronics, photocatalysis, and charge collection and transport at heterojunctions due to their unique optical, electrical, and piezoelectric properties [20–25]. ZnO is a wide bandgap semiconductor (~3.37 eV) with high exciton binding energy (~60 meV) that has also been shown to exhibit biocompatibility and low cytotoxicity [26,27], which makes it an excellent companion to 1D carbon nanotubes and even 2D graphene.

Many reported studies have investigated the unique properties of specific nanomaterials and have explored their myriad useful applications. However, fewer studies have focused on developing scalable methods for engineering heterostructures and interfaces with different nanomaterials to complement each other synergistically.

The flexibility, optical transparency, and electrical properties of 2D atomically thin graphene have generated interest in its use in photonics, transistors, and applications in energy production and storage [28] making it a valuable material to investigate for devices.

Preparing interfaces between two or more nanomaterials provides opportunities to harness properties inherent to each material and synergistically combine them to create a broader range of applications. Hybrid structures assembled from CNTs and ZnO NWs [20,29] and graphene and ZnO NWs [30–37] have been examined and shown promising results of the enhancement of their native properties when two materials are integrally combined. In a recent study, researchers have reported high-performance supercapacitor electrodes (~192 F/g capacitance) manufactured using hybrid structures of ZnO NWs on aligned MWCNTs [38]. Another study has shown that ZnO attached to MWCNTs can be useful for ultrafast nonlinear optical switching [17]. Investigations into hybrid systems containing CNTs and zinc oxide also include ZnO nanoparticles (NPs) beaded on CNTs [17], ZnO NPs dispersed in scaffolds containing CNTs [26], and ZnO nanorods grown on bundles of buckled multi-walled carbon nanotubes (MWCNTs) [29].

The heterostructure formed between graphene and ZnO NWs has also been a motivation for researchers. Previous studies have reported ZnO NWs grown on graphene via a solution-based hydrothermal method and have shown that graphene can be beneficial to the growth of ZnO NWs and can enhance their field electron emission [30]. At the same time, hybrid systems made of Gr/ZnO or graphene oxide/ZnO structures have also been shown to operate as hydrogen sensors [31], strain sensors [32], and for the photoinactivation of bacteria [33] effectively. Hybrid materials in these studies were formed through hydrothermal growth or proximity placement methods; we found that the piezoelectric, photoactive, and antibacterial properties of 1D ZnO NWs combine well with the conductive, transparent 2D support structure of the graphene sheet. This makes hybrid ZnO NWs/Gr structures a promising nanomaterial for use in sensors [31,32], photovoltaic cells [34], and energy storage [35].

With respect to the methodologies of producing hybrid ZnO NWs/CNTs systems, MWCNTs have been successfully implemented. For example, in the study by Ok et al., ZnO NWs were produced on the surface of vertically aligned MWCNT forests, exhibiting enhanced electrical responses while remaining mechanically robust [20]. The authors' method consisted in growing a vertically aligned MWCNT forest via CVD at 775 °C, followed by the growth of ZnO NWs on the outer surfaces of these tube clusters in a subsequent thermal evaporative method at ~420 °C. In their process, the entire MWCNTs forest was encapsulated in ZnO NWs, creating a structure with a MWCNT core and ZnO NW outer shell. They found that the intimate contact between the photoactive, semiconducting ZnO NWs with the conductive CNTs produced a rapid photoelectric response. Due to the unique organization of the MWCNT forest with overlying ZnO NWs, the materials have been shown to have anisotropic electrical conductivity. Another group produced horizontally aligned buckled MWCNTs forests through pyrolysis followed by

CVD growth of ZnO NWs on the sidewalls of these structures. They have shown the material to have Schottky-like behavior and p-type conductivity, which allows it to be used in applications of ultraviolet detectors, high current p-type field-effect transistors, and other multifunctional devices [29]. Despite these promising results, methodologies combining low-temperature ZnO NW synthesis, such as hydrothermal growth (70–200 °C), require extended growth periods (12–48 h) and do not produce high-quality crystal structures capable through CVD [37,39]. Additionally, the use of MWCNT bundles limits one's control over the electrical properties of the material unlike the use of SWCNTs, which allow greater flexibility for being tailored to either semiconductive or metallic performance.

There has also been considerable interest in developing hybrid structures of graphene and ZnO in various forms. Researchers have been successful in developing methods such as transferring as-grown ZnO NWs onto graphene [34], growing ZnO nanorods directly on graphene via a hydrothermal process [30,35], or even electrophoretic deposition of graphene oxide within ZnO NWs [33], among other methods [40]. Although effective in creating heterostructures, and some even at lower temperatures, these methods did not produce high-quality materials and are time-consuming (involve many more steps) unlike those observed with CVD.

In this study, we have presented a novel approach to demonstrate a selective, scalable two-step CVD method of growing high-quality heterostructure interfaces of ZnO NWs/SWCNTs and ZnO NWs/Gr. Our bottom-up synthesis approach produces nanomaterials with highly organized crystal structures. Here, for the first time, we developed a method of producing ZnO NWs uniformly on a flat network of CNTs in a two-step CVD process using our universal catalytic Fe ink to catalyze both materials. This creates an interface between the semiconducting ZnO NWs and the underlying network of semiconducting CNTs, which would provide future opportunities to engineer more complex architectures based on the properties of both materials given the ability to control electrical properties of the base layer, which consists of carbon nanostructures, by adjusting the expression of CNT characteristics.

Additionally, we realized a two-step CVD procedure whereby ZnO NWs are grown directly on top of CVD-grown graphene. Dense ZnO NW forests were grown on the surface of graphene, covering the entire area of the conductive 2D sheet with semiconducting ZnO NWs. This process is efficient and scalable, producing high-quality ZnO NWs/Gr interfaces with desirable morphologies, exploiting unique surface characteristics or graphene, its atomic flatness and significant hydrophobicity, thus enabling to tune the resulting morphology of the ZnO NWs/Gr system.

Finally, we showed that our catalytic ink precursor allows for selective heterostructure growth in predetermined areas demonstrated with our novel direct-write patterning (DWP) approach. The ability to create intimate interfaces between these unique materials provides platforms for generating an opportunity for the two or more materials to complement one another's physical and electrical properties or mechanical and thermal performance.

2. Materials and Methods

2.1. Materials

Iron (III) nitride nonahydrate ($Fe(NO_3)_3{\cdot}9H_2O$) (Sigma-Aldrich, St. Louis, MO, USA, 99.99% purity), N,N-dimethylformamide (Sigma-Aldrich, St. Louis, MO, USA, ≥99% purity), and glycerol (Sigma-Aldrich, St. Louis, MO, USA, ≥99.0% purity) were used as solvents in the custom composite iron-based (CCFe) ink. Graphite (Alfa Aesar, Haverhill, MA, USA, 99.999% purity) was mixed with ZnO powder (Alfa Aesar, Haverhill, MA, USA, 99.999% purity) in equal proportion (35 mg each) and used as ZnO/C source precursors during the growth process. Copper (Cu) foil (Alfa Aesar, Haverhill, MA, USA, 99.8% purity and thickness of 0.025 mm) was used as the catalyzing substrate for Gr synthesis. Si/SiO_2 wafers (UniversityWafer, South Boston, MA, P/B, (100), resistivity: 1–5 mΩ·cm) with a 285 nm wet thermal oxide layer were cut into pieces measuring 5×5 cm^2 and used as growth substrates in this study.

2.2. *Direct-Write Patterning (DWP) Method*

DWP was utilized for delivering catalytic ink directly into etched-in features. This was performed with a custom-built instrument equipped with three piezo-driven stages which aid precise XYZ manipulation/positioning [37]. Additionally, atomic force microscopy (AFM) cantilevers outfitted with multiple tips (12 pens) and custom ink reservoirs were utilized during patterning (cantilevers and ink reservoirs purchased from Advanced Creative Solutions Technology LLC, Des Plaines, IL, USA). A stock "master solution" (MS) was created by dissolving 2.8 mg of iron (III) nitride nonahydrate in 12 mL of deionized (DI) water (18.2 M·Ω cm) and sonicated for 10 min (BRANSON, Model # 2800, 40 kHz). To prepare the CCFe ink, the master stock solution was further diluted in solvents such as N,N-dimethylformamide (DMF) and glycerol at parts per volume ratio as follows: 6MS/3DI/2DMF/1Glycerol. Mixed CCFe precursor ink was used throughout the study either for dip coating and direct-write patterning [41].

2.3. *CVD Synthesis Methods*

2.3.1. CVD Synthesis of CNTs

A catalytic CVD method modified from previously established protocols [41] was modified and optimized to grow dense, randomly oriented SWCNT networks. The CVD system equipped with a three-zone furnace (Thermo Scientific Lindberg/Blue M, Waltham, MA, USA), a quartz tube (6 ft L; ID, 22 mm; OD, 25 mm) (Technical Glass Products, Painesville, OH, USA), and a digital mass flow controller (Sierra Instrument, Monterey, CA. USA) was used in all the experiments. The substrates were cleaned in DI water, acetone, and isopropanol by 10-min sonication (BRANSON, Model #2800, 40 kHz). They were further processed in a UV ozone system for 2 min (Novascan PSD Pro Series Digital UV Ozone System, Boone, IA, USA) to increase hydrophilicity immediately prior to coating with the CCFe ink (see Section 2.1). Several SiO_2/Si samples dip-coated in the CCFe ink were placed in a 5-inch quartz boat (Technical Glass Products). CVD growth of CNTs was performed under the following conditions: first, the system was purged with ultrahigh-purity Ar (500 sccm) for 10 min at 25 °C. The temperature was then increased to 365 °C under Ar/H_2 (300/150 sccm), and the samples were preconditioned for 65 min. H_2 gas (500 sccm) alone was run for 10 min at 900 °C before growth. A mix of H_2/Ar/CH_4 (140/60/900 sccm) was used during the growth for 17–20 min. The furnace was then cooled under the protection of H_2/Ar (100/200 sccm).

2.3.2. CVD Synthesis of ZnO NWs

For CVD growth of ZnO NWs, an additional, smaller quartz tube with one end closed (L, 600 mm; ID, 16 mm; OD, 18 mm) (Technical Glass Products, Painesville, OH, USA) contained the ZnO/C precursor and samples. The precursor was placed in a small quartz boat (L, 20 mm; W, 10 mm; H, 8 mm) (Technical Glass Products, Painesville, OH, USA) at the closed end of the quartz tube in a ratio of 35 mg/35 mg (m/m) (graphite/ZnO). The samples were placed near the open end of the tube, approximately 3–5 cm from the opening. The small-diameter quartz tube was placed inside of the CVD with the open end facing the gas inlet and positioned such that the ZnO/C precursor was located within the 930 °C zone of the furnace, while the samples were located within the 270 °C zone. The CVD growth procedure for ZnO NWs was modified from previously established CVD protocols [37,39]. It involved purging with Ar (600 sccm) at 25 °C, increasing the temperature under Ar (70 sccm) to 930 °C, at which point ZnO NWs growth occurred over 100 min in an Ar (150 sccm) atmosphere. The furnace was cooled under the protection of Ar (150 sccm).

2.3.3. CVD Synthesis of Graphene

A modified CVD method was employed to grow few-layer graphene using a previously established procedure [42]. Pieces of Cu foils were cleaned, annealed in a hydrogen rich environment at 550 °C for 60 min, and used as substrates. Prior to growth, the system was purged with argon/hydrogen (Ar/H_2) gases in a ratio of 450/50 sccm. Methane was used as the carbon source (20 sccm) to grow graphene sheets atop the prepared Cu foil substrates during the growth. High CVD temperatures (970 °C) were utilized to achieve a uniform and high-quality product. After 10 min of growth time, the CVD furnace cooled naturally under Ar/H_2 protection. A commonly used wet transfer method was employed to transfer graphene to a precleaned SiO_2/Si substrate. More details can be found in the Supplementary Materials.

2.4. Measurements and Characterization Tools

In this study, various characterization methods were used to characterize the samples, including scanning electron microscopy (SEM; FEI Inspect F50, Lausanne, Switzerland), atomic force microscopy (AFM; Park NX 10, Suwon, Korea), Raman spectroscopy (Renishaw, InVia, 532 nm, 100× objective, Wotton-under-Edge, Gloucestershire, England), X-ray photoelectron spectroscopy (XPS; PHI 5000 Versa Probe-II, Lafayette, LA, USA), energy-dispersive spectroscopy (EDS; JEOL JSM-7001 LVF Field Emission SEM), photoluminescent spectroscopy (PL; Nanolog spectrofluorometer, Horiba, Kyoto-shi, Japan), high-resolution transmission electron microscopy (HRTEM; JEOL JEM-2000 FX TEM, 200 kV; JEOL 2100F STEM, 200 kV; Peabody, MA, USA)

3. Results and Discussion

CVD syntheses of CNTs, ZnO NWs, and graphene, individually, are well-studied processes. However, the growth of these materials as heterostructures in consecutive CVD cycles has very few supporting studies despite being a preferred methodology due to higher quality of produced nanostructures, control and selectivity of the products, and overall scalability of the method. Earlier reported studies have shown that Au-catalyzed vapor–liquid–solid (VLS) growth of ZnO NWs on top of CNTs compromises CNT structure [20], and proximity placement of graphene can be a complicated process that limits placement selectivity [34].

Here, we developed novel procedures that allow for a streamlined direct two-step CVD process to grow either ZnO NWs/CNTs or ZnO NWs/Gr heterostructures in a controlled manner; and we showed that the integrity of an individual nanomaterial is maintained in each heterostructured assembly which has hardly any precedents in the field.

Figure 1 depicts a schematic of the flow process and the sequential steps to produce high-quality heterostructures of ZnO NWs/CNTs and ZnO NWs/Gr. First, a dense network of horizontally oriented CNTs (Figure 1a(i–iii)) or few-layer graphene (Figure 1b(i–iv)) were prepared via CVD as shown. While CNT networks were prepared on a SiO_2/Si substrate using our CCFe ink as a catalyst, graphene was prepared on Cu foil, which also acts as a catalyst. Complete CVD protocols can be found in Figure S6 (Supplementary Materials).

Our universal CCFe molecular ink deposited on the substrates catalyzes both CNTs and, subsequently, ZnO NWs, which is unique to our developed process, and is reported here for the first time. The advantage of the CCFe ink is that it can be effectively used for either dip-coating samples or selectively patterning regions of interest prior to growth, hence enabling a much more simplified and universal method for producing a desirable heterostructure interface. The CCFe ink is preferable to water-based inks as the low vapor pressure of DMF allows a better wetting of the substrate surface [41]. Likewise, this ink does not exhibit humidity-dependent constraints during DWP, which are commonly experienced with water-based inks. After CNT growth, the samples did not require recoating with catalytic ink for ZnO NW growth since our CCFe ink effectively catalyzes both materials, CNTs and ZnO NWs. CNTs were synthesized prior to ZnO NWs because the reduction process under H_2 during the CNT synthesis could etch ZnO NW introducing impurities and sometimes detrimental vacancies, compromising ZnO NWs [20]. Graphene, on the other hand, once it is grown on Cu foil, could be transferred to a variety of substrates, both rigid and flexible, providing great versatility in applications. To produce ZnO NWs/Gr heterostructures, graphene must be coated with CCFe prior to ZnO NW growth to serve as a catalyst (Figure 1c(ii)). It is important to note that the graphene surface is not modified by UV ozone before dip coating as this could damage graphene or introduce defects.

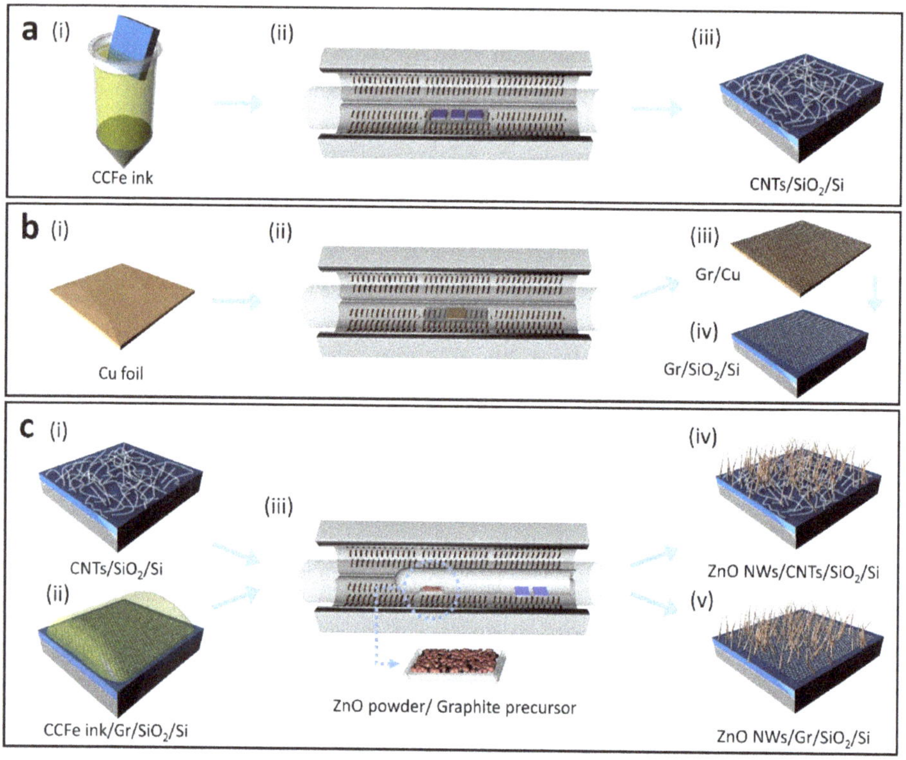

Figure 1. A schematic illustration of the process to grow ZnO NWs/CNTs and ZnO NWs/Gr heterostructures on SiO_2/Si substrates. (**a**) CNT growth including (**i**) dip coating in the catalytic CCFe ink precursor, (**ii**) CVD synthesis of CNTs, and (**iii**) a representative CNT sample. (**b**) Graphene growth on (**i**) Cu foil via (**ii**) CVD and subsequent transfer of graphene from (**iii**) the Cu substrate to (**iv**) SiO_2/Si. (**c**) Heterostructure growth beginning with (**i**) a CNT sample and (**ii**) a graphene sample dip-coated in the CCFe ink followed by (**iii**) CVD synthesis of ZnO NWs using the double-tube method, resulting in (**iv**) ZnO NWs/ CNTs and (**v**) ZnO NWs/Gr heterostructures.

The growth of both ZnO NWs/CNTs (Figure 1c(iv)) and ZnO NWs/Gr (Figure 1c(v)) heterostructures utilized the same unique (customized) ZnO NW CVD recipe modified from our previous studies [37,39]. The ZnO NW growth process consists of a double-tube arrangement where the open end of the inner tube faces the gas inlet. In our process, the inner tube allowed Zn^{2+} and O_2 vapors to achieve regional saturation at the open end of the tube where the samples were placed. Unlike some studies, our procedure did not require low pressures or furnace temperatures greater than 1000 °C providing a more simplified and time-efficient protocol. Substrate and source temperatures are important factors in controlling ZnO NW morphology, and here it was achieved through three individually controlled zones of the CVD furnace [43]. The versatility of the CCFe catalytic molecular ink allowed the ink to be patterned in a highly selective manner providing utility in applications requiring microscale or nanoscale precision. See the Section 2 for more details about DWP.

Our protocols for growing heterostructure interfaces start with the synthesis and characterization of the individual materials. To quantify changes in quality or morphology of the nanomaterials, they are characterized prior to and following their incorporation into the heterostructures. The CNTs grown for this study were optimized in density and uniformity (Figure 2a) over the entire sample (SiO_2/Si surface). While CNTs grow horizontally, as a monolayer, ZnO NWs tend to grow vertically in a forest-like arrangement (Figure 2d) with high density and relatively random organization. Representative (AFM) image of CNTs grown on SiO_2/Si (Figure 2c) shows the diameters of CNTs ranged from 1.7 nm to 2.4 nm, which indicates the CNTs grown were likely single-walled. Likewise, an AFM image of ZnO NWs grown on the SiO_2/Si substrate (Figure 2f) indicates an average diameter of ~24.5 nm. This AFM image was acquired from a reference sample of low-density ZnO NWs.

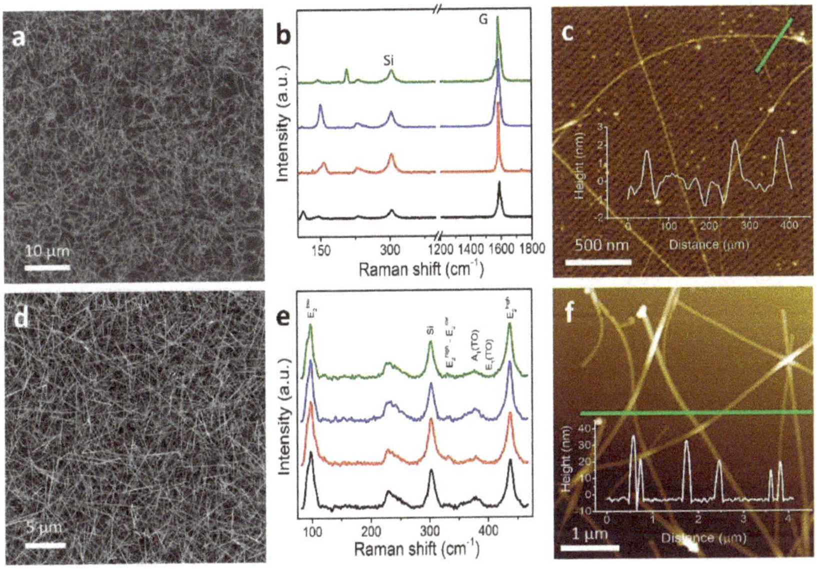

Figure 2. Characterization of as-grown 1D nanomaterials. (**a**) SEM image of CVD-grown CNTs on the SiO_2/Si substrate demonstrating uniformity and high density. (**b**) Raman spectra of four representative locations from a CNT sample highlighting radial breathing modes (RBM) and G-band regions of the spectra. (**c**) AFM image of a few CNTs with a line profile (inset). (**d**) SEM image of ZnO NWs grown randomly on SiO_2/Si. (**e**) Raman spectra of four representative spots of (**d**) with characteristic E_2^{high} and E_2^{low} peaks of ZnO NWs. (**f**) AFM topographic image of ZnO NWs grown on the reference sample, the SiO_2/Si substrate, and a line profile (inset).

Resonant Raman spectroscopy was employed to evaluate the quality of as-grown CNTs to determine their physical nature and electronic structure. The Raman characteristic spectrum for CNTs at four representative locations of the sample in (c) is shown in Figure 2b. This spectrum spanned the range of Raman shifts specific to radial breathing modes (RBMs) (~100–350 cm^{-1}), D-band (~1350 cm^{-1}), and G-band (~1580 cm^{-1}). From analyses of the RBM values, the diameters of the CNTs were confirmed to range from 0.7 to 2.3 nm, supporting our AFM results [44,45]. A narrow, intense G-band at 1583 cm^{-1}, full width at half-maximum (FWHM) of ~23.2 cm^{-1}, and a quality factor (intensity ratio $I_G/I_D > 100$) in Figure 2b confirm that the produced CNTs were of high quality and single-walled, with only a minute amount of defects. A number of as-grown CNT samples were measured (usually, an array of 25 spots tested per sample), and with examination of the G-band peak (its line shape and average FWHM), it was determined that our CVD growing process renders predominantly semiconducting CNTs.

ZnO NWs also exhibit a characteristic Raman spectrum. Of the twelve theoretical phonon branches in the wurtzite ZnO, nine are optically active [46]. Predictable modes of the lattice optical phonons are $\Gamma = 1A_1 + 2B_1 + 1E_1 + 2E_2$, of which A_1, E_1, and E_2 show Raman activity while B_1 is considered to be Raman-silent. The E_2 modes indicate the level of crystallinity of as-grown ZnO NWs, while A_1 and E_1 correspond to the common defects or vacancies. The Raman spectra (Figure 2e) at representative locations show high-intensity modes for E_2^{low} and E_2^{high} peaks (97.6 cm^{-1} and 436.9 cm^{-1}) which correlate with the Zn and O_2 sublattices [37,39,47,48]. The strongest peak in our Raman data was the E_2^{high} peak with the FWHM of ~10.87 cm^{-1}, indicating a high level of crystallinity. Notably, the E_1(TO) and E_1(LO) modes were insignificant in our ZnO NW samples [48,49]. To further confirm the crystal orientation and structural properties of ZnO NWs, high-resolution transmission electron microscopy (HRTEM) and selected area electron diffraction (SAED) were performed. Nickel TEM grids were prepared by scrapping ZnO NWs from ZnO NWs/CNTs and ZnO NWs/Gr samples onto the grid with flat-edged tweezers. HRTEM images showed crystal lattice spacing of ~0.53 nm and ~0.52 nm (Figure 3c,f); SAED images are shown in the insets. A wurtzite crystal structure was observed for our ZnO NWs, and the values were consistent with other reported HRTEM analyses [22,39]. TEM images of these samples indicated average diameters of ~89 nm and ~59 nm and minimum average lengths of ~988 nm and ~1055 nm for ZnO NWs from ZnO NWs/CNTs and ZnO NWs/Gr heterostructures, respectively (Figure S4, Table S2; Supplementary Materials).

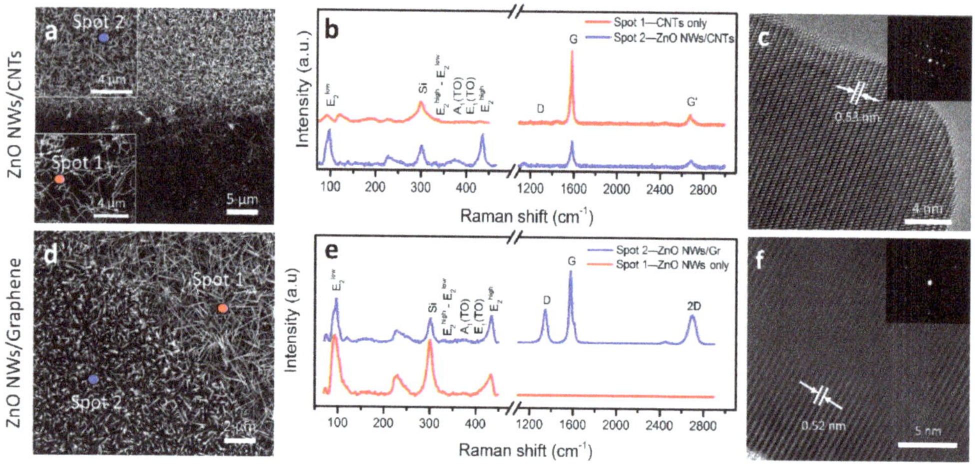

Figure 3. SEM and Raman data of the prepared heterostructures. (**a**) SEM image of ZnO NWs/CNTs heterostructure at a boundary location, where the upper half of the SEM image shows the presence of both materials while the lower half only contains CNTs. Insets show zoomed-in regions of both the top and the bottom features. (**b**,**e**) Raman spectra of ZnO NWs/CNTs and ZnO NWs/Gr heterostructures spanning both the ZnO NWs characteristic region (90 cm^{-1} to 460 cm^{-1}) and the CNTs/Gr-characteristic region (1100 cm^{-1} to 3000 cm^{-1}). The two spots correlate with the spots shown in (**a**) and in (**d**), respectively. (**c**,**f**) HRTEM images of the ZnO NWs lattice structure show lattice spacing of ~0.532 nm with SAED patterns (inset) from ZnO NWs/CNTs and ZnO NWs/Gr heterostructures, respectively. (**d**) SEM image of the ZnO NWs/Gr heterostructure at the boundary location, where the top right corner shows only the ZnO NWs grown on SiO_2/Si, while the bottom left corner shows the ZnO NWs grown on multilayered graphene.

To confirm that the materials' integrity post-heterostructure formation was preserved, we further evaluated the samples. The morphology of the ZnO NWs/CNTs heterostructure was observed via SEM (Figure 3a). Insets show dense, vertically aligned ZnO NWs grown on top of CNTs (top) and CNTs only (bottom). The ZnO NWs/Gr heterostructure (Figure 3d), however, showed distinct morphological changes when ZnO NWs were grown on the SiO_2/Si (top right corner) substrate versus graphene (bottom left corner). Clear morphological differences in the ZnO NWs grown on graphene were manifested in greater alignment and density, while the ZnO NWs grown on the SiO_2/Si surface appeared identical to the reference samples of ZnO NWs. The atomically flat crystal structure of graphene promotes greater alignment in the ZnO NWs grown directly on its surface, and with greater alignment, the density of ZnO NWs can increase. In contrast, while ZnO NWs still grow with vertical tendencies, they do not exhibit strong preferential alignment when grown on amorphous SiO_2 surface and appear at relatively lower densities. The wettability of graphene is lower compared to Si/SiO_2, thus creating more dense nucleation points and larger nanocluster aggregates from which ZnO NWs can grow, and that could lead to larger diameter of ZnO NWs as well as the higher density per unit surface area [35].

Additionally, Raman spectroscopy of both heterostructures confirmed that the quality and crystallinity of CNTs, graphene, and ZnO NWs was preserved in the heterostructure interfaces. In Figure 3b, spot 1 is selected from the region designed to contain only CNTs, so the characteristic E_2^{high} (~436.9 cm^{-1}) signal for ZnO NWs is negligible, while the G-band (1592 cm^{-1}) is intense and narrow (FWHM of ~15.9 cm^{-1}), with very few defects (I_D/I_G ~0.0149). These results further identified high-quality CNTs even where ZnO NWs did not grow to a complete crystal structure. Conversely, spot 2 was selected from a region where the ZnO NWs/CNTs heterostructure was formed, so the E_2^{high} and E_2^{low} (~436.9 cm^{-1} and ~97.6 cm^{-1}) modes were clearly observed, consistent with the results shown in Figure 2. Likewise, the CNTs' G-band (1593.24 cm^{-1}) was narrow (FWHM ~26.82 cm^{-1}), with minimal defects (I_D/I_G ~0.0192), further confirming the quality of our CNTs was not adversely affected by ZnO NW growth. The Raman spectra of the ZnO NWs/Gr heterostructure (Figure 3e) confirmed ZnO NWs were present at both spot 1 and spot 2 of the SEM image (Figure 3d), with intense peaks representing E_2^{high} and E_2^{low} (~438.9 cm^{-1} and ~99.7 cm^{-1}) consistent with the ZnO NW reference sample in Figure 2. The characteristic peaks of graphene were present only at spot 2 where the ZnO NWs/Gr heterostructure had formed but were absent at spot 1. The narrow G-band at 1583 cm^{-1} (FWHM of ~28.5 cm^{-1}) indicated the graphene was of high quality and the G-band/2D-band ratio of ~2.51 confirmed the few-layer nature of graphene, which was consistent with the Raman spectra of the graphene samples on Cu and SiO_2/Si prior to heterostructure formation (Figure S1, Supplementary Materials). This characterization confirms that the integrity of graphene was not damaged by the ZnO NWs grown on its surface.

Energy-dispersive X-ray spectroscopy (EDS) was performed on both heterostructures and on ZnO NW reference samples to confirm elemental compositions of the prepared samples. EDS spectra for each material with a representative SEM image are shown in Figure 4a–c. All the samples had signature peaks for oxygen, zinc, and the Si substrate near 0.5 KeV, 1.1 KeV, and 1.75 KeV, respectively. A Si peak was also observed, typical for samples prepared on SiO_2/Si substrates [50]. While the data were consistent within each sample (at three different locations/spots), there were small variations in wt% between samples, which was not surprising given that the representative spectral data were collected from random spots (~200 nm^2) within each sample. Interestingly, there was no peak representing residual iron traces for any of the samples, which indicated that this growth process was primarily base-growth (catalyst particles encapsulated by ZnO NWs) [37,51,52]. Specific details on the wt% composition of each element can be found in Table S1 (Supplementary Materials) which confirms the presence of Zn, O, and Si in the expected ratios consistent with previous studies [37]. Enlarged SEM images of Figure 4a–c may be found in Figure S5 (Supplementary Materials).

Photoluminescence (PL) spectroscopy was also employed to analyze the photoexcitation exhibited by ZnO NWs in the prepared samples and was conducted at room temperature. The PL spectra (Figure 4d) confirmed sharp peaks with full width at half-maxima (FWHM ~ 0.105 eV). The PL peaks' maxima are located at ~3.3 eV near band edge range excitonic emission (NBE) at the excitation wavelength ~276 nm laser light [37]. There were additional wide peaks (at 2.50, 2.50, and 2.54 eV) associated with defects in ZnO NWs/Gr, ZnO NWs/CNTs, and ZnO NWs, respectively. The data were consistent with other reported PL characterization of defects in ZnO NWs [20].

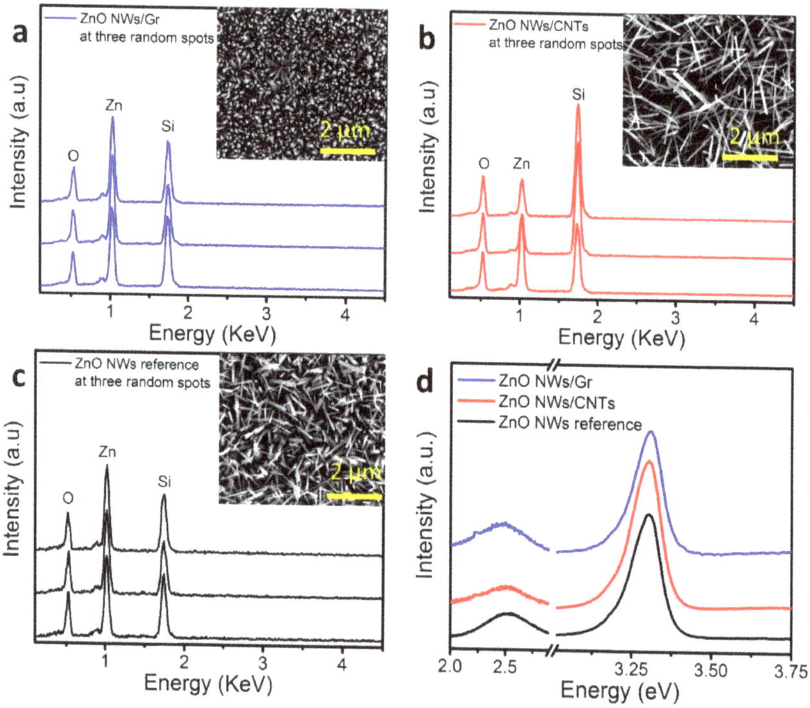

Figure 4. EDS data and PL spectroscopy. (**a–c**) EDS analysis demonstrated the expected Zn, O, and Si presence in the prepared ZnO NWs/Gr and ZnO NWs/CNTs heterostructures and in the ZnO NWs reference samples taken at three selected locations. The insets show SEM micrographs for each respective sample where EDS data were collected. (**d**) PL spectra of all the three prepared sample types had observed peak maxima of ~3.31 eV and defect peaks at ~2.50 eV.

X-ray photoelectron spectroscopy (XPS) was performed to characterize the elemental chemical composition and chemical states of the prepared heterostructures (ZnO NWs/CNTs and ZnO NWs/Gr) and the reference sample (ZnO NWs). Figure 5a illustrates the core-level spectra of Zn2p for the ZnO NW samples, ZnO NWs/CNTs, and ZnO NWs/Gr heterostructures, which are indicated by two notable peaks located at ≈1021.4 eV and ≈1044.5 eV representing $Zn2p_{3/2}$ and $Zn2p_{1/2}$ in the Zn^{2+} state, respectively [37]. The O1s energy-level spectra are represented in Figure 5b, where three distinct peaks (outlined by the Gaussian fit) are found at ≈530.1 eV, 531.7 eV, and 532.9 eV in all the three samples, the ZnO NWs/CNTs and ZnO NWs/Gr heterostructures and the ZnO NWs reference. The peak at lower binding energy (530.1 eV) confirmed the participation of O ions with Zn ions, while the other two peaks at higher binding energies (531.7 eV and 532.9 eV) implied oxygen vacancies/defects absorption with some contribution from Si–O bonding [52]. It is important to note here that the oxygen defects peak (represented by the higher binding energy ≈532.9 eV) showed higher intensity in the ZnO NWs/CNTs heterostructures than in the reference sample (ZnO NWs/SiO_2/Si) and the ZnO NWs/Gr samples. We speculate that the intensity of the relatively higher defect in the ZnO NWs/CNTs heterostructure could be due to additional residuals accumulated on sample surfaces during the two subsequent CVD runs. When subtracting a peak (~532.9 eV) from all the spectra, in Figure 5b (see Figure S2b (Supplementary Materials) post-subtraction), we observe near-identical spectral plots in all the three samples showing consistency between the sample types outside of their oxygen defect presence.

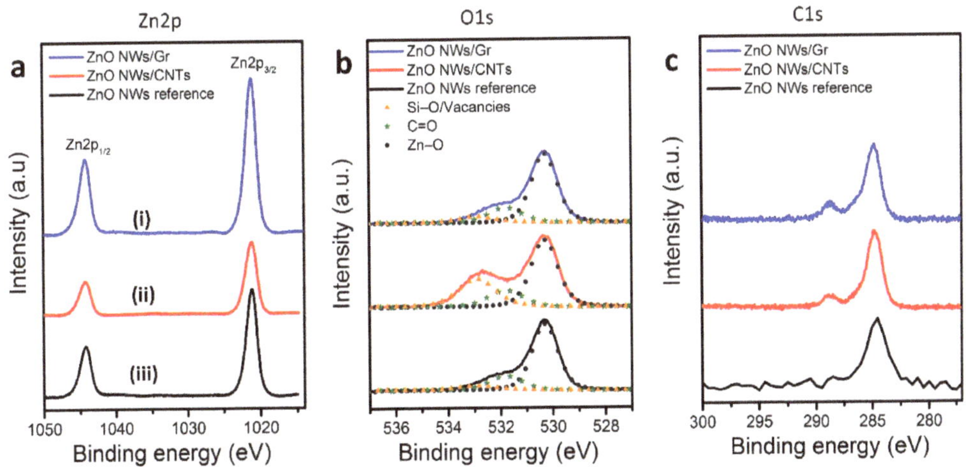

Figure 5. XPS analyses of the ZnO NWs, ZnO NWs/CNTs, and ZnO NWs/Gr heterostructure; (**a**–**c**) represent the binding energy positions of Zn2p, O1s, and C1s for the ZnO NWs (black), ZnO NWs/CNTs (red), and ZnO NWs/Gr (blue) heterostructures, respectively. Gaussian fits are applied to the O1s curves in (**b**), showing three-peak deconvolution in each sample. Survey data from each sample can be found in Figure S2a (Supplementary Materials).

A single CVD run was used to fabricate ZnO NW reference samples and ZnO NWs/Gr hybrids prepared on cleaner and smoother surfaces of SiO_2/Si and graphene substrates, respectively. Figure 5c shows the XPS analyses of the C1s corresponding to ZnO NWs (reference sample, black curve). At the same time, ZnO NWs/CNTs (red) and ZnO NWs/Gr (blue) represent heterostructured samples. Figure 5c clearly shows a sharp peak at ~284.7 eV, attributed to C=C bonds [53,54]. In addition to the prominent peak that was shown in all the fabricated specimens, we could notice a small hump localized at ≈288.4 eV only in the heterostructured samples, ZnO NWs/CNTs (red) and ZnO NWs/Gr (blue), which was expected due to the presence of the C=O bonding [53,54].

While growing quality materials and heterostructure interfaces consistently and reproducibly is essential, the ability to grow heterostructure interfaces selectively at desired locations is of great practical importance. A proof-of-concept experiment was conducted to demonstrate ZnO NWs/CNTs heterostructures formed on a SiO_2/Si sample with etched-in silicon features. Etched star-shaped geometrical features (trenches) were prepared by optical laser lithography following RIE in SiO_2/Si chips. Details of the lithography and RIE etching are described in the Supplementary Materials.

CNTs were first grown on these samples following dip coating with the CCFe ink (Figure 6a) and showed to grow densely only outside of the star-shaped features. While this conveys a well-documented notion of the importance of SiO_2 in the growth of CNTs, it also allows for selective exclusion of CNTs from the etched area. Thus, it allows for morphology control and placement of ZnO NWs in the next step of the growth as described below [55,56]. DWP was used to selectively fill star-shaped features with the CCFe catalytic ink (volumes of tens to hundreds of femtoliters (10^{-12} L)) using custom AFM cantilevers (Figure 6b,c). Depositing the CCFe ink into these star-shaped features increased the effective density of catalytic particles per unit area. It can be used to alter the resulting morphology of the grown ZnO NWs. Indeed, the diameters of the ZnO NWs originating from these features appeared larger—a result which has been observed in ZnO NW growth using high-density catalyst aggregation [37]. Hence, following ZnO NW CVD growth, we observed ZnO NWs grew more densely, exhibited greater alignment, and had larger diameters at the locations where additional ink was deposited as compared to those produced at the neighboring etched features with a lower concentration of the catalyst or on the surrounding substrate surface (Figure 6d).

Figure 6e–g shows Raman spectral plots and Raman intensity maps of a region surrounding a star-shaped feature. Two spots, inside and outside of the feature shown in Figure 6d, were analyzed at each respective location; the spectra are shown in Figure 6g. The Raman spectrum at spot 1 (ZnO NWs only) exhibited strong E_2^{high} and E_2^{low} (438.9 cm^{-1} and 99.7 cm^{-1}) peaks consistent with the discussion of Figure 2 for ZnO NWs while the G-band signal for CNTs was not present. Meanwhile, the Raman spectrum at spot 2 (ZnO NWs/CNTs) exhibited signature peaks of both ZnO NWs and CNTs with minor defects (D-band/G-band ~0.0191). Thus, Raman mapping of the region containing the feature allowed us to characterize the quality of materials present in the locations where they grew. The regions colored in red represent the relative intensity of the CNTs' G-band (1582 cm^{-1}) and the ZnO NWs' E_2^{high} mode (440 cm^{-1}) in Figure 6e,f, respectively. The Raman map in Figure 6e shows intense G-band signatures selectively outside of the star-shaped feature, which is reprehensive of CNT presence.

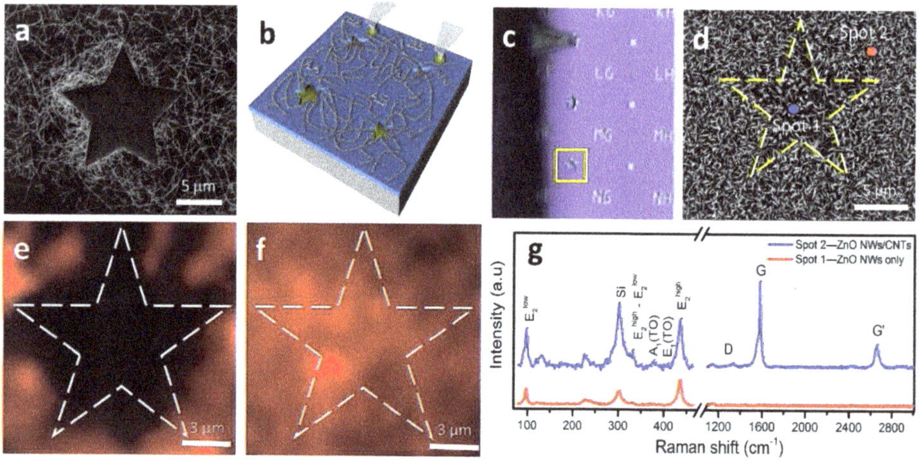

Figure 6. Direct-write patterning of the ZnO NWs/CNTs heterostructure. (**a**) SEM image of the etched SiO_2/Si substrate with CNTs grown uniformly. (**b**) Schematic of the patterning process used to fill selected features with the CCFe catalytic ink prior to ZnO NW growth. (**c**) Optical image of a filled star-shaped feature with as-grown CNTs present prior to ZnO NW growth. (**d**) SEM micrograph of the etched feature indicated in (**d**) following ZnO NWs growth. (**e**,**f**) Raman mapping of the star feature in (**d**) with the CNTs' G-band mapped at 1589 cm^{-1} (**e**) and the E_2^{high} peak mapped at 98.8 cm^{-1} (**f**). (**g**) Representative Raman spectra taken at the corresponding spots indicated in (**d**).

Meanwhile, the Raman map in Figure 6f shows high-intensity signals for E_2^{high} over the entire region, which is representative of ZnO NW presence. This map also shows a slightly higher relative intensity of E_2^{high} signal inside of the star-shaped feature, where we observed a greater alignment and density of ZnO NWs. ZnO NWs were also grown selectively inside of the other etched-in geometrical features on the SiO_2/Si reference samples. As it can be seen in Figure S3 (Supplementary Materials), ZnO NWs only grew in the features where catalytic ink was directly deposited. CNTs grow selectively on the exterior of the feature because the SiO_2 layer is completely removed during the RIE process, leaving bare Si substrate lacking necessary SiO_2, which is critical for optimal growth of CNTs [55–58]. This example demonstrates flexibility in fabrication (lithography and subsequent CVD growth) of various micro- and nanoarchitectures; it can also be used to design hybrid heterostructures in a selective/predefined fashion, thus providing greater control over material placement, which is paramount for interface and device engineering.

4. Conclusions

In this study, we developed unique and highly efficient two-step CVD processes of synthesizing ZnO NWs/CNTs and ZnO NWs/Gr heterostructure interfaces to demonstrate a reproducible, scalable method of producing high-quality nanomaterials. This method allows for the direct growth and formation of 1D–1D or 1D–2D interfaces. The resulting materials exhibit high levels of crystallinity and desirable morphologies and targeted electronic and optical properties as confirmed by various characterization methods such as Raman spectroscopy, SEM/EDS, XPS, TEM, PL, and AFM. The CNT networks prepared for the ZnO NWs/CNTs heterostructure interfaces were characterized as high-quality and predominately semiconducting single-walled, with average diameters ranging between 0.7 and 2.4 nm as determined by AFM and Raman spectroscopy. We also showed that the ZnO NWs formed in our heterostructured systems were highly crystalline and had a wurtzite crystal structure with lattice spacing ~0.52–0.53 nm as confirmed by HRTEM, SAED, and Raman spectroscopy. Our universal catalytic ink greatly simplified the formation of ZnO NWs/CNTs heterostructures through optimized reactivation of the catalysts' nanoparticles derived from the CCFe ink during the growth processes because it catalyzes both CNTs and ZnO NWs, providing a more efficient protocol for catalyst usage. The graphene used in the heterostructure was shown to promote more densely packed arrangements of ZnO NWs, greater vertical alignment, with controlled diameters of individual ZnO NWs, due to graphene's unique surface properties.

Additionally, we demonstrated that it is possible to selectively grow ZnO NWs/CNTs heterostructures with our direct-write patterning technique, which can tailor an interface based on the precise predefined location and morphological specifications. Such heterostructures could offer a foundation for developing electronic, optical, and energy-related devices fabricated in customizable vertical or horizontal arrangements in the templated multiplexed architectures, thus allowing for many promising future applications, including biomedical applications.

Supplementary Materials: The following are available online at https://www.mdpi.com/2079-4991/11/7/1836/s1, Transfer method of Gr from Cu to SiO_2/Si. Figure S1: Raman spectroscopic data of graphene on Cu (red) before transfer and graphene on SiO_2/Si (black) after transfer, Table S1: EDS composition data for the ZnO NW reference sample and both ZnO NWs/Gr and ZnO NWs/CNTs heterostructures, Figure S2: XPS survey data and O1s with a subtracted peak at ~532.9 eV. Laser lithography and RIE etching. Figure S3: Controlled selective synthesis of ZnO NWs. Figure S4: TEM image of ZnO NWs. Table S2: ZnO NW diameters measured from the TEM. Figure S5: Enlarged SEM images of ZnO NWs/Gr, ZnO NWs/CNTs, and the ZnO NW reference. Figure S6: CVD protocols for growing carbon nanotubes, graphene, and ZnO NWs.

Author Contributions: Conceptualization, I.K.; methodology, N.S. and I.K.; investigation, N.S., Y.K., B.T., K.M., M.C., R.D., D.J.G. and Y.L.; validation, N.S., Y.K. and K.M.; visualization, N.S. and Y.K.; formal analysis, N.S., D.A., Y.K. and I.K.; writing—original draft preparation, N.S., D.A., Y.K. and I.K.; writing—review and editing, N.S., D.A., Y.K. and I.K.; project administration, I.K.; funding acquisition, I.K. All authors have read and agreed to the published version of the manuscript.

Funding: This research received no external funding.

Acknowledgments: The use of the Center for Nanoscale Materials, an Office of Science user facility, was supported by the US Department of Energy, Office of Science, Office of Basic Energy Sciences, under contract No. DE-AC02 06CH11357. The authors acknowledge the use of the Washington University in St. Louis user facility Institute of Materials and Engineering. I.K. acknowledges the support of Saint Louis University, President's Research Fund (PRF).

Conflicts of Interest: The authors declare no conflict of interest.

References

1. Pitkethly, M.J. Nanometerials—The Driving Force. *Mater. Today* **2004**, *7*, 20–29. [CrossRef]
2. Iijima, S.; Ichihashi, T. Single-Shell Carbon Nanotubes of 1-nm Diameter. *Nature* **1993**, *363*, 603–605. [CrossRef]
3. Bethune, D.S.; Kiang, C.H.; de Vries, M.S.; Gorman, G.; Savoy, R.; Vazquez, J.; Beyers, R. Cobalt-Catalysed Growth of Carbon Nanotubes with Single-Atomic-Layer Walls. *Nature* **1993**, *363*, 605–607. [CrossRef]
4. Ruddaraju, L.K.; Pammi, S.V.N.; Pallela, P.N.V.K.; Padavala, V.S.; Kolapalli, V.R.M. Antibiotic Potentiation and Anti-Cancer Competence through Bio-Mediated ZnO Nanoparticles. *Mater. Sci. Eng. C* **2019**, *103*, 109756. [CrossRef] [PubMed]

5. Li, Y.; Zhang, W.; Niu, J.; Chen, Y. Mechanism of Photogenerated Reactive Oxygen Species and Correlation with the Antibacterial Properties of Engineered Metal-Oxide Nanoparticles. *ACS Nano* **2012**, *6*, 5164–5173. [CrossRef] [PubMed]
6. Modugno, G.; Ménard-Moyon, C.; Prato, M.; Bianco, A. Carbon Nanomaterials Combined with Metal Nanoparticles for Theranostic Applications. *Br. J. Pharm.* **2015**, *172*, 975–991. [CrossRef]
7. Pandurangan, M.; Enkhtaivan, G.; Kim, H.D. Anticancer Studies of Synthesized ZnO Nanoparticles against Human Cervical Carcinoma Cells. *J. Photochem. Photobiol. B Biol.* **2016**, *158*, 206–211. [CrossRef]
8. Hanley, C.; Layne, J.; Punnoose, A.; Reddy, K.M.; Coombs, I.; Coombs, A.; Feris, K.; Wingett, D. Preferential Killing of Cancer Cells and Activated Human T Cells Using ZnO Nanoparticles. *Nanotechnology* **2008**, *19*, 295103. [CrossRef]
9. Rasmussen, J.W.; Martinez, E.; Louka, P.; Wingett, D.G. Zinc Oxide Nanoparticles for Selective Destruction of Tumor Cells and Potential for Drug Delivery Applications. *Expert Opin. Drug Deliv.* **2010**, *7*, 1063–1077. [CrossRef] [PubMed]
10. Ahmed, N.; Fessi, H.; Elaissari, A. Theranostic Applications of Nanoparticles in Cancer. *Drug Discov.* **2012**, *17*, 928–934. [CrossRef]
11. Wang, J.; Lee, J.S.; Kim, D.; Zhu, L. Exploration of Zinc Oxide Nanoparticles as a Multitarget and Multifunctional Anticancer Nanomedicine. *ACS Appl. Mater. Interfaces* **2017**, *9*, 39971–39984. [CrossRef]
12. Keefer, E.W.; Botterman, B.R.; Romero, M.I.; Rossi, A.F.; Gross, G.W. Carbon Nanotube Coating Improves Neuronal Recordings. *Nat. Nanotechnol.* **2008**, *3*, 434–439. [CrossRef]
13. Schwartz, A.B.; Cui, X.T.; Weber, D.J.; Moran, D.W. Brain-Controlled Interfaces: Movement Restoration with Neural Prosthetics. *Neuron* **2006**, *52*, 205–220. [CrossRef]
14. Xie, Y.; Martini, N.; Hassler, C.; Kirch, R.D.; Stieglitz, T.; Seifert, A.; Hofmann, U.G. In Vivo Monitoring of Glial Scar Proliferation on Chronically Implanted Neural Electrodes by Fiber Optical Coherence Tomography. *Front. Neuroeng.* **2014**, *7*, 00034. [CrossRef] [PubMed]
15. Kordzadeh, A.; Amjad-Iranagh, S.; Zarif, M.; Modarress, H. Adsorption and Encapsulation of the Drug Doxorubicin on Covalent Functionalized Carbon Nanotubes: A Scrutinized Study by Using Molecular Dynamics Simulation and Quantum Mechanics Calculation. *J. Mol. Graph. Mod.* **2019**, *88*, 11–22. [CrossRef] [PubMed]
16. Kocaman, S.; Gursoy, M.; Karaman, M.; Ahmetli, G. Synthesis and Plasma Surface Functionalization of Carbon Nanotubes for Using in Advanced Epoxy-Based Nanocomposites. *Surf. Coat. Technol.* **2020**, *399*, 126144. [CrossRef]
17. Zhu, Y.; Elim, H.I.; Foo, Y.L.; Yu, T.; Liu, Y.; Ji, W.; Lee, J.Y.; Shen, Z.; Wee, A.T.S.; Thong, J.T.L.; et al. Multiwalled Carbon Nanotubes Beaded with ZnO Nanoparticles for Ultrafast Nonlinear Optical Switching. *Adv. Mater.* **2006**, *18*, 587–592. [CrossRef]
18. Zhao, T.; She, S.; Ji, X.; Jin, W.; Dang, A.; Li, H.; Li, T.; Shang, S.; Zhou, Z. In-Situ Growth Amorphous Carbon Nanotube on Silicon Particles as Lithium-Ion Battery Anode Materials. *J. Alloys Compd.* **2017**, *708*, 500–507. [CrossRef]
19. Baughman, R.H.; Zakhidov, A.A.; de Heer, W.A. Carbon Nanotubes-the Route toward Applications. *Science* **2002**, *297*, 787–792. [CrossRef] [PubMed]
20. Ok, J.G.; Tawfick, S.H.; Juggernauth, K.A.; Sun, K.; Zhang, Y.; Hart, A.J. Electrically Addressable Hybrid Architectures of Zinc Oxide Nanowires Grown on Aligned Carbon Nanotubes. *Adv. Funct. Mater.* **2010**, *20*, 2470–2480. [CrossRef]
21. Caglar, M.; Ilican, S.; Caglar, Y.; Yakuphanoglu, F. Electrical Conductivity and Optical Properties of ZnO Nanostructured Thin Film. *Appl. Surf. Sci.* **2009**, *255*, 4491–4496. [CrossRef]
22. Xu, F.; Qin, Q.; Mishra, A.; Gu, Y.; Zhu, Y. Mechanical Properties of ZnO Nanowires under Different Loading Modes. *Nano Res.* **2010**, *3*, 271–280. [CrossRef]
23. Bettini, S.; Pagano, R.; Bonfrate, V.; Maglie, E.; Manno, D.; Serra, A.; Valli, L.; Giancane, G. Promising Piezoelectric Properties of New ZnO@Octadecylamine Adduct. *J. Phys. Chem. C* **2015**, *119*, 20143–20149. [CrossRef]
24. Ta, H.Q.; Zhao, L.; Pohl, D.; Pang, J.; Trzebicka, B.; Rellinghaus, B.; Pribat, D.; Gemming, T.; Liu, Z.; Bachmatiuk, A.; et al. Graphene-like ZnO: A Mini Review. *Crystals* **2016**, *6*, 100. [CrossRef]
25. Wojnarowicz, J.; Chudoba, T.; Lojkowski, W. A Review of Microwave Synthesis of Zinc Oxide Nanomaterials: Reactants, Process Parameters and Morphologies. *Nanomaterials* **2009**, *10*, 1086. [CrossRef] [PubMed]
26. Shrestha, B.K.; Shrestha, S.; Tiwari, A.P.; Kim, J.I.; Ko, S.W.; Kim, H.J.; Park, C.H.; Kim, C.S. Bio-Inspired Hybrid Scaffold of Zinc Oxide-Functionalized Multi-Wall Carbon Nanotubes Reinforced Polyurethane Nanofibers for Bone Tissue Engineering. *Mater. Des.* **2017**, *133*, 69–81. [CrossRef]
27. Xie, J.; Frachioni, A.; Williams, D.S.; White, B.E. Thermal Conductivity of a ZnO Nanowire/Silica Aerogel Nanocomposite. *Appl. Phys. Lett.* **2013**, *102*, 193101. [CrossRef]
28. Zhang, Y.; Zhang, L.; Zhou, C. Review of Chemical Vapor Deposition of Graphene and Related Applications. *Acc. Chem. Res.* **2013**, *46*, 2329–2339. [CrossRef]
29. Singh, L.T.; Sugavaneshwar, R.P.; Nanda, K.K. Carbon Nanotube-ZnO Nanowire Hybrid Architectures as Multifunctional Devices. *AIP Adv.* **2013**, *3*, 082106. [CrossRef]
30. Song, Z.; Wei, H.; Liu, Y.; Wang, J.; Long, H.; Wang, H.; Qin, P.; Zeng, W.; Fang, G. Enhanced Field Emission from Aligned ZnO Nanowires Grown on a Graphene Layer with Hydrothermal Method. *IEEE Trans. Nanotechnol.* **2014**, *13*, 167–171. [CrossRef]
31. Anand, K.; Singh, O.; Singh, M.P.; Kaur, J.; Singh, R.C. Hydrogen Sensor Based on Graphene/ZnO Nanocomposite. *Sens. Actuators B Chem.* **2014**, *195*, 409–415. [CrossRef]
32. Panth, M.; Cook, B.; Zhang, Y.; Ewing, D.; Tramble, A.; Wilson, A.; Wu, J. High-Performance Strain Sensors Based on Vertically Aligned Piezoelectric Zinc Oxide Nanowire Array/Graphene Nanohybrids. *ACS Appl. Nano Mater.* **2020**, *3*, 6711–6718. [CrossRef]

33. Nourmohammadi, A.; Rahighi, R.; Akhavan, O.; Moshfegh, A. Graphene Oxide Sheets Involved in Vertically Aligned Zinc Oxide Nanowires for Visible Light Photoinactivation of Bacteria. *J. Alloys Compd.* **2014**, *612*, 380–385. [CrossRef]
34. Fu, X.W.; Liao, Z.M.; Zhou, Y.B.; Wu, H.C.; Bie, Y.Q.; Xu, J.; Yu, D.P. Graphene/ZnO Nanowire/Graphene Vertical Structure Based Fast-Response Ultraviolet Photodetector. *Appl. Phys. Lett.* **2012**, *100*, 223114. [CrossRef]
35. Lee, J.M.; Pyun, Y.B.; Yi, J.; Choung, J.W.; Park, W.I. ZnO Nanorod-Graphene Hybrid Architectures for Multifunctional Conductors. *J. Phys. Chem. C* **2009**, *113*, 19134–19138. [CrossRef]
36. Alameri, D.; Ocola, L.E.; Kuljanishvili, I. Mask-Free Fabrication and Chemical Vapor Deposition Synthesis of Ultrathin Zinc Oxide Microribbons on Si/SiO_2 and 2D Substrates. *J. Vac. Sci. Technol. A* **2018**, *36*, 05G506. [CrossRef]
37. Alameri, D.; Ocola, L.E.; Kuljanshvili, I. Controlled Selective CVD Growth of ZnO Nanowires Enabled by Mask-Free Fabrication Approach Using Aqueous Fe Catalytic Inks. *Adv. Mater. Interfaces* **2017**, *4*, 1700950. [CrossRef]
38. Al-Asadi, A.S.; Henley, L.A.; Wasala, M.; Muchharla, B.; Perea-Lopez, N.; Carozo, V.; Lin, Z.; Terrones, M.; Mondal, K.; Kordas, K.; et al. Aligned Carbon Nanotube/Zinc Oxide Nanowire Hybrids as High Performance Electrodes for Supercapacitor Applications. *J. Appl. Phys.* **2017**, *121*, 124303. [CrossRef]
39. Geng, C.; Jiang, Y.; Yao, Y.; Meng, X.; Zapien, J.A.; Lee, C.S.; Lifshitz, Y.; Lee, S.T. Well-Aligned ZnO Nanowire Arrays Fabricated on Silicon Substrates. *Adv. Funct. Mater.* **2004**, *14*, 589–594. [CrossRef]
40. Little, D.J.; Pfund, J.D.; Mclain, A.A.; Sporie, J.A.; Lantvit, S.M.; King, S.T. Synthesis of a Zinc Oxide/Graphene Hybrid Material by the Direct Thermal Decomposition of Oxalate. *Mater. Res. Express* **2020**, *7*, 65005. [CrossRef]
41. Kuljanishvili, I.; Dikin, D.A.; Rozhok, S.; Mayle, S.; Chandrasekhar, V. Controllable Patterning and CVD Growth of Isolated Carbon Nanotubes with Direct Parallel Writing of Catalyst Using Dip Pen Nanolithography. *Small* **2009**, *5*, 2523–2527. [CrossRef]
42. Dong, R.; Moore, L.; Ocola, L.E.; Kuljanishvili, I. Enabling Quality Interfaces with Mask-Free Approach to Selective Growth of MoS2/Graphene Stacked Structures. *Adv. Mater. Interfaces* **2016**, *3*, 1600098. [CrossRef]
43. Wongchoosuk, C.; Subannajui, K.; Menzel, A.; Burshtein, I.A.; Tamir, S.; Lifshitz, Y.; Zacharias, M. Controlled Synthesis of ZnO Nanostructures: The Role of Source and Substrate Temperatures. *J. Phys. Chem. C* **2011**, *115*, 757–761. [CrossRef]
44. Jorio, A.; Pimenta, M.A.; Souza Filho, A.G.; Saito, R.; Dresselhaus, G.; Dresselhaus, M.S. Characterizing Carbon Nanotube Samples with Resonance Raman Scattering. *New J. Phys.* **2003**, *5*, 139. [CrossRef]
45. Dresselhaus, M.S.; Jorio, A.; Souza Filho, A.G.; Saito, R. Defect Characterization in Graphene and Carbon Nanotubes Using Raman Spectroscopy. *Philos. Trans. R. Soc.* **2010**, *368*, 5355–5377. [CrossRef]
46. Arguello, C.A.; Rousseau, D.L.; Porto, S.P.S.; Reynolds, D.C.; Collins, T.C.; Bauman, R.P. First-Order Raman Effects in Wurtzite-Type Crystals. *Phys. Rev.* **1969**, *181*, 1351–1363. [CrossRef]
47. Cheng, A.J.; Tzeng, Y.; Xu, H.; Alur, S.; Wang, Y.; Park, M.; Wu, T.H.; Shannon, C.; Kim, D.J.; Wang, D. Raman Analysis of Longitudinal Optical Phonon-Plasmon Coupled Modes of Aligned ZnO Nanorods. *J. Appl. Phys.* **2009**, *105*, 073104. [CrossRef]
48. Calzolari, A.; Nardelli, M.B. Dielectric Properties and Raman Spectra of ZnO from a First Principles Finite-Differences/Finite-Fields Approach. *Sci. Rep.* **2013**, *3*, 2999. [CrossRef]
49. Xu, L.; Li, X.; Zhan, Z.; Wang, L.; Feng, S.; Chai, X.; Lu, W.; Shen, J.; Weng, Z.; Sun, J. Catalyst-Free, Selective Growth of ZnO Nanowires on SiO_2 by Chemical Vapor Deposition for Transfer-Free Fabrication of UV Photodetectors. *ACS Appl. Mater. Interfaces* **2015**, *7*, 20264–20271. [CrossRef]
50. Zhu, G.; Zhou, Y.; Wang, S.; Yang, R.; Ding, Y.; Wang, X.; Bando, Y.; Wang, Z.L. Synthesis of Vertically Aligned Ultra-Long ZnO Nanowires on Heterogeneous Substrates with Catalyst at the Root. *Nanotechnology* **2012**, *23*, 055604. [CrossRef] [PubMed]
51. Zhu, Z.; Chen, T.-L.; Gu, Y.; Warren, J.; Osgood, R.M. Zinc Oxide Nanowires Grown by Vapor-Phase Transport Using Selected Metal Catalysts: A Comparative Study. *Chem. Mater.* **2005**, *17*, 4227–4234. [CrossRef]
52. Lim, K.; Hamid, M.A.A.; Shamsudin, R.; Al-Hardan, N.H.; Mansor, I.; Chiu, W. Temperature-Driven Structural and Morphological Evolution of Zinc Oxide Nano-Coalesced Microstructures and Its Defect-Related Photoluminescence Properties. *Materials* **2016**, *9*, 300. [CrossRef] [PubMed]
53. Hossain, M.M.; Shima, H.; Son, S.; Hahn, J.R. In Situ Fabrication of a Thermally Stable and Highly Porous Conductive Solar Light-Driven ZnO-CNT Fiber Photocatalyst. *RSC Adv.* **2016**, *6*, 71450–71460. [CrossRef]
54. Sharma, M.; Joshi, M.; Nigam, S.; Shree, S.; Avasthi, D.K.; Adelung, R.; Srivastava, S.K.; Kumar Mishra, Y. ZnO Tetrapods and Activated Carbon Based Hybrid Composite: Adsorbents for Enhanced Decontamination of Hexavalent Chromium from Aqueous Solution. *Chem. Eng. J.* **2019**, *358*, 540–551. [CrossRef]
55. Cao, A.; Ajayan, P.M.; Ramanath, G.; Baskaran, R.; Turner, K. Silicon Oxide Thickness-Dependent Growth of Carbon Nanotubes. *Appl. Phys. Lett.* **2004**, *84*, 109–111. [CrossRef]
56. Simmons, J.M.; Nichols, B.M.; Marcus, M.S.; Castellini, O.M.; Hamers, R.J.; Eriksson, M.A. Critical Oxide Thickness for Efficient Single-Walled Carbon Nanotube Growth on Silicon Using Thin SiO_2 Diffusion Barriers. *Small* **2006**, *2*, 902–909. [CrossRef] [PubMed]
57. Jung, Y.J.; Wei; Vajtai, R.; Ajayan, P.M.; Homma, Y.; Prabhakaran, K.; Ogino, T. Mechanism of Selective Growth of Carbon Nanotubes on SiO_2/Si Patterns. *Nano Lett.* **2003**, *3*, 561–564. [CrossRef]
58. Xiang, R.; Einarsson, E.; Okabe, H.; Chiashi, S.; Shiomi, J.; Maruyama, S. Patterned Growth of High-Quality Single-Walled Carbon Nanotubes from Dip-Coated Catalyst. *Jpn. J. Appl. Phys.* **2010**, *49*, 02BA03. [CrossRef]

2.0 eV and a type II band alignment with ZnO NWs [3]. These core–shell heterostructures benefit from a large number of assets, including (i) efficient light absorption phenomena through radiated and guided optical modes and (ii) efficient charge carrier management through charge carrier separation and collection [10–13]. The first efficient ETA solar cell integrating ZnO NWs was reported in 2005 with the use of a p-type CdSe shell and of CuSCN as the hole transporting material (HTM), leading to the photovoltaic conversion efficiency (PCE) of 2.3% [14]. A wide variety of shells, including CdTe [15–17], CdSe [14,18,19], CdS [20], CdS/CdTe [21], In_2S_3 [22–24], TiO_2/$CuInSe_2$ [25,26], and Cu_2O [27,28], as well as the wide bandgap ZnSe [29] and ZnS [30], through the type II interfacial transition, have been developed in ETA solar cells, along with different HTMs, such as CuSCN or iodide/triiodide and poly-sulfur electrolytes, resulting in PCEs lying in the range of 1.5–5%.

Following pioneering works in 1990s [31,32], antimony tri-sulfide (Sb_2S_3) as a p-type V/VI semiconductor with a high optical absorption coefficient of 7.5×10^4 cm^{-1} at 550 nm and a direct bandgap energy of 1.7 eV at room temperature [33,34] has recently emerged as a highly promising optical absorber in semiconductor-sensitized solar cells [35–37]. The combination of Sb_2S_3 with nanoporous/mesoporous TiO_2 has led to the fabrication of semiconductor-sensitized solar cells with a record PCE of 7.5% [38], following the optimization of HTMs and post-deposition treatments [39–43]. A special emphasis has recently been placed on the defect reduction and interface optimization to boost the fairly low open-circuit voltage (V_{OC}) in Sb_2S_3-sensitized solar cells [44]. Alternative approaches for increasing efficiencies have included the development of $Sb_2(S,Se)_3$-sensitized solar cells, using in situ hydrothermal growth/post-selenization [45] or vapor transport deposition [46], leading to a PCE of 6.14 and 7.31%, respectively.

The combination of a Sb_2S_3 shell with TiO_2 nanostructures in the form of nanofibers [47], nanotubes [48], or NWs [49–52] has also been explored, resulting in the fabrication of semiconductor-sensitized solar cells with a PCE in the typical range of 2–5%. Recently, the use of surface modifiers with different functional groups and carbon numbers has led to the fabrication of Sb_2S_3-sensitized solar cells, reaching a PCE of 5.37%, which showed the capability of getting a high photovoltaic performance, using oxide nanorods [52]. Alternatively, there has been an increasing interest in coupling an Sb_2S_3 shell with ZnO NWs [53–60]. First, ZnO NWs with a very short length of 100 nm have been gap-filled by Sb_2S_3, using vacuum evaporation, leading to the fabrication of a nanocomposite cell with a PCE of 2.9% [53]. An immersion into thioacetamide, followed by a metal cation exchange process, was further employed to transform the surface layers of ZnO NWs into ZnS/Sb_2S_3 shells, resulting in the fabrication of ETA solar cells with a PCE of 1.32% [54]. Later on, Parize et al. reported the first ETA solar cells integrating ZnO/TiO_2/Sb_2S_3 core–shell NW heterostructures by using chemical bath deposition (CBD), atomic layer deposition (ALD), and ultrasonic chemical spray pyrolysis (CSP), respectively [55]. The TiO_2 conformal shell with the anatase phase was used to act as a protective, passivating layer [61]. A PCE of 2.3%, along with a V_{OC} of 656 mV and a short-circuit current density (J_{SC}) of 7.5 mA/cm^2, was achieved by using a Sb_2S_3 shell of approximately 10 nm, in which the presence of a senarmontite Sb_2O_3 phase was clearly revealed [55]. No optimization for the Sb_2S_3 shell purity was shown, although it is well-known to significantly influence the photovoltaic performance. The oxidation of the bulk of the Sb_2S_3 absorber layer has been proven to be detrimental to the photovoltaic performance of solar cells in both planar and nanostructured configurations, because Sb_2O_3 creates deep traps that cause the recombination of generated charge carriers [38,62]. Surface oxidation is by some accounts a valid technique for passivating the surface of the Sb_2S_3 absorber to boost the PCE [39], although the record PCE of 7.5% was reached with a post-sulfurized absorber [38]. More recently, significant efforts have been achieved on the development of a Sb_2S_3 shell by standard successive ionic layer adsorption and reaction (SILAR) [56] and spin-coating-assisted SILAR [58–60] techniques. The fabrication of ZnO/TiO_2/Sb_2S_3 core–shell NW heterostructures with a higher purity of the Sb_2S_3 shell was shown by standard SILAR process through the formation of three-dimensional clusters [56]. The use of spin-coating-assisted SILAR to form a Cu-doped

Sb_2S_3 shell modulating the bandgap alignment with ZnO NWs further resulted in the fabrication of sensitized-solar cells with a PCE of 3.14% [58]. The verticality and length of ZnO NWs were also found to strongly affect the photovoltaic performances of Sb_2S_3-sensitized solar cells, resulting in a typical PCE of around 2% after optimization [59,60]. The SILAR-related techniques have a high potential in the field of Sb_2S_3-sensitized solar cells, but typically lead to the formation of three-dimensional clusters or dots [56,58–60], which, in contrast to the CSP technique, results in the formation of a continuous thin shell for Sb_2S_3-based ETA solar cells [55]. Interestingly, the recent development of the CSP of Sb_2S_3 thin films by using a two-step process has further improved the deposit continuity as compared to the one-step process, while drastically increasing its purity by suppressing the formation of the senarmontite Sb_2O_3 phase, in turn increasing the PCE to 5.5% [63]. Actually, the highest efficiencies in Sb_2S_3-sensitized solar cells when grown by chemical deposition techniques have, to date, been only achieved by using an amorphous Sb_2S_3 phase as the first step.

In this study, the $ZnO/TiO_2/Sb_2S_3$ core–shell NW heterostructures grown by low-cost chemical deposition techniques, along with P3HT as the HTM and Au as the top electrode, are developed for ETA solar cells. We optimize the thickness of the Sb_2S_3 shell, whereas the ZnO NW/TiO_2 stack below the absorber is intentionally kept constant to decouple the physical phenomena at work. We further modify the growth conditions of the Sb_2S_3 shell by revising the sulfur content in the CSP solution to suppress the oxidation of the bulk of the Sb_2S_3 shell during deposition. The effect of the Sb_2S_3 shell thickness on the structural and optical properties of $ZnO/TiO_2/Sb_2S_3$ core–shell NW heterostructures, as well as on the performances of the resulting ETA solar cells, is investigated in detail by field-emission scanning electron microscopy (FESEM) imaging, in-plane X-ray diffraction (XRD), Raman and cathodoluminescence spectroscopy, UV–visible absorption and current density (J)–voltage (V) measurements under dark and air mass (AM) 1.5 G illumination conditions. Our investigation aims at identifying and clarifying the materials and technological issues limiting the performances of ZnO NW-based ETA solar cells, while proposing innovative solutions.

2. Materials and Methods

The $ZnO/TiO_2/Sb_2S_3$ core–shell NW heterostructures were grown by low-cost and easily scalable techniques on indium tin oxide (ITO)–glass substrates. The 150 nm–thick ITO layer (Delta Technologies, Loveland, CO, USA) had a sheet resistance in the range of 5–15 Ω/sq. and an optical transmittance larger than 85%. Its sheet resistance after a thermal treatment at a temperature higher than 400 °C was increased to a typical value in the range of 40–80 Ω/sq. ZnO NW arrays were grown by a two-step wet chemical route. First, the polycrystalline ZnO seed layers were deposited by sol–gel process. An equimolar solution of zinc acetate dihydrate ($Zn(CH_3OOH)_2 \cdot 2H_2O$, Sigma-Aldrich, St. Louis, MO, USA) and monoethanolamine (MEA, Sigma-Aldrich, St. Louis, MO, USA) was mixed in pure ethanol. The sol was stirred for 12 h at 60 °C and then for 12 h at room temperature to complete the dissolution of $Zn(CH_3OOH)_2 \cdot 2H_2O$. The xerogel film was formed by dipping the ITO–glass substrates into the sol and by withdrawing them at the speed of 3.3 mm/s in ambient atmosphere with a relative humidity below 15%. The xerogel film was put for 10 min on a hot plate kept at 300 °C to evaporate the organic compounds, and then annealed for 1 h in an oven kept at 500 °C to crystallize the ZnO seed layer. Second, the ZnO-seed-layer-coated ITO–glass substrates were placed face down in a sealed reactor to form ZnO NWs by CBD. A 30 mM equimolar solution of zinc nitrate hexahydrate ($Zn(NO_3)_2 \cdot 6H_2O$, Sigma-Aldrich, St. Louis, MO, USA) and hexamethylenetetramine (HMTA, Sigma-Aldrich, St. Louis, MO, USA) was prepared in deionized water. The sealed reactor was kept at 90 °C for 3 h in a regular oven. Subsequently, ZnO NW arrays were covered by ALD with an amorphous 10 nm–thick TiO_x layer grown at 200 °C from *tetrakis*-dimethylamino titanium (TDMAT) and H_2O in a F200 Fiji reactor from Cambridge Nanotech (Cambridge, MA, USA). A post-deposition thermal treatment in air was performed for 3 h in a regular oven kept at 300 °C to crystallize the anatase TiO_2 shell. Eventually, after purifying the sample surface under UV–ozone treatment for 30 min at room temperature in ambient air, an amorphous Sb_2S_3 shell was deposited by ultrasonic CSP in air at 210 °C from a solution of 60 mM antimony chloride ($SbCl_3$, Sigma-Aldrich, St. Louis, MO, USA) and 180 mM thiourea (Alfa Aesar, Ward Hill, MA, USA) dissolved in 99.8 vol% methanol (Honeywell, Charlotte, NC, USA) [63]. The deposition time was varied in cycles, where one cycle lasts 20 s, as described previously in Reference [63]. Afterward, Sb_2S_3 was crystallized in a tubular furnace in flowing 99.999% N_2 at 300 °C for 5 min. To deposit the HTM layer for the ETA solar cell, the samples were immersed in a solution of 2.0 wt% of regioregular poly(3-hexylthiophene-2,5-diyl) (P3HT, Carbosynth, Newbury, UK), dissolved in chlorobenzene (Sigma-Aldrich, St. Louis, MO, USA), and were ultrasonicated for 15 min. Thereafter, the samples were withdrawn from the solution and dried in air at 50 °C for 10 min, followed by a thermal treatment in vacuum ($5.3 \cdot 10^{-4}$ Pa) at 170 °C for 5 min. The ETA solar cells made of ZnO NW heterostructures were completed by thermally evaporating a layer of 99.999% Au through a mask to form an array of contacts, each 1.7 mm^2 in area.

FESEM images in top-view and cross-sectional view configurations of the incomplete structure were collected with a ZEISS GeminiSEM 300 instrument (Carl Zeiss, Oberkochen, Germany) operating at an accelerating voltage of 3 kV. FESEM images of the complete structure were collected with a ZEISS HR Ultra 55 instrument (Carl Zeiss, Oberkochen, Germany) operating at an accelerating voltage of 4 kV, and the FESEM–EDX data were therein recorded at an accelerating voltage of 7 kV, using a Bruker ESPRIT 1.8 EDX detector (Bruker, Billerica, MA, USA). In-plane XRD patterns were recorded with a RIGAKU Smartlab diffractometer (Rigaku, Tokyo, Japan) equipped with a 9 kW rotating anode, using the K_α(Cu) radiation and operating at 45 kV and 200 mA. The in-plane configuration was run on a 5 circle-goniometer that was specifically designed for this type of acquisitions. The X-ray beam was about 2 mm/0.05 nm parallel/perpendicular to the sample surface, respectively. The samples were placed in a horizontal position on a double tilt stage during the acquisition. Moreover, 2Theta-Chi/Phi XRD measurements were performed in the range of 20° to 60°, with a step of 0.04° and a speed of 1.0°/min, while setting the Omega incidence to 0.5°. The bixbyite In_2O_3, wurtzite ZnO, anatase TiO_2, and stibnite Sb_2S_3 phases were identified by using the 00-006-0416, 00-036-1451, 00-021-1272, and 00-042-1393 files of the International Center for Diffraction Data (ICDD), respectively. Raman spectra were recorded with a HORIBA/JOBIN YVON Labram spectrometer (Jobin Yvon, Palaiseau, France) equipped with a liquid-nitrogen-cooled CCD detector. The 632.8 nm excitation line of a Ne/He laser was used with a power on the sample surface close to 10 μW. The laser beam was focused to a spot size of 1 μm^2, using a 100× objective, leading to a power density of 10 $\mu W/\mu m^2$. The spectra were calibrated in wavenumber at room temperature by using a silicon reference sample where the theoretical position of the silicon Raman line was set to 520.7 cm^{-1}. The acquisition time for Raman scattering measurements was 600 s. Then 5 K cathodoluminescence measurements were performed on $ZnO/TiO_2/Sb_2S_3$ core–shell NW heterostructures with an FEI Inspect F50 FESEM instrument (FEI, Hillsboro, OR, USA) equipped with a liquid-helium-cooled stage. The cathodoluminescence signal was collected through a parabolic mirror and analyzed with a 550 mm focal length monochromator equipped with 600 grooves/mm diffraction grating. Cathodoluminescence spectra were recorded with a thermoelectric-cooled silicon CCD detector over a 3 × 3 μm^2 area, using a low acceleration voltage of 10 kV and a small spot size (i.e., less than 10 nm), along with an acquisition time of 30 s. Optical total transmittance and total reflectance were measured by using a JASCO V-670 UV–visible–NIR spectrophotometer (Jasco Applied Sciences, Halifax, NS, Canada) equipped with a 40 mm integrating sphere. ETA solar cell *J*–*V* curves under dark and AM 1.5 G illumination conditions were measured by using a factory-calibrated solar simulator (Xe light source, Newport Oriel Sol3A class AAA, Newport, Irvine, CA, USA) and a shadow mask with an aperture larger than the cell area defined by the back contact. The intensity of the solar simulator was calibrated to 100 mW/cm^2 at AM 1.5 G illumination conditions with a reference silicon solar cell. External quantum efficiency (EQE) spectra were measured at room temperature, using a light source (Newport 300 W Xenon lamp, 69911 equipped with a monochromator Newport Cornerstone 260, Newport, Irvine, CA, USA), a digital lock-in detector (Merlin, Newport, Irvine, CA, USA), and a factory-calibrated silicon reference detector. EQE-integrated J_{SC} was calculated in AM 1.5 G illumination conditions with the online tool Open Photovoltaics Analysis Platform (http://web.archive.org/web/20191216042337if_/http://opvap.com/eqe.php, accessed on 15 November 2021).

3. Results and Discussion

3.1. Structural and Optical Properties

The structural morphology of ZnO/TiO_2 core–shell NW heterostructures covered with a Sb_2S_3 shell grown by CSP, using 30, 50, 70, and 90 cycles, is presented in Figure 1 through FESEM images.

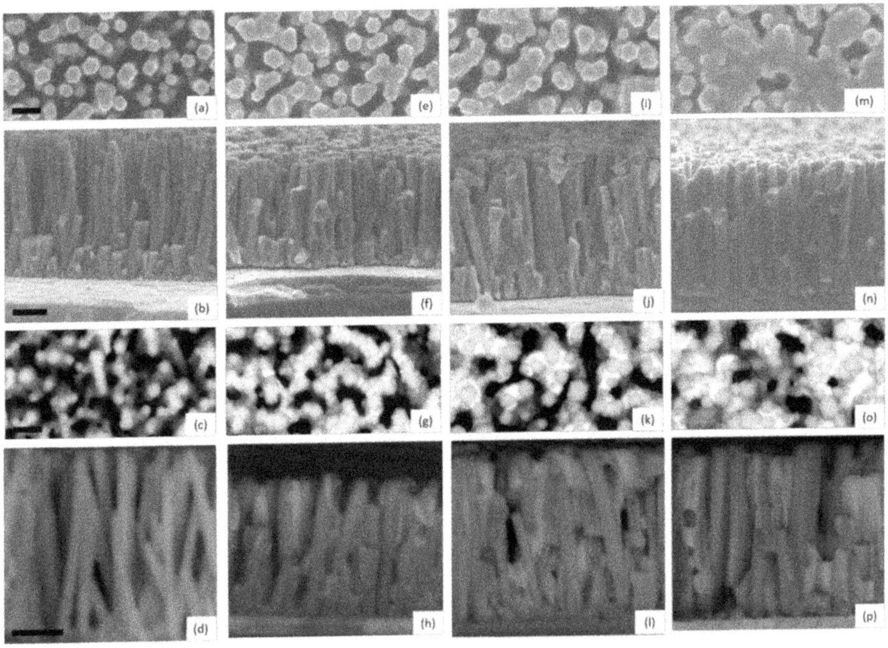

Figure 1. Top-view and cross-sectional view FESEM images of ZnO/TiO_2 core–shell NW heterostructures covered with a Sb_2S_3 shell grown by CSP, using (**a–d**) 30, (**e–h**) 50, (**i–l**) 70, and (**m–p**) 90 cycles. FESEM images were collected by using the detectors of secondary electrons (**a,b,e,f,i,j,m,n**) and backscattered electrons (**c,d,g,h,k,l,o,p**), respectively. The scale bars denote 200 and 300 nm for the top-view and cross-sectional view FESEM images, respectively.

Vertically aligned ZnO NWs are homoepitaxially grown on the grains with polar orientations composing the *c*-axis oriented ZnO seed layer [64,65]. They exhibit a mean length and diameter of 998 ± 115 nm and 92 ± 21 nm, respectively, when using an equimolar concentration of 30 mM [66]. The TiO_2 shell has a mean thickness of 8.5 nm on the sidewalls of ZnO NWs and of around 10 nm on their top *c*-face, and exhibits a pure anatase phase as determined by HRTEM imaging in Reference [61], which is desirable to get a type II band alignment with ZnO NWs. The ZnO/TiO_2 core–shell NW heterostructures are, on average, separated from each other by a distance of about 100 nm, forming a gap. A typical top-view FESEM image of ZnO/TiO_2 core–shell NW heterostructures is shown in Supplementary Materials Figure S1. These heterostructures are well covered by the Sb_2S_3 shell from bottom to top, regardless of the cycle number, as seen in Supplementary Materials Figure S2. Overall, the thickness of the Sb_2S_3 shell increases as the cycle number is increased. When using 30 cycles, the Sb_2S_3 shell appears to be very thin, as indicated by the hexagonal section of ZnO/TiO_2/Sb_2S_3 NWs and by the absence of any connections between them in Figure 1a–d. In contrast, in the cross-sectional view of ZnO/TiO_2/Sb_2S_3 NWs, contours are more rounded when the cycle number is increased to 50 as seen in Figure 1e–h, revealing the thickening of the Sb_2S_3 shell through the development of the thin, conformal layer. Moreover, a couple of connections start to be established at that cycle number between closely spaced ZnO/TiO_2 NWs. The connections between the ZnO/TiO_2 NWs are even more pronounced at the cycle number of 70, as seen in Figure 1i–l. The gaps between ZnO/TiO_2 NWs are almost filled by the Sb_2S_3 shell at the cycle number of 90, as seen in Figure 1m–p, switching from the core–shell configuration typically used in ETA solar cells to the fully impregnated configuration usually employed in bulk-heterojunction quantum dot solar cells and organic/hybrid solar cells [3]. From the present FESEM data, it is inferred that the thickness of the Sb_2S_3 shell varies, on average, from a couple of nanometers to several tens of nanometers as the cycle number is increased from 30 to 90.

The crystallinity and purity of the Sb_2S_3 shell, as well as its dependence on the cycle number, were assessed by XRD and Raman spectroscopy. The in-plane XRD patterns of ZnO/TiO_2 core–shell NW heterostructures covered with a Sb_2S_3 shell grown by CSP, using 30, 50, 70 and 90 cycles, are presented in Figure 2.

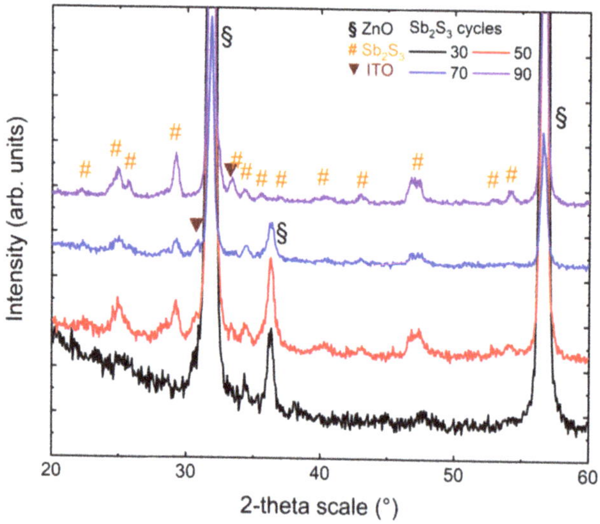

Figure 2. In-plane XRD patterns of ZnO/TiO_2 core–shell NW heterostructures covered with a Sb_2S_3 shell grown by CSP, using 30, 50, 70, and 90 cycles.

The in-plane configuration is well-designed to measure the diffracting planes of the Sb_2S_3 shell that are perpendicular to the sample surface, namely along the growth axis on the sidewalls of ZnO NWs. The XRD patterns are dominated by the $(10\bar{1}0)$ and $(11\bar{2}0)$ diffraction peaks located at 31.7° and 56.6°, respectively, corresponding to the wurtzite phase of vertically aligned ZnO NWs exhibiting the six-fold sidewalls with the nonpolar *m*-planes. The weaker $(10\bar{1}1)$ diffraction peak at 36.3° indicates that some of the ZnO NWs are slightly tilted with respect to the normal to the sample surface, following their nucleation over some grains with the same semipolar orientation composing the ZnO seed layer. The even weaker diffraction peak at 30.6° is attributed to the (222) planes of the bixbyite phase of the ITO layer. Its weak intensity originates from the in-plane configuration used, in which the X-ray beam was centered on the $ZnO/TiO_2/Sb_2S_3$ core–shell NW heterostructures. More importantly, the remaining diffraction peaks are all attributed to the stibnite phase of the Sb_2S_3 shell. The XRD pattern of the ZnO/TiO_2 core–shell NW heterostructures covered by the Sb_2S_3 shell grown with 30 cycles does not show any intense peaks. The weak diffraction peaks at 25.0°, 34.3°, and around 47.5° correspond to the (130)/(310), (131)/(311), and (002)/(151)/(511) planes. The very thin Sb_2S_3 shell grown with 30 cycles was, thus, partially crystallized. As the cycle number is increased to 90, the XRD patterns of the ZnO/TiO_2 core–shell NW heterostructures covered by the Sb_2S_3 shell exhibit a larger number of diffraction peaks with a higher intensity coming from the stibnite phase. Regardless of the cycle number beyond 30, the diffraction peaks at 24.93°, 29.20°, and around 46.80° corresponding to the (130)/(310), (121)/211), and (530)/(002)/(151)/(511) planes, respectively, dominate. A list of all of the diffraction peaks, along with their exact position and nature, is given in Supplementary Materials Table S1. Correlatively, the intensity of the ITO- and ZnO-related diffraction peaks decreases. Interestingly, no sign of the presence of the senarmontite Sb_2O_3 phase occurs in the XRD patterns, indicating the absence of this minor phase that was typically detected when Sb_2S_3 has been grown by CBD [67]. Additionally, no significant shifts in the position of the diffraction peaks of the ZnO NWs and of the TiO_2 and Sb_2S_3 shells are detected, indicating that the $ZnO/TiO_2/Sb_2S_3$ core–shell NW heterostructures are fully relaxed. The local epitaxy between the ZnO NWs and TiO_2 shell grown by ALD is plastically accommodated, such that the epitaxial strain is totally relieved [61]. The Sb_2S_3 shell grown by CSP follows a Volmer–Weber growth mode, where the intrinsic stress can be relieved by different processes occurring at grain boundaries [68].

The Raman spectra of ZnO/TiO_2 core–shell NW heterostructures covered with a Sb_2S_3 shell grown by CSP using 30, 50, 70, and 90 cycles are presented in Figure 3. A very small laser power density of 10 $\mu W/\mu m^2$ was used to avoid any photo-induced degradation of the Sb_2S_3 phase, as reported in Reference [69]. The stibnite Sb_2S_3 phase belongs to the *Pbnm* (centrosymmetric) space group. According to the factor group analysis at the Γ point, the orthorhombic stibnite structure exhibiting 20 atoms per Sb_2S_3 primitive cell has 30 active Raman modes: $\Gamma_{Raman} = 10\ A_g + 5\ B_{1g} + 10\ B_{2g} + 5\ B_{3g}$ [70–72].

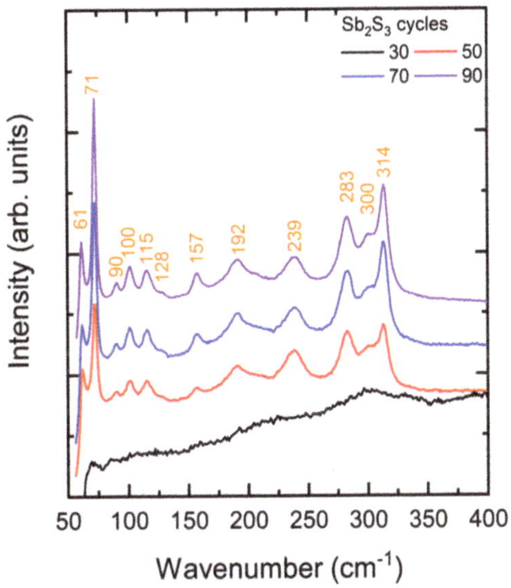

Figure 3. Raman spectra of ZnO/TiO_2 core–shell NW heterostructures covered with a Sb_2S_3 shell grown by CSP using 30, 50, 70, and 90 cycles.

The Raman spectrum of the ZnO/TiO_2 core–shell NW heterostructures covered by the Sb_2S_3 shell grown with 30 cycles does not show any intense narrow lines. Rather, a broad Raman band centered at around 300 cm^{-1} is observed and is typical of the amorphous Sb_2S_3 phase [67,73]. This confirms that the very thin Sb_2S_3 shell grown with 30 cycles is mainly amorphous and has only been partially crystallized. Any role of chlorine residues in that process due to the lower deposition temperature is ruled out by the fact that chlorine is known to improve the crystallinity of Sb_2S_3 thin films [74]. Instead, more and more energy (i.e., higher annealing temperature) is required to crystallize the Sb_2S_3 shell at nanoscale dimensions, below a typical thickness of 10 nm. The need for crystallizing layers with a thickness of less than 10 nm at a substantially higher annealing temperature than the same layers with a thickness of several tens of nanometers has been related to an increase in the activation energy for the crystallization process in oxides and phase change materials [75–77]. For extremely thin layers, the surface-over-volume ratio is even so high that the surface energy predominantly contributes to the total free energy hampering the amorphous-to-crystalline state transformation below a critical thickness [77]. As the cycle number is increased to 90, the Raman spectra of the ZnO/TiO_2 core–shell NW heterostructures covered by the Sb_2S_3 shell show intense narrow lines originating from the stibnite phase. Regardless of the cycle number beyond 30, the Raman lines at 61 cm^{-1} (B_{1g}/B_{3g}), 71 cm^{-1} (A_g), 283 cm^{-1} (A_g), and 314 cm^{-1} (B_{2g}) dominate [70–72,78]. A list of all of the Raman lines, along with their exact position and nature, is given in Supplementary Materials Table S2. The Raman line intensity generally increases as the cycle number is increased. This confirms that the Sb_2S_3 shell was well-crystallized and that its thickness increased as the cycle number does, which is in agreement with the FESEM and in-plane XRD measurements. Additionally, it should be noted that two Raman lines at 90 and 115 cm^{-1} are detected when the cycle number lies in the range of 50–90. The assignment of these Raman lines is still under debate. On the one hand, they can be attributed to the senarmontite Sb_2O_3 phase through the B_2 and E modes, respectively [79]. The formation of that senarmontite phase takes place when the Sb_2S_3 phase is in contact with air, usually passivating the surface layer and thus being beneficial for the heterojunction interface. On the other hand, it was reported in Reference [72], from density-functional theory calculations, that two B_{2g} modes originating from the stibnite phase lie in the same range of wavenumber. In the present case, on the basis of the in-plane XRD patterns and in the absence of the usually dominant Raman lines at around 190 and 255 cm^{-1} belonging to the senarmontite Sb_2O_3 phase [67,73], it is deduced that the two Raman lines at 90 and 115 cm^{-1} are attributed to the B_{2g} modes of the Sb_2S_3 phase. The absence of the Sb_2O_3 phase can be explained by the likely oxidation suppression, because the deposition temperature of the Sb_2S_3 shell was decreased from 220 °C in our previous study [55] to 210 °C in this study. Furthermore, the concentration of the Sb and S precursors as $SbCl_3$ and thiourea in the CSP process was doubled. Increasing the concentration of the sulfur source in the spray solution is known to suppress the oxidation of metal sulfide layers during the deposition [80].

The 5 K cathodoluminescence spectra of ZnO/TiO_2 core–shell NW heterostructures covered with a Sb_2S_3 shell grown by CSP using 30, 50, 70, and 90 cycles are presented in Figure 4 following an analysis over a fixed surface area of 3×3 μm^2.

These spectra are composed of three typical emission bands located at 3.36 eV, in the range of 2.00–2.25 eV and at around 1.80 eV, as seen in Figure 4a. The 3.36 eV line corresponds to the near-band edge (NBE) emission of ZnO NWs, which is assigned to donor-bound A exciton transitions involving hydrogen-related defects [81]. From the *I* nomenclature used to label the excitonic transitions [82], the I_4 and I_5 lines assigned to substitutional hydrogen on the oxygen lattice site [83] and zinc vacancy/hydrogen defect complexes [84], along with radiative transitions involving interstitial hydrogen in bond centered sites, significantly contribute to the NBE emission of ZnO NWs grown by CBD [81]. The yellow-green emission band at around 2.25 eV is attributed to the (V_{Zn}-2H) defect complex, while the red-orange emission band at around 1.86 eV is assigned to the (V_{Zn}-H) defect complex. No green-blue emission band centered at around 2.65 eV and associated with hydrogen-related defects on the surfaces of ZnO NWs occurs, indicating, as expected, that the TiO_2 shell efficiently passivates the surface defects [61]. Furthermore, it should be noted that the visible emission band lies in the range of 1.50–2.25 eV when the Sb_2S_3 shell is grown for 30 cycles, and thus is much broader than the visible emission band lying in the range of 1.50–1.90 eV when the Sb_2S_3 shell is grown for 50, 70, and 90 cycles, as seen in Figure 4b. The much lower intensity of the radiative transitions extending beyond 1.90 eV when the Sb_2S_3 shell is grown for 30 cycles reveals that the NBE emission of amorphous Sb_2S_3 takes place in the yellow-green emission band that is close to its bandgap energy of 2.2 eV [85]. The contribution of the amorphous Sb_2S_3 shell to the yellow-green emission band vanishes at a cycle number beyond 30, which is in agreement with in-plane XRD and Raman scattering measurements. Moreover, the NBE emission of the crystallized Sb_2S_3 shell occurs in the red emission band that is close to its bandgap energy lying in the range of 1.70–1.80 eV [33,34,85]. As the cycle number is increased from 30 to 90, the ratio of the NBE emission of ZnO NWs over the red-orange emission band significantly decreases. A prominent contribution around 1.78 eV corresponding to the red emission is mainly attributed to the NBE emission of the crystallized Sb_2S_3 shell. As the shell thickness is increased with a higher cycle number in the CSP process, the red emission band becomes more and more intense. However, the increase of around 20% in the intensity of the red emission band when the Sb_2S_3 shell is grown with a cycle number of 90, as compared to 70, is less than the estimated 30% increase in the thickness of the Sb_2S_3 shell. This is likely the sign of a small increase in the density of defects in the bulk of the Sb_2S_3 shell when grown with a cycle number of 90.

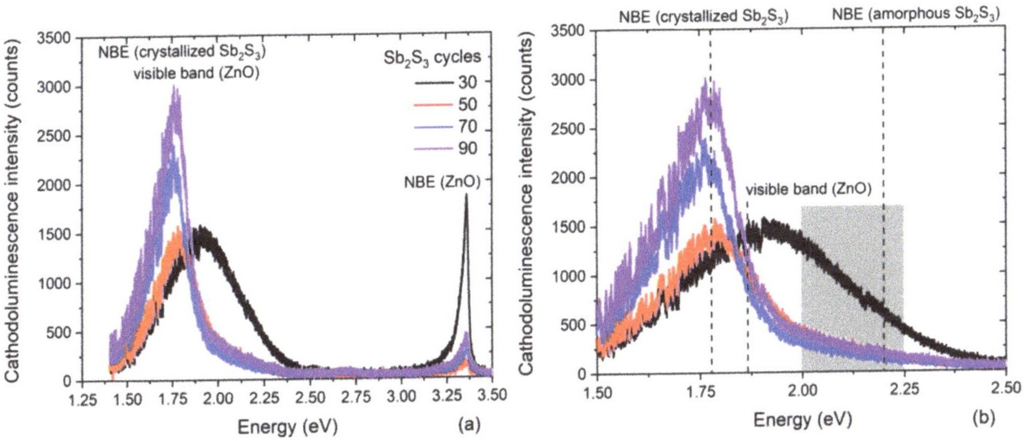

Figure 4. (**a**) 5 K cathodoluminescence spectra collected on an ensemble of single ZnO/TiO_2 core–shell NW heterostructures covered with a Sb_2S_3 shell grown by CSP using 30, 50, 70, and 90 cycles. A fixed surface area of 3 × 3 μm^2 was chosen for the acquisition. (**b**) Zoom-in in the area of interest corresponding to the visible energy range from 1.5 to 2.5 eV.

Correlatively, the sample color switches from translucent orange to opaque dark brown by increasing the cycle number, which reveals an increase in the absorption of the visible light by the Sb_2S_3 shell. The optical bandgap energy of the Sb_2S_3 shell was extracted from EQE measurements following the $(EQE \times h\nu)^2$ method [86] and is reported in Figure 5. The $(EQE \times h\nu)^2$, as a function of photon energy, was plotted, and a linear regression fitting provided the value of the optical bandgap energy. The optical bandgap energy obtained for 30, 50, 70, and 90 cycles is 1.78, 1.76, 1.74, and 1.72 eV, respectively, thus converging to the theoretical value of around 1.7 eV in bulk Sb_2S_3 when the shell thickness is increased [33,34,85].

3.2. Photovoltaic Performances

The architecture of $ZnO/TiO_2/Sb_2S_3$ core–shell NW heterostructure-based ETA solar cells and the corresponding diagram of energy levels are presented in Figure 6. Under AM 1.5 G illumination through the glass substrate, visible photons are absorbed by the Sb_2S_3 shell with a bandgap energy of around 1.7 eV to generate electron–hole pairs. The successive type II band alignments are favorable for the separation and collection of electrons and holes to generate the photocurrent as follows: electrons migrate to the TiO_2 shell and then to the ZnO NWs, acting as the ETM toward their collection with the ITO contact as the topside electrode, while holes migrate to P3HT, acting as the HTM toward their collection with the Au contact as the backside electrode. The TiO_2 shell acts as a protective, passivating layer [61], further improving the quality of the interface between the ZnO NWs and Sb_2S_3 shell. The absence of the TiO_2 shell very strongly degrades the photovoltaic performances in the present ETA solar cells [55].

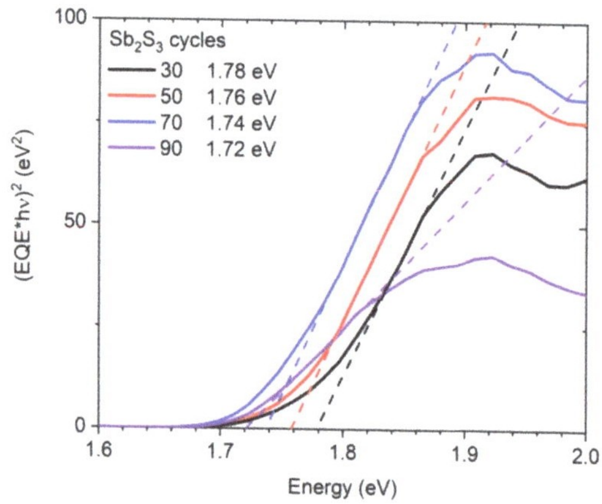

Figure 5. (EQE × hν)^2 vs. photon energy (solid line) of ZnO/TiO_2 core–shell NW heterostructures covered with a Sb_2S_3 shell grown by CSP, using 30, 50, 70, and 90 cycles. The fitting dashed lines reveal the optical bandgap energy as the intercept of the linear part with the photon energy axis.

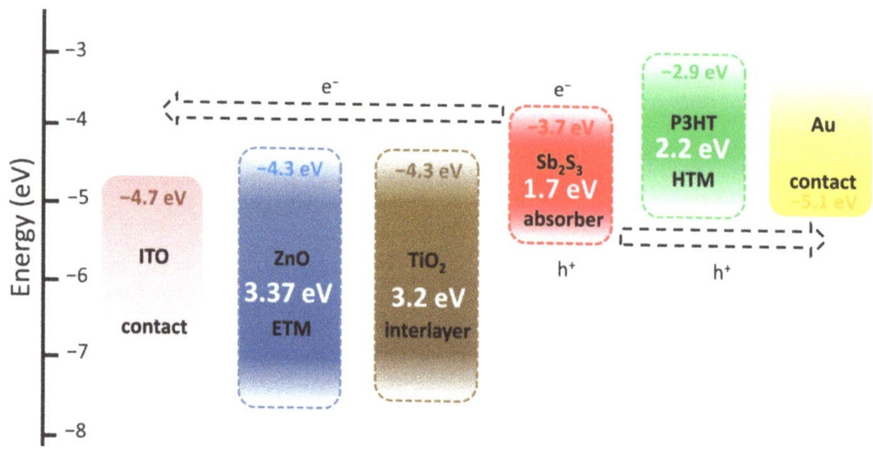

Figure 6. Architecture and corresponding diagram of energy levels of $ZnO/TiO_2/Sb_2S_3$ core–shell NW heterostructure-based ETA solar cells.

The ZnO/TiO_2 core–shell NW heterostructures covered with a Sb_2S_3 shell grown by CSP, using 30, 50, 70, and 90 cycles, were filled by P3HT, using immersion and thereafter covered with a thin layer of Au by thermal evaporation to form the complete ETA solar cell structure of $ITO/ZnO/TiO_2/Sb_2S_3$/P3HT/Au. A cross-sectional view FESEM image of the entire ETA solar cell, as presented in Figure 7, clearly indicates the efficient penetration of the Sb_2S_3 shell and P3HT when using the cycle number of 70.

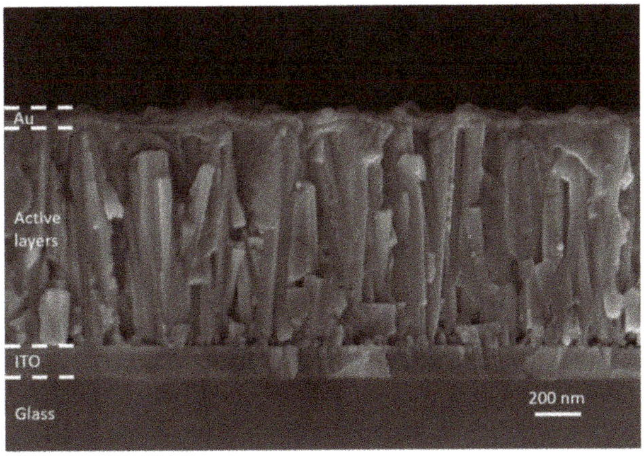

Figure 7. Cross-sectional view FESEM image of the complete ETA solar cell structure of $ITO/ZnO/TiO_2/Sb_2S_3$/P3HT/Au when using the cycle number of 70.

The *J–V* curves collected in dark and AM 1.5 G illumination conditions are presented in Figure 8, and the photovoltaic properties are reported in Table 1 and plotted in Figure 9.

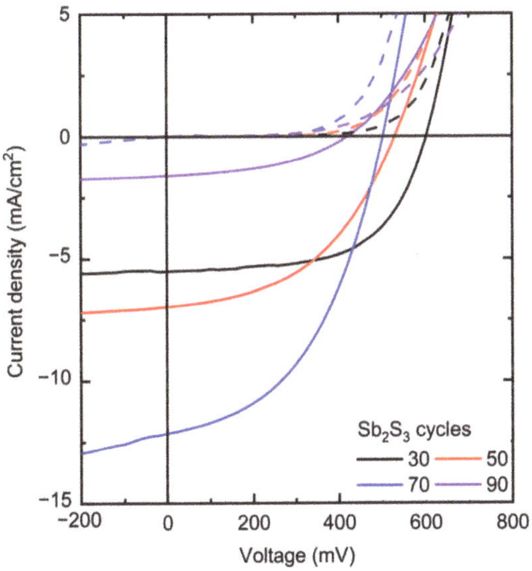

Figure 8. *J–V* curves of the best ETA solar cells made of ZnO/TiO_2 core–shell NW heterostructures covered with a Sb_2S_3 shell grown by CSP, using 30, 50, 70, and 90 cycles, collected under dark (dashed lines) and AM 1.5 G illumination (solid lines) conditions.

Table 1. Photovoltaic properties of the ETA solar cells involving ZnO/TiO_2 core–shell NW heterostructures covered with a Sb_2S_3 shell grown by CSP, using 30, 50, 70, and 90 cycles. The mean values and standard deviations are given in brackets.

Cycle Number	V_{OC} (mV)	J_{SC} (mA/cm^2)	R_s ($\Omega\cdot$cm^2)	R_{sh} ($\Omega\cdot$cm^2)	FF (%)	PCE (%)	No. of Cells
30	605 (481 ± 49)	5.48 (4.80 ± 0.87)	16.0 (24.8 ± 5.3)	2784 (1394 ± 789)	60.2 (51.7 ± 3.2)	2.00 (1.22 ± 0.38)	12
50	530 (499 ± 32)	6.97 (6.51 ± 0.32)	22.9 (26.8 ± 3.9)	980 (681 ± 442)	46.9 (44.6 ± 4.3)	1.73 (1.46 ± 0.25)	9
70	502 (484 ± 56)	12.08 (10.90 ± 0.76)	15.2 (15.3 ± 1.5)	221 (343 ± 107)	46.7 (43.7 ± 2.2)	2.83 (2.32 ± 0.42)	10
90	399 (323 ± 94)	1.57 (1.38 ± 0.30)	84.9 (127 ± 38)	818 (630 ± 719)	43.8 (33.4 ± 8.5)	0.27 (0.16 ± 0.08)	16

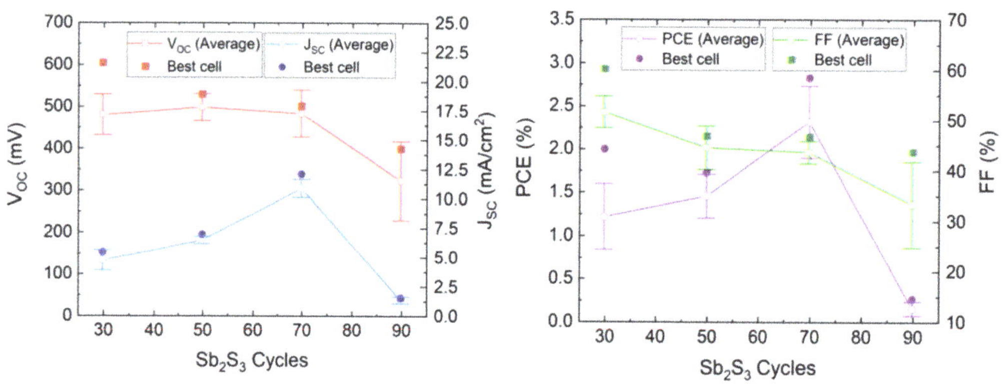

Figure 9. Evolution of the V_{OC}, J_{SC}, FF, and PCE values of the ETA solar cells made of ZnO/TiO_2 core–shell NW heterostructures covered with a Sb_2S_3 shell grown by CSP as a function of the cycle number. Horizontal bars denote standard deviation.

The photovoltaic performances of the ETA solar cells involving the $ZnO/TiO_2/Sb_2S_3$ core–shell NW heterostructures offer a clear trend of the dependence of its characteristics on the Sb_2S_3 shell thickness. V_{OC} is quite high, with a mean value above 400 mV when the Sb_2S_3 shell is grown for 30, 50, and 70 cycles. The V_{OC} mean value initially increases from 481 to 499 mV as the cycle number is increased from 30 to 50, and then it decreases continuously to 323 mV as the cycle number reaches a value of 90. The initial increase in the V_{OC} mean value is related to a more continuous and better crystallinity of the Sb_2S_3 shell when grown with a cycle number of 50, in turn decreasing the density of defects in its bulk. The further decrease in the V_{OC} mean value likely results from the two following major reasons. First, we expect that the growth of a thicker Sb_2S_3 shell over high-aspect-ratio ZnO NWs in an ETA solar cell results in the increase in the density of defects in its bulk, as revealed in the cathodoluminescence spectra. Second, the spatial variance of the thickness of the Sb_2S_3 shell is expected to appear as the spacings get narrower, which further leads to the issue of incomplete immersion of P3HT into the gaps between the ZnO NWs in the case of the largest Sb_2S_3 thickness at 90 cycles, as supported in Supplementary Materials Figure S3. By performing an FESEM–EDX analysis over a given rectangular area located on the cross-section of the ETA solar cells, it is revealed that the Sb/Zn element ratio continuously increases as the Sb_2S_3 shell is thickened and, more importantly, that the S/Sb element ratio gradually decreases toward a value of 1.7 as the cycle number is increased to 90. This indicates that the amount of excess sulfur coming from P3HT at a given thickness of the entire ETA solar cells strongly decreases down to a S/Sb element ratio value close to 1.5, which is expected from the sulfur coming only from the Sb_2S_3 shell. As such, the penetration depth of the P3HT is reduced by the thickening of the Sb_2S_3 shell, raising the issue of a progressive incomplete immersion of P3HT as the cycle number is increased from 30 to 90. Furthermore, it is well-known from ETA solar cell studies that the uniformity of the thickness of the absorber shell is a critically important performance factor [3,55]. Consequently, the heterojunction quality worsens as the cell conceptually changes by gradually transforming from the ETA cell type to a 3D cell type in which the active component is not a conformal shell anymore, leading to the observed drop in the V_{OC}. In contrast, the J_{SC} mean value significantly increases from 4.80 to 10.90 mA/cm^2 as the cycle number is increased to 70, which can be attributed to an enhancement of the optical absorption that increases the charge carrier generation rate thanks to the increase in the thickness of the Sb_2S_3 shell. However, the J_{SC} mean value then falls to 1.38 mA/cm^2 as the cycle number reaches 90. In particular, as discussed above, at larger cycle numbers to deposit the Sb_2S_3 shell, the increased density of defects in its bulk induces more recombination while the ZnO NWs exhibit less gap space to accommodate HTM, with both of them leading to poor charge carrier collection and extraction. In fact, in order to reach the HTM in the ETA solar cells with a thick Sb_2S_3 shell and a poor HTM coverage, the holes are expected to move along the Sb_2S_3 shell for as much as the length of the core–shell NWs, in excess of 1000 nm, which is five times longer than the diffusion length of charge carriers reported for Sb_2S_3 [87]. Recombination thus occurs before the holes reach the HTM.

The calculated fill factor (FF) also decreases continuously from the mean values of 51.7 to 33.4% as the cycle number is increased up to 90; hence, its trend generally follows the trend of the V_{OC}. The series resistance (R_s) mean values are 24.8, 26.8, 15.3 and 127 $\Omega \cdot cm^2$ for a cycle number of 30, 50, 70, and 90, respectively. This indicates that, at a cycle number of 70, the thickness of the Sb_2S_3 shell yields the lowest R_s, and increasing to a cycle number of 90 immediately causes an eightfold increase in R_s, as is apparent from the trend in J_{SC}. The shunt resistance (R_{sh}) mean values are 1394, 681, 343, and 630 $\Omega \cdot cm^2$ for a cycle number of 30, 50, 70, and 90, respectively. Both R_{sh} and V_{OC} decrease as the cycle number is increased. Thus, the poor penetration of P3HT into the space between core–shell NWs combined with the uneven thickness of the Sb_2S_3 coating could possibly cause a reduction in charge carrier transfer from Sb_2S_3 to P3HT. Thereby, interface recombination is increased, leading to a reduction in R_{sh}. In addition, Sb_2S_3 shells interconnecting at closely positioned core–shell NWs could in turn cause shunting (i.e., short-circuits). Furthermore, the reduction of R_{sh} as the cycle number is increased could also be due to the increased density of defects in the bulk of the Sb_2S_3 shell. The PCE mean value of the ETA solar cells increases from 1.22 to 2.32% as the cycle number is increased from 30 to 70, and then it decreases drastically to 0.16% as the cycle number is further increased to 90. The evolution of the PCE mean value is thus driven strongly by the evolution of the J_{SC} mean value, along with that of the V_{OC} mean value to a lesser extent. The best ETA solar cell is obtained when the Sb_2S_3 shell is grown for 70 cycles, with a maximum PCE of 2.83% (V_{OC} = 502 mV, J_{SC} = 12.08 mA/cm^2, and FF = 46.7%). The increase in the thickness of the Sb_2S_3 shell is favorably related to an improvement of the optical absorption of visible photons generating more charge carriers, but a balance must be found, as its further increase is liable to increase the density of defects in its bulk and lead to the poor penetration depth of the HTM, both of which likely cause the reduced hole collection. As the distance between bare core–shell NWs is constant, as seen in Figure 1, a thicker absorber inevitably leads to a narrower and shallower space between the core–shell NWs for the HTM to occupy. As a result, the penetration depth of the HTM decreases at a larger Sb_2S_3 thickness, which is evident in the performance drop above 70 cycles. The optimization of the Sb_2S_3 shell thickness thereby results in an increase in the PCE value of over 0.5%, as compared to our first ETA solar cells reported in Reference [55].

The EQE measurements of the best ETA solar cells made of ZnO/TiO_2 core–shell NW heterostructures covered with a Sb_2S_3 shell grown by CSP, using 30, 50, 70, and 90 cycles, are reported in Figure 10, along with the absorptance and absorption coefficient of each layer in Figure 11. By integrating the EQE spectra over the wavelength, the ideal J_{SC} values of 10.2, 9.7, 15.6, and 4.2 mA/cm^2 for a cycle number of 30, 50, 70, and 90, respectively, corresponding to the best samples reported in Table 1, were obtained, and, as a trend, they are in relative agreement with the J_{SC} values deduced from the *J*–*V* measurements under AM 1.5 G illumination condition. The J_{SC} value calculated from EQE measurements is overall higher than that calculated from *J–V* curves, probably due to fundamental differences in illumination intensities for both techniques. There exists a clear positive offset in favor of the values calculated from the EQE data, which were collected by using monochromatic low-intensity light scans. At low illumination intensities, the PCE value (and, thus, the J_{SC} value when normalized to the light intensity) is expected to be higher, as already demonstrated for a planar analogue of this type of cell [88].

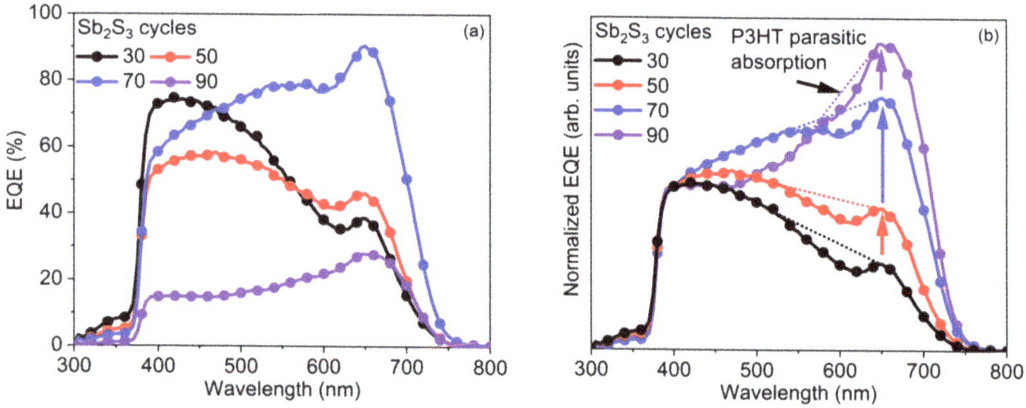

Figure 10. (**a**) EQE vs. wavelength and (**b**) normalized EQE to its value at 400 nm vs. wavelength of the best ETA solar cells made of ZnO/TiO_2 core–shell NW heterostructures covered with a Sb_2S_3 shell grown by CSP, using 30, 50, 70, and 90 cycles. The dotted lines are a guide-to-the-eye, lying roughly between the absorption onset of P3HT at 650 nm on one side, and the absorption maximum of P3HT, at 525 nm, on the other side.

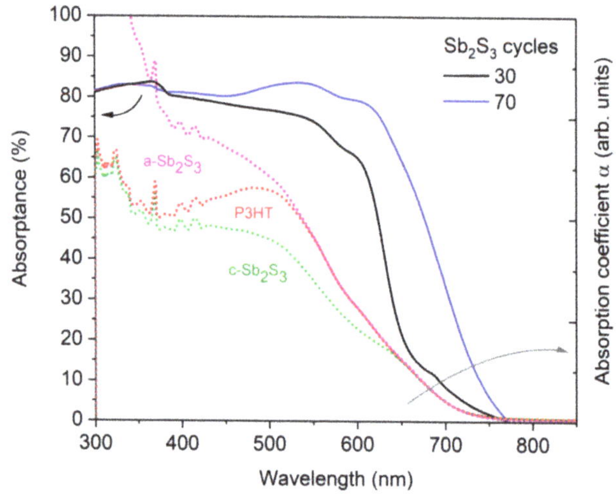

Figure 11. Absorptance (solid lines) of the best ETA solar cells made of ZnO/TiO_2 core–shell NW heterostructures covered with a Sb_2S_3 shell grown by CSP, using 30 and 70 cycles, and topped by P3HT. Absorption coefficient, α (dashed lines), of each layer and of the expected additional phase (a-Sb_2S_3) in the ETA cell. The absorption coefficients of P3HT and amorphous Sb_2S_3 (a-Sb_2S_3) were elevated above crystalline Sb_2S_3 (c-Sb_2S_3) and P3HT, respectively, to illustrate the cumulative effect of each added layer on the optical density of the ETA cell.

The EQE value at around 650 nm, attributed solely to Sb_2S_3, as neither P3HT nor any of the other layers in the core–shell NWs absorb light in this region [89], as further shown in Figure 10, increases from 38.7% to 46.1%, peaking at 90.6% as the Sb_2S_3 shell is grown for a cycle number increasing from 30 to 70, and then falls to 28.0% for a cycle number of 90. As the cycle number is increased, the EQE at 380–500 nm steadily decreases, as seen in Figure 10a. EQE could decrease in this region because the holes generated by visible photons with a low penetrating depth (25–50 nm based on the absorption coefficient of Sb_2S_3 in this range [88]) have to move through a thicker Sb_2S_3 shell to reach the HTM. However, as the EQE in this region is higher at a cycle number of 70 vs. 50, it is more likely that the trend is related to the shift in the position at which photons are absorbed in the stack relative to the ITO and Au contacts. The issue of hole collection and extraction is further amplified when the Sb_2S_3 shell is grown for a cycle number larger than 70, due to a non-conformal HTM coating, or even the absence of any HTM coating originating from the decreased gap size as discussed earlier. In contrast, the EQE at 550–720 nm increases as the cycle number is increased, peaking at 70 cycles, thanks to the increased optical absorption in the thicker Sb_2S_3 shell, whereas enough space still remains for P3HT to penetrate to the deepest parts of the ETA solar cell. Finally, at 90 cycles, the Sb_2S_3 shell is already so thick that recombination dominates owing to the poor P3HT penetration, and the entire EQE spectrum has plummeted in intensity. The EQE normalized to the high-energy-absorption edge in Figure 10b illustrates the proportional increase of optical absorption in the low-energy region as the cycle number is increased. The dotted line shows the perceived EQE loss due to the parasitic absorption in the P3HT, as well-established in earlier reports [43], fitting the shape of its absorption that peaks at around 600 nm. As the cycle number is increased, the EQE loss due to the parasitic absorption seems to decrease and span across a smaller wavelength range. This can be explained by a decreasing amount of light reaching the P3HT, as a thicker Sb_2S_3 shell absorbs more incident light at the 550–720 nm wavelength range. Evidently, further gains in J_{SC} and PCE values could be achieved for this type of ETA solar cell by making use of a UV–vis transparent HTM to prevent parasitic absorption.

The recapitulated photovoltaic performances of the nanostructured solar cells involving the heterostructures made of ZnO NWs and Sb_2S_3, as a comparison with this work, are presented in Table 2. Most of the reported data involve the core–shell configuration at the expense of the fully impregnated configuration, as defined in Reference [3], owing to its higher potential to benefit from the advantages of integrating ZnO NWs in the architecture. In comparison with our previous results in Reference [55], the J_{SC} value of the best device significantly increased from 7.5 to 12.08 mA/cm^2, while the V_{OC} value decreased from 656 to 502 mV, at a similar FF value of 47%, resulting in an increase in the PCE from 2.3% to 2.83% (Tables 1 and 2). The V_{OC} and J_{SC} values thus govern the PCE variation in the present study. As shown in Reference [55], the V_{OC} value primarily stems from the use of the TiO_2 passivating layer and P3HT HTM, with minor contribution from the added Sb_2S_3 shell. In contrast, the J_{SC} value is mainly driven by the Sb_2S_3 absorber shell. Moreover, it is known that longer ZnO NWs as the ETM generally result in a proportionally larger J_{SC} value, with a small decrease in the V_{OC} value due to added interfacial recombination [90]. Thus, the increase in the J_{SC} value in the current development, as compared to Reference [55], originates from the use of slightly longer ZnO NWs (998 ± 115 nm vs. 900 nm), as well as the enhanced quality and increased thickness of the Sb_2S_3 absorber shell. In contrast, the decrease in the V_{OC} value can partially stem from the small increase in the length of ZnO NWs. It should be noted that the recent development of nanostructured solar cells with liquid electrolyte as the HTM has emerged as an alternative to the ETA solar cells with P3HT as the HTM. The direct comparison shows that the J_{SC} values are so far less in these solar cells, which, however, reveal higher FF values. They further raise the problem of stability with time, as in the case of dye-sensitized solar cells, for which the quality of the interface between the inorganic materials and liquid electrolyte still represents a major issue. In that respect, the introduction of an interlayer, such as ZnS [54] and TiO_2 [55]

between the ZnO NWs and Sb_2S_3 shell when grown by chemical deposition techniques, appears as a typical route to further optimize the architecture of these ETA solar cells.

Table 2. Recapitulated photovoltaic properties of the nanostructured solar cells involving the heterostructures made of ZnO NWs and Sb_2S_3 in the fully impregnated or core–shell configurations according to the definition used in Reference [3].

Materials	Architecture	Sb_2S_3 shell	HTM	V_{OC} (mV)	J_{SC} (mA/cm^2)	FF (%)	PCE (%)	Reference
ZnO/Sb_2S_3	Full impregnation	Thermal evaporation	P3HT	450	16.0	40	2.9	[53]
$ZnO/ZnS/Sb_2S_3$	Core–shell	Chemical conversion	P3HT	440	5.57	54	1.32	[54]
$ZnO/TiO_2/Sb_2S_3$	Core–shell	CSP	P3HT	656	7.5	47	2.3	[55]
ZnO/Sb_2S_3	Core–shell	CBD	Electrolyte	438	1.46	31	0.20	[57]
ZnO/Sb_2S_3:Cu	Core–shell	SILAR	Electrolyte	580	9.18	59	3.14	[58]
ZnO/Sb_2S_3	Core–shell	SILAR	Electrolyte	586	7.02	59	2.43	[59]
ZnO/Sb_2S_3	Core–shell	SILAR	Electrolyte	582	5.91	59	2.04	[60]
$ZnO/TiO_2/Sb_2S_3$	Core–shell	CSP	P3HT	502	12.08	46.7	2.83	This Work

These findings show that the dimensional optimization of all the components in the $ZnO/TiO_2/Sb_2S_3$ core–shell NW heterostructures is crucial to further improve the photovoltaic performance of the related ETA solar cells, while the issue of the HTM is capital to properly collect the charge carriers, specifically the holes. It is foreseen that, by carefully varying the length and aspect ratio of ZnO NWs, as well as the spacing between them, a larger amount of absorber could be incorporated in the core–shell NW heterostructures in a way that the gaps between the core–shells are not absorber-filled, thus promoting a more uniform HTM coating and preserving the integrity of the ETA concept. Moreover, the application of an HTM, which is more transparent in the visible range, could further increase the J_{SC} value to boost the PCE. To accommodate the thicker Sb_2S_3 shell, detrimental interconnections between the core–shell NW individual heterostructures must be eliminated by growing each ZnO NW perpendicular to the substrate and parallel to adjacent ZnO NWs in a more optimized arrangement. On the other end, at the extremely low absorber thicknesses, elevated crystallization temperatures are liable to cause materials and technological issues when the substrate cannot handle the elevated thermal load and should be taken into account when modeling and designing experiments.

4. Conclusions

In summary, we investigated the effect of the thickness of the Sb_2S_3 shell by varying the cycle number from 30 to 90 during the CSP process on the structural and optical properties of the $ZnO/TiO_2/Sb_2S_3$ NW heterostructures, along with the photovoltaic performance of ETA solar cells, using P3HT as the HTM. By growing the Sb_2S_3 shell at the moderate temperature of 210 °C, using the CSP process, the Sb_2S_3 shell was found to be of high purity and free of the unwanted senarmontite Sb_2O_3 phase. This represents a strong advantage of the CSP process over the CBD process. The limitations at both the low and high end of Sb_2S_3 shell thicknesses from a couple of nanometers to several tens of nanometers were discussed in detail. The low end is limited by challenges in the crystallization of the Sb_2S_3 shell that is amorphous at nanoscale dimensions, resulting in the low optical absorption of visible photons. In contrast, the high end is limited by the increased density of defects in the bulk of the Sb_2S_3 shell, degrading charge carrier dynamics, and by the incomplete immersion of the P3HT in the structure, resulting in the poor hole collection. The best ETA solar cell with the intermediate optimum thickness of the Sb_2S_3 shell shows a J_{SC} of 12.1 mA/cm^2, a V_{OC} of 502 mV, and a PCE of 2.83%. These findings deepen the knowledge of the advantages and limitations of the architecture of ETA solar cells through materials and technological issues, further emphasizing the intricate nature of any dimensional optimization in the structure as a major challenge to boost their overall photovoltaic performance.

Supplementary Materials: The following supporting information can be downloaded at https://www.mdpi.com/article/10.3390/nano12020198/s1. Figure S1: Typical top-view FESEM image of ZnO/TiO_2 core–shell NW heterostructures without any Sb_2S_3 shell, Figure S2: High-magnification cross-sectional view FESEM image of ZnO/TiO_2 core–shell NW heterostructures covered by a Sb_2S_3 shell grown by CSP, using 70 cycles, Figure S3: (a) Cross-sectional view FESEM image of the complete ETA solar cell structure of ITO/ZnO/TiO_2/Sb_2S_3/P3HT/Au when using the cycle number of 70, (b) FESEM-EDX spectra collected on the yellow rectangular area as denoted in (a), (c) Element ratio as a function of the cycle number ranging from 30 to 90, Table S1: In-plane XRD data recapitulating the nature and position of the diffraction peaks in this work, as compared to the 00-042-1393 ICDD file, Table S2: Raman scattering data recapitulating the nature and position of the Raman lines in this work.

Author Contributions: Conceptualization, E.K. and V.C.; methodology, A.K., H.R., and O.C.-P.; validation, E.K. and V.C.; formal analysis, G.H. and J.S.E.; investigation, G.H., J.S.E., A.K., H.R., O.C.-P., F.D., and E.A.; resources, I.O.A. and V.C.; data curation, G.H. and J.S.E.; writing—original draft preparation, G.H., J.S.E., E.K., and V.C.; writing—review and editing, G.H., J.S.E., O.C.-P., E.A., E.K., and V.C.; supervision, E.K. and V.C.; project administration, E.K. and V.C.; funding acquisition, I.O.A. and V.C. All authors have read and agreed to the published version of the manuscript.

Funding: The authors acknowledge the Estonian Research Council, as well as MENESR and MAEDI French Ministries, for their financial support through the Parrot program (n°42401QH). This work was also supported by the French National Research Agency through the DOSETTE project (ANR-17-CE24-0004). G.H. held a doctoral fellowship from the DOSETTE project. This work was also supported by the Estonian Research Council project PRG627, Estonian Centre of Excellence project TAR16016EK, and the European Commission project 952509 (5GSOLAR). The authors further acknowledge the facilities and the scientific and technical assistance of the CMTC characterization platform of Grenoble INP, which is supported by the Centre of Excellence of Multifunctional Architectured Materials (LabEx CEMAM) under the contract ANR-10-LABX-44-01 funded by the "Investments for the Future" Program. Funding by the French National Research Agency in the framework of the "Investments for the Future" Program (ANR-15-IDEX-02) through the CDP Eco-SESA project is also acknowledged.

Institutional Review Board Statement: Not applicable.

Informed Consent Statement: Not applicable.

Data Availability Statement: The data that support the findings of this study are available from the corresponding authors upon reasonable request.

Conflicts of Interest: The authors declare no conflict of interest.

References

1. Schmidt-Mende, L.; MacManus-Driscoll, J.L. ZnO—Nanostructures, defects, and devices. *Mater. Today* **2007**, *10*, 40–48. [CrossRef]
2. Xu, S.; Wang, Z.L. One-dimensional ZnO nanostructures: Solution growth and functional properties. *Nano Res.* **2011**, *4*, 1013–1098. [CrossRef]
3. Consonni, V.; Briscoe, J.; Karber, E.; Li, X.; Cossuet, T. ZnO nanowires for solar cells: A comprehensive review. *Nanotechnology* **2019**, *30*, 362001. [CrossRef] [PubMed]
4. Ozgur, U.; Alivov, Y.I.; Liu, C.; Teke, A.; Reshchikov, M.A.; Dogan, S.; Avrutin, V.; Cho, S.J.; Morkoc, H. A comprehensive review of ZnO materials and devices. *J. Appl. Phys.* **2005**, *98*, 041301. [CrossRef]
5. O'Regan, B.; Schwartz, D.T.; Zakeeruddin, S.M.; Gratzel, M. Electrodeposited nanocomposite n-p heterojunctions for solid-state dye-sensitized photovoltaics. *Adv. Mater.* **2000**, *12*, 1263–1267. [CrossRef]
6. Mora-Sero, I.; Bisquert, J. Breakthroughs in the Development of Semiconductor-Sensitized Solar Cells. *J. Phys. Chem. Lett.* **2010**, *1*, 3046–3052. [CrossRef]
7. Dittrich, T.; Belaidi, A.; Ennaoui, A. Concepts of inorganic solid-state nanostructured solar cells. *Sol. Energy Mater. Sol. Cells* **2011**, *95*, 1527–1536. [CrossRef]
8. Briscoe, J.; Dunn, S. Extremely thin absorber solar cells based on nanostructured semiconductors. *Mater. Sci. Technol.* **2011**, *27*, 1741–1756. [CrossRef]
9. Hodes, G.; Cahen, D. All-Solid-State, Semiconductor-Sensitized Nanoporous Solar Cells. *Acc. Chem. Res.* **2012**, *45*, 705–713. [CrossRef]
10. Taretto, K.; Rau, U. Modeling extremely thin absorber solar cells for optimized design. *Prog. Photovolt.* **2004**, *12*, 573–591. [CrossRef]
11. Mora-Sero, I.; Gimenez, S.; Fabregat-Santiago, F.; Azaceta, E.; Tena-Zaera, R.; Bisquert, J. Modeling and characterization of extremely thin absorber (eta) solar cells based on ZnO nanowires. *Phys. Chem. Chem. Phys.* **2011**, *13*, 7162–7169. [CrossRef]
12. Michallon, J.; Bucci, D.; Morand, A.; Zanuccoli, M.; Consonni, V.; Kaminski-Cachopo, A. Light trapping in ZnO nanowire arrays covered with an absorbing shell for solar cells. *Opt. Express* **2014**, *22*, A1174–A1189. [CrossRef]

13. Michallon, J.; Bucci, D.; Morand, A.; Zanuccoli, M.; Consonni, V.; Kaminski-Cachopo, A. Light absorption processes and optimization of ZnO/CdTe core-shell nanowire arrays for nanostructured solar cells. *Nanotechnology* **2015**, *26*, 075401. [CrossRef]
14. Levy-Clement, C.; Tena-Zaera, R.; Ryan, M.A.; Katty, A.; Hodes, G. CdSe-Sensitized p-CuSCN/nanowire n-ZnO heterojunctions. *Adv. Mater.* **2005**, *17*, 1512–1515. [CrossRef]
15. Levy-Clement, C.; Katty, A.; Bastide, S.; Zenia, F.; Mora, I.; Munoz-Sanjose, V. A new CdTe/ZnO columnar composite film for Eta-solar cells. *Physica E* **2002**, *14*, 229–232. [CrossRef]
16. Consonni, V.; Renet, S.; Garnier, J.; Gergaud, P.; Artus, L.; Michallon, J.; Rapenne, L.; Appert, E.; Kaminski-Cachopo, A. Improvement of the physical properties of ZnO/CdTe core-shell nanowire arrays by $CdCl_2$ heat treatment for solar cells. *Nanoscale Res. Lett.* **2014**, *9*, 13. [CrossRef]
17. Akbarnejad, E.; Nayeri, F.D.; Ghoranneviss, M. Core-shell solar cell fabrication using heterostructure of ZnO-nanowires arrays decorated with sputtered CdTe-nanoparticles. *J. Phys. D -Appl. Phys.* **2018**, *51*, 7. [CrossRef]
18. Xu, J.; Yang, X.; Wang, H.K.; Chen, X.; Luan, C.Y.; Xu, Z.X.; Lu, Z.Z.; Roy, V.A.L.; Zhang, W.J.; Lee, C.S. Arrays of ZnO/$Zn_xCd_{1-x}Se$ Nanocables: Band Gap Engineering and Photovoltaic Applications. *Nano Lett.* **2011**, *11*, 4138–4143. [CrossRef]
19. Levy-Clement, C.; Elias, J. Optimization of the Design of Extremely Thin Absorber Solar Cells Based on Electrodeposited ZnO Nanowires. *ChemPhysChem* **2013**, *14*, 2321–2330. [CrossRef] [PubMed]
20. Tak, Y.; Hong, S.J.; Lee, J.S.; Yong, K. Fabrication of ZnO/CdS core/shell nanowire arrays for efficient solar energy conversion. *J. Mater. Chem.* **2009**, *19*, 5945–5951. [CrossRef]
21. Kartopu, G.; Turkay, D.; Ozcan, C.; Hadibrata, W.; Aurang, P.; Yerci, S.; Unalan, H.E.; Barrioz, V.; Qu, Y.; Bowen, L.; et al. Photovoltaic performance of CdS/CdTe junctions on ZnO nanorod arrays. *Sol. Energy Mater. Sol. Cells* **2018**, *176*, 100–108. [CrossRef]
22. Dittrich, T.; Kieven, D.; Rusu, M.; Belaidi, A.; Tornow, J.; Schwarzburg, K.; Lux-Steiner, M. Current-voltage characteristics and transport mechanism of solar cells based on ZnO nanorods/In_2S_3/CuSCN. *Appl. Phys. Lett.* **2008**, *93*, 053113. [CrossRef]
23. Belaidi, A.; Dittrich, T.; Kieven, D.; Tornow, J.; Schwarzburg, K.; Lux-Steiner, M. Influence of the local absorber layer thickness on the performance of ZnO nanorod solar cells. *Phys. Status Solidi-Rapid Res. Lett.* **2008**, *2*, 172–174. [CrossRef]
24. Belaidi, A.; Dittrich, T.; Kieven, D.; Tornow, J.; Schwarzburg, K.; Kunst, M.; Allsop, N.; Lux-Steiner, M.C.; Gavrilov, S. ZnO-nanorod arrays for solar cells with extremely thin sulfidic absorber. *Sol. Energy Mater. Sol. Cells* **2009**, *93*, 1033–1036. [CrossRef]
25. Krunks, M.; Katerski, A.; Dedova, T.; Acik, I.O.; Mere, A. Nanostructured solar cell based on spray pyrolysis deposited ZnO nanorod array. *Sol. Energy Mater. Sol. Cells* **2008**, *92*, 1016–1019. [CrossRef]
26. Krunks, M.; Karber, E.; Katerski, A.; Otto, K.; Acik, I.O.; Dedova, T.; Mere, A. Extremely thin absorber layer solar cells on zinc oxide nanorods by chemical spray. *Sol. Energy Mater. Sol. Cells* **2010**, *94*, 1191–1195. [CrossRef]
27. Izaki, M.; Ohta, T.; Kondo, M.; Takahashi, T.; Mohamad, F.B.; Zamzuri, M.; Sasano, J.; Shinagawa, T.; Pauporte, T. Electrodeposited ZnO-Nanowire/Cu_2O Photovoltaic Device with Highly Resistive ZnO Intermediate Layer. *ACS Appl. Mater. Interfaces* **2014**, *6*, 13461–13469. [CrossRef]
28. Chen, X.; Lin, P.; Yan, X.Q.; Bai, Z.M.; Yuan, H.G.; Shen, Y.W.; Liu, Y.C.; Zhang, G.J.; Zhang, Z.; Zhang, Y. Three-Dimensional Ordered ZnO/Cu2O Nanoheterojunctions for Efficient Metal-Oxide Solar Cells. *ACS Appl. Mater. Interfaces* **2015**, *7*, 3216–3223. [CrossRef]
29. Zhang, Y.; Wu, Z.M.; Zheng, J.J.; Lin, X.A.; Zhan, H.H.; Li, S.P.; Kang, J.Y.; Bleuse, J.; Mariette, H. ZnO/ZnSe type II core-shell nanowire array solar cell. *Sol. Energy Mater. Sol. Cells* **2012**, *102*, 15–18. [CrossRef]
30. Wang, K.; Chen, J.J.; Zeng, Z.M.; Tarr, J.; Zhou, W.L.; Zhang, Y.; Yan, Y.F.; Jiang, C.S.; Pern, J.; Mascarenhas, A. Synthesis and photovoltaic effect of vertically aligned ZnO/ZnS core/shell nanowire arrays. *Appl. Phys. Lett.* **2010**, *96*, 123105. [CrossRef]
31. Savadogo, O.; Mandal, K.C. Studies on new chemically deposited photoconducting antimony trisulfide thin-films. *Sol. Energy Mater. Sol. Cells* **1992**, *26*, 117–136. [CrossRef]
32. Savadogo, O.; Mandal, K.C. Characterizations of antimony trisulfide chemically deposited with silicotungstic acid. *J. Electrochem. Soc.* **1992**, *139*, L16–L18. [CrossRef]
33. Nair, M.T.S.; Pena, Y.; Campos, J.; Garcia, V.M.; Nair, P.K. Chemically deposited Sb_2S_3 and Sb_2S_3-CuS thin films. *J. Electrochem. Soc.* **1998**, *145*, 2113–2120. [CrossRef]
34. Ben Nasr, T.; Maghraoui-Meherzi, H.; Ben Abdallah, H.; Bennaceur, R. Electronic structure and optical properties of Sb_2S_3 crystal. *Physica B* **2011**, *406*, 287–292. [CrossRef]
35. Kondrotas, R.; Chen, C.; Tang, J. Sb_2S_3 Solar Cells. *Joule* **2018**, *2*, 857–878. [CrossRef]
36. Wang, Q.; Chen, Z.; Wang, J.; Xu, Y.; Wei, Y.; Wei, Y.; Qiu, L.; Lu, H.; Ding, Y.; Zhu, J. Sb_2S_3 solar cells: Functional layer preparation and device performance. *Inorg. Chem. Front.* **2019**, *6*, 3381–3397. [CrossRef]
37. Shah, U.A.; Chen, S.; Khalaf, G.M.G.; Jin, Z.; Song, H. Wide Bandgap Sb_2S_3 Solar Cells. *Adv. Funct. Mater.* **2021**, *31*, 2100265. [CrossRef]
38. Choi, Y.C.; Lee, D.U.; Noh, J.H.; Kim, E.K.; Seok, S.I. Highly Improved Sb_2S_3 Sensitized-Inorganic-Organic Heterojunction Solar Cells and Quantification of Traps by Deep-Level Transient Spectroscopy. *Adv. Funct. Mater.* **2014**, *24*, 3587–3592. [CrossRef]
39. Itzhaik, Y.; Niitsoo, O.; Page, M.; Hodes, G. Sb_2S_3-Sensitized Nanoporous TiO_2 Solar Cells. *J. Phys. Chem. C* **2009**, *113*, 4254–4256. [CrossRef]
40. Moon, S.J.; Itzhaik, Y.; Yum, J.H.; Zakeeruddin, S.M.; Hodes, G.; Gratzel, M. Sb_2S_3-Based Mesoscopic Solar Cell using an Organic Hole Conductor. *J. Phys. Chem. Lett.* **2010**, *1*, 1524–1527. [CrossRef]

41. Chang, J.A.; Rhee, J.H.; Im, S.H.; Lee, Y.H.; Kim, H.J.; Seok, S.I.; Nazeeruddin, M.K.; Gratzel, M. High-Performance Nanostructured Inorganic-Organic Heterojunction Solar Cells. *Nano Lett.* **2010**, *10*, 2609–2612. [CrossRef] [PubMed]
42. Im, S.H.; Lim, C.S.; Chang, J.A.; Lee, Y.H.; Maiti, N.; Kim, H.J.; Nazeeruddin, M.K.; Gratzel, M.; Seok, S.I. Toward Interaction of Sensitizer and Functional Moieties in Hole-Transporting Materials for Efficient Semiconductor-Sensitized Solar Cells. *Nano Lett.* **2011**, *11*, 4789–4793. [CrossRef]
43. Chang, J.A.; Im, S.H.; Lee, Y.H.; Kim, H.J.; Lim, C.S.; Heo, J.H.; Seok, S.I. Panchromatic Photon-Harvesting by Hole-Conducting Materials in Inorganic-Organic Heterojunction Sensitized-Solar Cell through the Formation of Nanostructured Electron Channels. *Nano Lett.* **2012**, *12*, 1863–1867. [CrossRef] [PubMed]
44. Dong, J.; Liu, Y.; Wang, Z.; Zhang, Y. Boosting V_{OC} of antimony chalcogenide solar cells: A review on interfaces and defects. *Nano Sel.* **2021**, *2*, 1818–1848. [CrossRef]
45. Wang, W.; Wang, X.; Chen, G.; Yao, L.; Huang, X.; Chen, T.; Zhu, C.; Chen, S.; Huang, Z.; Zhang, Y. Over 6% Certified $Sb_2(S,Se)_3$ Solar Cells Fabricated via In Situ Hydrothermal Growth and Postselenization. *Adv. Electron. Mater.* **2019**, *5*, 1800683. [CrossRef]
46. Hu, X.; Tao, J.; Wang, R.; Wang, Y.; Pan, Y.; Weng, G.; Luo, X.; Chen, S.; Zhu, Z.; Chu, J.; et al. Fabricating over 7%-efficient $Sb_2(S,Se)_3$ thin-film solar cells by vapor transport deposition using Sb_2Se_3 and Sb_2S_3 mixed powders as the evaporation source. *J. Power Sources* **2021**, *493*, 229737. [CrossRef]
47. Zhong, J.; Zhang, X.J.; Zheng, Y.J.; Zheng, M.; Wen, M.J.; Wu, S.J.; Gao, J.W.; Gao, X.S.; Liu, J.M.; Zhao, H.B. High Efficiency Solar Cells as Fabricated by Sb_2S_3-Modified TiO_2 Nanofibrous Networks. *ACS Appl. Mater. Interfaces* **2013**, *5*, 8345–8350. [CrossRef]
48. Sharma, V.; Das, T.K.; Ilaiyaraja, P.; Dakshinamurthy, A.C.; Sudakar, C. Growth of Sb_2S_3 semiconductor thin film on different morphologies of TiO_2 nanostructures. *Mater. Res. Bull.* **2020**, *131*, 110980. [CrossRef]
49. Cardoso, J.C.; Grimes, C.A.; Feng, X.J.; Zhang, X.Y.; Komarneni, S.; Zanoni, M.V.B.; Bao, N.Z. Fabrication of coaxial TiO_2/Sb_2S_3 nanowire hybrids for efficient nanostructured organic- inorganic thin film photovoltaics. *Chem. Commun.* **2012**, *48*, 2818–2820. [CrossRef] [PubMed]
50. Meng, X.Q.; Wang, X.Z.; Zhong, M.Z.; Wu, F.M.; Fang, Y.Z. Sb_2S_3 surface modification induced remarkable enhancement of TiO_2 core/shell nanowries solar cells. *J. Solid State Chem.* **2013**, *201*, 75–78. [CrossRef]
51. Ying, C.; Shi, C.; Lv, K.; Ma, C.; Guo, F.; Fu, H. Fabrication of Sb_2S_3 sensitized TiO_2 nanorod array solar cells using spin-coating assisted successive ionic layer absorption and reaction. *Mater. Today Commun.* **2019**, *19*, 393–395. [CrossRef]
52. Ying, C.; Guo, F.; Wu, Z.; Lv, K.; Shi, C. Influence of Surface Modifier Molecular Structures on the Photovoltaic Performance of Sb_2S_3-Sensitized TiO_2 Nanorod Array Solar Cells. *Energy Technol.* **2020**, *8*, 1901368. [CrossRef]
53. Liu, C.P.; Chen, Z.H.; Wang, H.E.; Jha, S.K.; Zhang, W.J.; Bello, I.; Zapien, J.A. Enhanced performance by incorporation of zinc oxide nanowire array for organic-inorganic hybrid solar cells. *Appl. Phys. Lett.* **2012**, *100*, 243102. [CrossRef]
54. Han, J.H.; Liu, Z.F.; Zheng, X.R.; Guo, K.Y.; Zhang, X.Q.; Hong, T.T.; Wang, B.; Liu, J.Q. Trilaminar $ZnO/ZnS/Sb_2S_3$ nanotube arrays for efficient inorganic-organic hybrid solar cells. *RSC Adv.* **2014**, *4*, 23807–23814. [CrossRef]
55. Parize, R.; Katerski, A.; Gromyko, I.; Rapenne, L.; Roussel, H.; Karber, E.; Appert, E.; Krunks, M.; Consonni, V. $ZnO/TiO_2/Sb_2S_3$ Core-Shell Nanowire Heterostructure for Extremely Thin Absorber Solar Cells. *J. Phys. Chem. C* **2017**, *121*, 9672–9680. [CrossRef]
56. Parize, R.; Cossuet, T.; Appert, E.; Chaix-Pluchery, O.; Roussel, H.; Rapenne, L.; Consonni, V. Synthesis and properties of $ZnO/TiO_2/Sb_2S_3$ core-shell nanowire heterostructures using the SILAR technique. *CrystEngComm* **2018**, *20*, 4455–4462. [CrossRef]
57. Sharma, V.; Das, T.K.; Ilaiyaraja, P.; Sudakar, C. Oxygen non-stoichiometry in TiO_2 and ZnO nano rods: Effect on the photovoltaic properties of dye and Sb_2S_3 sensitized solar cells. *Sol. Energy* **2019**, *191*, 400–409. [CrossRef]
58. Sun, Z.; Peng, Z.; Liu, Z.; Chen, J.; Li, W.; Qiu, W.; Chen, J. Band energy modulation on Cu-doped Sb_2S_3-based photoelectrodes for charge generation and transfer property of quantum dot–sensitized solar cells. *J. Nanopart. Res.* **2020**, *22*, 282. [CrossRef]
59. Li, W.; Peng, Z.; Sun, Z.; Liu, Z.; Chen, J.; Qiu, W.; Chen, J.; Zhang, S. Orientation modulation of ZnO nanorods on charge transfer performance enhancement for Sb_2S_3 quantum dot sensitized solar cells. *J. Alloys Compd.* **2020**, *816*, 152628. [CrossRef]
60. Peng, Z.; Ning, Z.; Liu, Z.; Chen, J.; Li, W.; Qiu, W.; Chen, J.; Tang, Y. Enhancement on Charge Generation and Transfer Properties of Sb_2S_3 Quantum-Dot-Sensitized Solar Cells by ZnO Nanorods Optimization. *J. Electron. Mater.* **2021**, *50*, 100–107. [CrossRef]
61. Cossuet, T.; Appert, E.; Chaix-Pluchery, O.; Roussel, H.; Rapenne, L.; Renou, G.; Sauvage, F.; Consonni, V. Epitaxial TiO_2 Shell Grown by Atomic Layer Deposition on ZnO Nanowires Using a Double-Step Process and Its Beneficial Passivation Effect. *J. Phys. Chem. C* **2020**, *124*, 13447–13455. [CrossRef]
62. Kim, D.-H.; Lee, S.-J.; Park, M.S.; Kang, J.-K.; Heo, J.H.; Im, S.H.; Sung, S.-J. Highly reproducible planar Sb_2S_3-sensitized solar cells based on atomic layer deposition. *Nanoscale* **2014**, *6*, 14549–14554. [CrossRef]
63. Eensalu, J.S.; Katerski, A.; Kärber, E.; Oja Acik, I.; Mere, A.; Krunks, M. Uniform Sb_2S_3 optical coatings by chemical spray method. *Beilstein J. Nanotechnol.* **2019**, *10*, 198–210. [CrossRef]
64. Guillemin, S.; Consonni, V.; Appert, E.; Puyoo, E.; Rapenne, L.; Roussel, H. Critical Nucleation Effects on the Structural Relationship Between ZnO Seed Layer and Nanowires. *J. Phys. Chem. C* **2012**, *116*, 25106–25111. [CrossRef]
65. Guillemin, S.; Appert, E.; Roussel, H.; Doisneau, B.; Parize, R.; Boudou, T.; Bremond, G.; Consonni, V. Controlling the Structural Properties of Single Step, Dip Coated ZnO Seed Layers for Growing Perfectly Aligned Nanowire Arrays. *J. Phys. Chem. C* **2015**, *119*, 21694–21703. [CrossRef]
66. Parize, R.; Garnier, J.; Chaix-Pluchery, O.; Verrier, C.; Appert, E.; Consonni, V. Effects of Hexamethylenetetramine on the Nucleation and Radial Growth of ZnO Nanowires by Chemical Bath Deposition. *J. Phys. Chem. C* **2016**, *120*, 5242–5250. [CrossRef]

67. Parize, R.; Cossuet, T.; Chaix-Pluchery, O.; Roussel, H.; Appert, E.; Consonni, V. In situ analysis of the crystallization process of Sb_2S_3 thin films by Raman scattering and X-ray diffraction. *Mater. Des.* **2017**, *121*, 1–10. [CrossRef]
68. Thompson, C.V. Structure evolution during processing of polycrystalline films. *Annu. Rev. Mater. Sci.* **2000**, *30*, 159–190. [CrossRef]
69. Makreski, P.; Petrusevski, G.; Ugarkovic, S.; Jovanovski, G. Laser-induced transformation of stibnite (Sb_2S_3) and other structurally related salts. *Vib. Spectrosc.* **2013**, *68*, 177–182. [CrossRef]
70. Sereni, P.; Musso, M.; Knoll, P.; Blaha, P.; Schwarz, K.; Schmidt, G. Polarization-Dependent Raman Characterization of Stibnite (Sb_2S_3). *AIP Proc.* **2010**, *1267*, 1131–1132. [CrossRef]
71. Liu, Y.; Chua, K.T.E.; Sum, T.C.; Gan, C.K. First-principles study of the lattice dynamics of Sb_2S_3. *Phys. Chem. Chem. Phys.* **2014**, *16*, 345–350. [CrossRef]
72. Ibanez, J.; Sans, J.A.; Popescu, C.; Lopez-Vidrier, J.; Elvira-Betanzos, J.J.; Cuenca-Gotor, V.P.; Gomis, O.; Manjon, F.J.; Rodriguez-Hernandez, P.; Munoz, A. Structural, Vibrational, and Electronic Study of Sb_2S_3 at High Pressure. *J. Phys. Chem. C* **2016**, *120*, 10547–10558. [CrossRef]
73. Karber, E.; Katerski, A.; Acik, I.O.; Mere, A.; Mikli, V.; Krunks, M. Sb_2S_3 grown by ultrasonic spray pyrolysis and its application in a hybrid solar cell. *Beilstein J. Nanotechnol.* **2016**, *7*, 1662–1673. [CrossRef] [PubMed]
74. Tang, R.; Wang, X.; Jiang, C.; Li, S.; Liu, W.; Ju, H.; Yang, S.; Zhu, C.; Chen, T. n-Type Doping of Sb_2S_3 Light-Harvesting Films Enabling High-Efficiency Planar Heterojunction Solar Cells. *ACS Appl. Mater. Interfaces* **2018**, *10*, 30314–30321. [CrossRef]
75. Zacharias, M.; Streitenberger, P. Crystallization of amorphous superlattices in the limit of ultrathin films with oxide interfaces. *Phys. Rev. B* **2000**, *62*, 8391–8396. [CrossRef]
76. Raoux, S.; Jordan-Sweet, J.L.; Kellock, A.J. Crystallization properties of ultrathin phase change films. *J. Appl. Phys.* **2008**, *103*, 114310. [CrossRef]
77. Zhang, L.; Zhang, J.; Jiao, H.; Bao, G.; Wang, Z.; Cheng, X. Thickness-dependent surface morphology and crystallization of HfO_2 coatings prepared with ion-assisted deposition. *Thin Solid Films* **2017**, *642*, 359–363. [CrossRef]
78. Kharbish, S.; Libowitzky, E.; Beran, A. Raman spectra of isolated and interconnected pyramidal XS_3 groups (X = Sb,Bi) in stibnite, bismuthinite, kermesite, stephanite and bournonite. *Eur. J. Mineral.* **2009**, *21*, 325–333. [CrossRef]
79. Mestl, G.; Ruiz, P.; Delmon, B.; Knozinger, H. Sb_2O_3/Sb_2O_4 in reducing/oxidizing environments—An In-situ Raman-spectroscopy study. *J. Phys. Chem.* **1994**, *98*, 11276–11282. [CrossRef]
80. Dedova, T.; Krunks, M.; Volobujeva, O.; Oja, I. ZnS thin films deposited by spray pyrolysis technique. *Phys. Status Sol. C* **2005**, *2*, 1161–1166. [CrossRef]
81. Villafuerte, J.; Donatini, F.; Kioseoglou, J.; Sarigiannidou, E.; Chaix-Pluchery, O.; Pernot, J.; Consonni, V. Zinc Vacancy–Hydrogen Complexes as Major Defects in ZnO Nanowires Grown by Chemical Bath Deposition. *J. Phys. Chem. C* **2020**, *124*, 16652–16662. [CrossRef]
82. Meyer, B.K.; Alves, H.; Hofmann, D.M.; Kriegseis, W.; Forster, D.; Bertram, F.; Christen, J.; Hoffmann, A.; Strassburg, M.; Dworzak, M.; et al. Bound exciton and donor-acceptor pair recombinations in ZnO. *Phys. Status Solidi B* **2004**, *241*, 231–260. [CrossRef]
83. Lavrov, E.V.; Herklotz, F.; Weber, J. Identification of two hydrogen donors in ZnO. *Phys. Rev. B* **2009**, *79*, 165210. [CrossRef]
84. Heinhold, R.; Neiman, A.; Kennedy, J.V.; Markwitz, A.; Reeves, R.J.; Allen, M.W. Hydrogen-related excitons and their excited-state transitions in ZnO. *Phys. Rev. B* **2017**, *95*, 054120. [CrossRef]
85. Versavel, M.Y.; Haber, J.A. Structural and optical properties of amorphous and crystalline antimony sulfide thin-films. *Thin Solid Films* **2007**, *515*, 7171–7176. [CrossRef]
86. Carron, R.; Andres, C.; Avancini, E.; Feurer, T.; Nishiwaki, S.; Pisoni, S.; Fu, F.; Lingg, M.; Romanyuk, Y.E.; Buecheler, S.; et al. Bandgap of thin film solar cell absorbers: A comparison of various determination methods. *Thin Solid Films* **2019**, *669*, 482–486. [CrossRef]
87. Darga, A.; Mencaraglia, D.; Longeaud, C.; Savenije, T.J.; O'Regan, B.; Bourdais, S.; Muto, T.; Delatouche, B.; Dennler, G. On Charge Carrier Recombination in Sb_2S_3 and Its Implication for the Performance of Solar Cells. *J. Phys. Chem. C* **2013**, *117*, 20525–20530. [CrossRef]
88. Eensalu, J.S.; Katerski, A.; Kärber, E.; Weinhardt, L.; Blum, M.; Heske, C.; Yang, W.; Oja Acik, I.; Krunks, M. Semitransparent Sb_2S_3 thin film solar cells by ultrasonic spray pyrolysis for use in solar windows. *Beilstein J. Nanotechnol.* **2019**, *10*, 2396–2409. [CrossRef]
89. Zimmermann, E.; Pfadler, T.; Kalb, J.; Dorman, J.A.; Sommer, D.; Hahn, G.; Weickert, J.; Schmidt-Mende, L. Toward High-Efficiency Solution-Processed Planar Heterojunction Sb_2S_3 Solar Cells. *Adv. Sci.* **2015**, *2*, 1500059. [CrossRef]
90. Kieven, D.; Dittrich, T.; Belaidi, A.; Tornow, J.; Schwarzburg, K.; Allsop, N.; Lux-Steiner, M. Effect of internal surface area on the performance of $ZnO/In_2S_3/CuSCN$ solar cells with extremely thin absorber. *Appl. Phys. Lett.* **2008**, *92*, 153107. [CrossRef]

nanomaterials

MDPI

Article

Low-Temperature Hydrothermal Growth of ZnO Nanowires on AZO Substrates for $FACsPb(IBr)_3$ Perovskite Solar Cells

Karthick Sekar [1,2,3,*], Rana Nakar [1], Johann Bouclé [2,3], Raphaël Doineau [1], Kevin Nadaud [1], Bruno Schmaltz [4] and Guylaine Poulin-Vittrant [1,*]

1 GREMAN UMR 7347, Université de Tours, CNRS, INSA Centre Val de Loire, 37071 Tours, CEDEX 2, France; rana.nakar@univ-tours.fr (R.N.); raphael.doineau@univ-tours.fr (R.D.); kevin.nadaud@univ-tours.fr (K.N.)
2 Univ. Limoges, XLIM, UMR 7252, 87000 Limoges, France; johann.boucle@unilim.fr
3 CNRS, XLIM, UMR 7252, 87000 Limoges, France
4 PCM2E EA 6299, Université de Tours, Parc de Grandmont, 37200 Tours, France; bruno.schmaltz@univ-tours.fr
* Correspondence: karthick.sekar@univ-tours.fr (K.S.); guylaine.poulin-vittrant@univ-tours.fr (G.P.-V.)

Abstract: Electron and hole transport layers (ETL and HTL) play an essential role in shaping the photovoltaic performance of perovskite solar cells. While compact metal oxide ETL have been largely explored in planar *n-i-p* device architectures, aligned nanowires or nanorods remain highly relevant for efficient charge extraction and directional transport. In this study, we have systematically grown ZnO nanowires (ZnO NWs) over aluminum-doped zinc oxide (AZO) substrates using a low-temperature method, hydrothermal growth (HTG). The main growth parameters were varied, such as hydrothermal precursors concentrations (zinc nitrate hexahydrate, hexamethylenetetramine, polyethylenimine) and growing time, in order to finely control NW properties (length, diameter, density, and void fraction). The results show that ZnO NWs grown on AZO substrates offer highly dense, well-aligned nanowires of high crystallinity compared to conventional substrates such as FTO, while demonstrating efficient $FACsPb(IBr)_3$ perovskite device performance, without the requirement of conventional compact hole blocking layers. The device performances are discussed based on NW properties, including void fraction and aspect ratio (NW length over diameter). Finally, AZO/ZnO NW-based devices were fabricated with a recent HTL material based on a carbazole moiety (Cz–Pyr) and compared to the spiro-OMeTAD reference. Our study shows that the Cz–Pyr-based device provides similar performance to that of spiro-OMeTAD while demonstrating a promising stability in ambient conditions and under continuous illumination, as revealed by a preliminary aging test.

Keywords: ZnO; nanowires; AZO; hydrothermal growth; perovskite solar cell

Citation: Sekar, K.; Nakar, R.; Bouclé, J.; Doineau, R.; Nadaud, K.; Schmaltz, B.; Poulin-Vittrant, G. Low-Temperature Hydrothermal Growth of ZnO Nanowires on AZO Substrates for $FACsPb(IBr)_3$ Perovskite Solar Cells. *Nanomaterials* **2022**, *12*, 2093. https://doi.org/10.3390/nano12122093

Academic Editors: Efrat Lifshitz and Vincent Consonni

Received: 20 May 2022
Accepted: 14 June 2022
Published: 17 June 2022

1. Introduction

In recent years, several emerging photovoltaic (PV) technologies raised a growing interest in terms of light-weight, cheap process, and high efficiency. Among them, hybrid solar cells using organic–inorganic metal halide perovskites has become one of the most promising topics in material science in the past few years [1,2]. Perovskite photovoltaics studies were triggered by the report on the 9.7% efficient solid-state perovskite solar cell (PSC) in 2012 [2,3]. PSCs composed of organic–inorganic metal halide perovskite have made impressive progress in only a few years, surpassing in efficiencies some well-established thin-film technologies such as CIGS or amorphous silicon [4,5]. In all cases, an electron transport layer (ETL) is used to selectively collect photogenerated electrons from the perovskite absorber layer while blocking holes [2]. The nature of the ETL is also crucial for the perovskite crystallization in *n-i-p* architectures. Considering the optimum specifications of any ETL (high optical transmission for the solar spectrum, suitable electronic configuration with regard to the perovskite for efficient electron extraction, suitable photo- and thermal stability, cheap processing at low temperature < 150 °C, etc.), many organic and inorganic materials have been proposed as a powerful lever to improve the electrical behavior of PSC under working conditions [6–9]. Among inorganic compounds, titanium dioxide (TiO_2) is a best-seller ETL material, widely used for efficient PSCs [10–13]. The appropriate bandgap, high transmittance, and low toxicity of TiO_2 guarantee a high selectivity for electrons, despite a relatively high processing temperature required to achieve highly crystalline layers. This high-temperature annealing step limits its application for flexible devices and increases the production cost [14]. On the other hand, ZnO was largely considered as a good alternative to TiO_2 for PSCs since it shows energy levels and physical properties similar to those of TiO_2, but is more easily processed from solution and at low temperature to achieve structures of different morphologies. Moreover, the intrinsic properties of ZnO thin films and nanostructures can be adjusted by manipulating their morphology, doping, and structural composition [6,15–18].

Recent studies on ZnO-based PSCs have demonstrated high efficiencies and provided many new concepts [11,14]. The increased interest aroused by nanomaterials in ZnO has been largely motivated by its excellent electrical and optoelectronic properties in the bulk, in particular a wide direct forbidden band (3.37 eV) [19], high exciton binding energy (60 meV) [20], and high electronic mobility (200–300 cm^2/Vs) [21]. A wide variety of ZnO-based structures, such as nanoparticles [11,22], nanowires [23], nanotubes [24], and nanobeads [25], were demonstrated. Nevertheless, ZnO nanoparticles suffer from relatively modest electron transport and high charge recombination due to the presence of numerous grain boundaries and surface defects. In this context, monocrystalline ZnO nanorods provide an easy path for charge transfer due to the absence of grain boundaries, which can delay or inhibit charge recombination [26].

Consequently, in ongoing research efforts for the miniaturization of electronic devices, near-dimensional (1D) ZnO nanowires (NWs) have proven to be potential candidates for ETL due to their unique properties, such as a high electromechanical coupling factor and a better charge injection/extraction at the metal level [27]. Varieties of bottom-up approaches, including pulsed laser ablation [28], flame transport approach [29], vapor-liquid–solid process [30], hydrothermal deposition [31–33], were exploited for the synthesis of 1D ZnO NWs. However, most techniques are limited by a high temperature, which cannot be scaled up at a very low cost. The need for an industrially scalable low-temperature method has led to important developments of the hydrothermal growth (HTG) process. HTG is a simplistic, environmentally friendly, cost-effective, straightforward, and low-temperature process (i.e., below 100 °C) in which a 1D monocrystalline material can be produced on various substrates, including plastics and textile fibers [34]. High-density ZnO NWs oriented perpendicularly to the growth substrate have been often reported using the HTG method, for a broad range of applications [35–37]. However, only a few papers have explored the complex relationships between NW density, aspect ratio (NW length over diameter), and alignment on their electronic properties, especially for applications in the field of PSC [36,37].

Coming back to the field of perovskite solar cells, the use of ZnO nanostructures as ETL is associated with detrimental degradation mechanisms. Perovskite-based materials are known to be sensitive to water and oxygen in the air but thermally stable under certain conditions [38]. However, according to early reports [24], perovskite materials such as methylammonium lead iodide ($CH_3NH_3PbI_3$ or MAPI) deposited on ZnO is found to decompose rapidly during heat treatment. Such a phenomenon is not observed between TiO_2 and MAPI. This behavior is mainly attributed to the different surface properties of ZnO and TiO_2 nanomaterials. The TiO_2 surface exhibits a low acidity, while the ZnO surface usually exhibits a basic behavior. Once perovskite is brought into contact with ZnO, there is a deprotonation reaction against the methylammonium cation, which is the basis of instability, and this hypothesis has been proven by theoretical calculations [24].

Another important reason for instability is the presence of chemical residues from manufacturing processes. Materials based on ZnO, in particular nanomaterials used for PSCs, are generally synthesized from chemical methods in solution, which tend to inevitably leave chemical residuals at the final material surface after synthesis. According to

Yang et al. and Cheng et al. [24,39], hydroxyl groups and residual acetate in the growth solution exacerbate the degradation of the subsequent deposited perovskite layer. Indeed, the ZnO materials synthesized by the solution process are not completely oxidized, and the surface of the obtained ZnO nanomaterials are covered with oxygenated chemisorbed species such as hydroxide, which could break the ionic interaction between $CH_3NH_3^+$ and PbI_3^- and destroy the crystal structure of $MAPbI_3$ sequentially. This process can be explained by the formula of the following reaction:

$$OH^- + CH_3NH_3I \rightarrow CH_3NH_3OH + I^-$$

$$CH_3NH_3OH \rightarrow CH_3NH_2 + H_2O \text{ (heating)}$$

Snaith et al. revealed that the cause of instability of the ZnO/perovskite interface is deprotonation of the methylammonium cation, leading to the formation of zinc hydroxide [40]. Based on this discovery, they replaced the MA cation with formamidinium (FA) of lower acidity, mixed with cesium cations in devices containing ZnO treated at low temperature (120 °C). The stability of the ZnO/perovskite was greatly improved, and an overall power conversion efficiency (PCE) of 21.1% was obtained for the corresponding devices.

In this picture, it is clear that the morphology of the photoanode material plays an important role on the PV performance of ZnO-based PSCs. Achieving well-aligned ZnO nanowires of optimized NW density (number of NWs per unit area) that can efficiently interact with the active layers by preventing charge recombination is a relevant objective to improve device operation. The perovskite infiltration, which should also ensure an optimal light-harvesting ability, is also a crucial parameter. Several groups have demonstrated perovskite solar cells based on such ZnO NW electrodes, which shows that the main growth parameters, such as hydrothermal precursor ratios and growing time, and growth solution concentration, significantly influence NWs' properties (i.e., NW length, diameter, density, and morphology) [41–46] and the device PV performances [17,23,47].

In this paper, ZnO nanowires grown on aluminum-doped ZnO (AZO) substrates by HTG method are carefully investigated with regard to the specific application as ETL in *n-i-p* perovskite solar cells. While the application drives specific features for efficient charge extraction, we first demonstrate that the use of AZO as growing substrates does not require any additional compact or seed layer for their growth, leading to a simplified fabrication process and device architecture. We then systematically investigate the influence of the main ZnO NW growth parameters, such as hydrothermal precursor concentration and growing time, on the NW properties, before discussing the impact on the performance of perovskite solar cells fully prepared under ambient conditions. Finally, optimized ZnO NW arrays deposited on AZO are combined with a novel carbazole hole transport material derivative (Cz–Pyr as HTL), demonstrating promising performance compared to the reference spiro-OMeTAD.

2. Experimental Details

2.1. Materials

Commercial AZO-coated glass substrates were purchased from MSE Supplies (Tucson, AZ, USA). Zinc nitrate hexahydrate $Zn(NO_3)_2.6H_2O$ (98%), hexamethylenetetramine (HMTA) $(CH_2)_6N_4$ (>99.5%) purchased from Sigma-Aldrich (St. Louis, MI, USA), and polyethylenimine branched (PEI) (average Mw ~25,000) from Sigma-Aldrich, also used as received. These materials were used for the preparation of ZnO NWs. The deposition of ZnO seed layer was done with physical vapor deposition (PVD) equipment (Plassys MP 550 S, Marolles-en-Hurepoix, France). A tubular furnace (Thermolyne 79300, Dubuque, IA, USA) was used for the ZnO seed layer annealing treatment. A stainless-steel autoclave was used to operate the synthesis of ZnO NWs. For perovskite solar cells: Formamidinium iodide (FAI) (≥90%, GreatCell Solar, Queanbeyan, Australia), cesium iodide (CsI) (99.9%, trace metals basis, Sigma Aldrich, St. Louis, MO, USA), lead iodide (PbI_2) (99%, Sigma Aldrich), lead bromide ($PbBr_2$) (≥90%, Sigma Aldrich), tin oxide nanoparticle colloidal solution (SnO_2) (15% in H_2O colloidal dispersion, Alfa Aesar (Haverhill, MA, USA)), spiro-OMeTAD (99% HPLC, Sigma Aldrich), 4-tert-butylpyridine (96%, Sigma Aldrich), chlorobenzene (CB, anhydrous, 99.8%, Sigma Aldrich), N-N dimethylformamide (DMF, anhydrous, 99.8%, Sigma Aldrich), dimethyl sulfoxide (DMSO anhydrous, ≥99.9%, Sigma Aldrich), acetonitrile (anhydrous, 99.8%, Sigma Aldrich), diethyl ether (DE, ≥99.5%, GC, Sigma Aldrich).

2.2. ZnO Nanowire Growth

ZnO nanowires were synthesized on all kinds of substrates via the hydrothermal growth route [48]. All the substrates, except AZO, were precoated by a ZnO seed layer (50 nm) prepared by radio-frequency sputtering during 16 min at 65 Watts and under a pressure of 5 mTorr in an argon atmosphere. Then, a thermal treatment was performed at 400 °C for 1 h in air through a horizontal quartz tubular furnace. The hydrothermal synthesis of ZnO NWs was carried out in the stainless-steel autoclave. Two clear and transparent fresh stock solutions of $Zn(NO_3)_2$ and HMTA were separately prepared in distilled water (DI), as well as a PEI solution. The concentrations of each solution were adjusted for each specific case, as described in the next sections. To hydrothermally grow the ZnO NWs, the autoclave was filled with the solutions following this order: zinc nitrate hexahydrate and HMTA, and PEI before stirring. The substrates were tilted against the walls to prevent the precipitation of homogenous nucleated ZnO on the seed layer surface. After hermetically sealing the autoclave in order to avoid evaporation, the temperature was maintained at 90 °C during the growth process. Finally, the samples were rinsed with distilled water and dried under airflow. Then, thermal treatment was performed on the ZnO NW coated samples at 450 °C for 30 min in air through a horizontal quartz tubular furnace.

2.3. Perovskite Solar Cells Fabrication

The $FA_{0.85}Cs_{0.15}Pb(I_{0.85}Br_{0.15})_3$ precursor solution was prepared by mixing FAI (0.1463 g), PbI_2 (0.392 g), $PbBr_2$ (0.0554 g), CsI (0.039 g), and DMSO (78 μL) in the DMF (600 μL), which was spread on the substrate and spun at 4000 rpm for 30 s [49]. While spinning, diethyl ether was dripped to induce adduct intermediate by eliminating DMF. The as-coated perovskite film was annealed at 100 °C for 20 min. The spiro-OMeTAD solution was prepared by dissolving 36.6 mg of spiro-OMeTAD in 0.500 mL chlorobenzene. Classical additives were added to the spiro-OMeTAD solution, namely 8.75 μL LiTFSI (520 mg in 1 mL acetonitrile) and 14.4 μL *t*BP. Then, the spiro-OMeTAD layer was formed by spin coating at 3000 rpm for 30 s on top of the perovskite active layer. For comparison, an alternative hole transport layer made of 3,6-Bis[N,N′-di(4-methoxyphenyl)amino]-9-(1-pyrenyl)carbazole (Cz–Pyr) was prepared by dissolving 49 mg Cz–Pyr in 0.500 mL chlorobenzene, where 8.75 μL LiTFSI and 14.4 μL *t*BP were added. Cz–Pyr HTL was recently demonstrated as a relevant alternative to spiro-OMeTAD, leading to similar photovoltaic performance [50]. Finally, an Au electrode (~100 nm) was deposited by using thermal evaporation under secondary vacuum through a shadow mask, defining the active area of the cell (>0.22 cm^2).

2.4. Characterization Techniques

The surface morphology of the ZnO NWs was measured using a field emission scanning electron microscope (SEM, Hitachi S-4800). Lengths were measured using cross-sectional SEM images, while diameters were estimated using top-view images. The average NW dimensions (length and diameter), the NWs′ density, and the void fraction were calculated from simple SEM image analysis through the ratio of the surface occupied by ZnO to the total surface of the substrate. Average values were obtained from measurements made on a large number of ZnO nanowires on a single AZO substrate. The current density–voltage (J–V) characteristics of photovoltaic devices were measured in ambient air (25–35 °C) on unencapsulated devices using a solar simulator (1600 W NEWPORT) equipped with an AM 1.5 G filter and a source meter (Keithley 2400). The active area of the devices was defined by a masked aperture of approximately 0.22 cm^2, and spectral mismatch correction was systematically applied, and all the characterizations were carried out in ambient air without encapsulation (moisture level of 55–60%). Several devices (>3) presenting comparable features have been characterized in each case, both in forward and reverse scans (simply FS and RS, voltage scan in the order of 50 mV/s), in order to extract the dispersion of performance. The shunt and series resistances (simply—R_{Sh} and R_S) were extracted from the simple estimation of the J–V curve slopes in the open- and short-circuit conditions, assuming $R_{Sh} \gg R_S$. The external quantum efficiency (EQE or IPCE) was measured under steady-state conditions using a continuous monochromated xenon lamp and a calibrated pico-amperemeter (Keithley 485). A reference silicon photodiode of known spectral sensitivity was used for EQE calibration.

3. Results and Discussion

Hydrothermal growth was used to synthesize ZnO NWs electrodes subsequently integrated into PSCs (see experimental section). The growth parameters (growing time, precursor's concentration, and growth temperature) were varied to identify the optimal features in the context of the photovoltaic application. We remind below the specific mechanisms involved during the ZnO NW formation on a ZnO-based growing substrate (either on AZO or ZnO seed layer), which can be summarized by the following equations [51,52]:

$$C_6H_{12}N_4 + 6H_2O \longrightarrow 4NH_3 + 6HCHO \quad (1)$$

$$NH_3 + H_2O \longrightarrow NH_4{}^+ + OH^- \quad (2)$$

$$Zn^{2+} + 4OH^- \longrightarrow Zn(OH)_4{}^{2-} \quad (3)$$

$$Zn(OH)_4{}^{2-} \longrightarrow ZnO + H_2O + 2OH^- \quad (4)$$

During hydrothermal growth, ZnO NWs develop following a series of reactions (Equations (1)–(4)). HMTA is initially hydrolyzed with heat, forming formaldehyde (HCHO) and NH_3 (Equation (1)). HCHO does not participate directly in the growth of NWs but reacts indirectly, as will be explained in the following parts. Then, NH_3 protonates, producing OH^- ions (Equation (2)). The Zn^{2+} ions formed following the solubilization of $Zn(NO_3)$ react with the OH^- ions to form $Zn(OH)_4{}^{2-}$ (Equation (3)), which dehydrates, leading to the direct crystallization of ZnO (Equation (4)). Certain milky white precipitation from homogeneous nucleation is present in the reaction medium. This will be explained in the following parts. Finally, the samples are rinsed with distilled water and dried under airflow.

In the following sections, we first discuss the influence of the growing substrate before focusing on ZnO NW forest grown on AZO to systematically point out the influence of the growth parameters on the NW properties and associated photovoltaic performance of devices.

3.1. Photovoltaic Performance as a Function of Growing Substrate and ETL

We all know that the solar cell performance does not solely depend on any layer (i.e., TCO, ETL, Absorber, HTL, or electrode) because in order to get a high power conversion efficiency, each layer arrangement in the device is crucial without any defects. Usually, most state-of-the-art PSCs are based on SnO_2-ETL on which the perovskite can be grown [53]. The ETL plays a role in the extraction of the negative charge carriers to the cathode and also strongly impacts the perovskite active layer morphology as well as the quality of the ETL/perovskite interface. The ETL also brings a specific processing step that has to be taken into account for large-scale developments. In the case of ZnO NWs, the hydrothermal growth required a starting ZnO seed layer, which can be formed by various techniques on different substrates, including conventional FTO-coated glass [41,54–56]. However, zinc–oxide-based transparent electrodes such as AZO have been suggested to potentially act as efficient seed layers [48,57,58]. For these reasons, the performance of devices based on NW grown on different substrates have been compared: FTO/compact ZnO (referred as ZnOc) or bare AZO (both without ETL). We also include comparison with planar devices based on various ETL without NW: FTO/SnO_2, AZO/ZnOc, or AZO/SnO_2. NW-based devices were processed from NWs grown using a set of reference experimental parameters (precursor concentration ratio of $Zn(NO_3)_2$ to HMTA and PEI fixed to $[Zn^{2+}]/[HMTA] = 1$ and $[Zn^{2+}]/[PEI] = 300$, growing time of 6 h and 15 h), which will be subsequently tuned in the next sections of this article on the best performing substrate. After NW-based ETL deposition, the mixed cation halide perovskite active layer $FA_{0.85}Cs_{0.15}Pb(I_{0.85}Br_{0.15})_3$ was deposited from the solution before the spiro-OMeTAD HTL and gold top electrode (see full procedure in the experimental section). Avoiding methylammonium was necessary in order to prevent the rapid degradation of the ZnO/perovskite interface, as discussed in the introduction section. The photovoltaic parameters, extracted from current–voltage characteristics recorded under simulated standard illumination (Figure S1), are given as supporting information in Table S1. First, we observe that the classical planar architecture based on the FTO/SnO_2 ETL is associated with the highest performance (up to nearly 14% efficiency, considering that no passivation strategies are used), as reported in previous work [49,59]. Besides, a clear S-shape in the J–V curve (reverse scan) is observed for AZO/SnO_2 device (Figure S1b), which is not the case of any devices based on ZnO ETLs (neither compact layer nor NWs). This observation is consistent with previous reports made in the literature, emphasizing that the AZO/ZnO interface can easily generate an ohmic contact due to the inherent nature of the materials in both cases (zinc oxide), which facilitates charge extraction [60]. The electrical characteristics of the AZO/SnO_2-based device suggest a strong limitation for charge extraction at this interface, which is not the case of the reference devices fabricated on fluorinated tin oxide (FTO) substrates. In Figure S1b, we clearly observe that the presence of the ZnO NW is required to achieve reasonable current densities up to ~20 mA cm^{-2}, which remain reduced with regard to the FTO/SnO_2 reference. This observation can be associated with a better charge extraction efficiency generally attributed to NWs compared to compact layers, which agrees with previous reports [61,62]. However, a much lower open-circuit voltage (V_{oc}) has been observed for ZnO-based devices compared with SnO_2 ETL devices (Figure S1b). This could be due to the inherent recombination losses associated with the ZnO/perovskite interface. Overall, on AZO substrate better device performances are displayed with ZnO NWs electrode rather than compact layers made of ZnO or SnO_2 (see Figure S1b and Table S1), especially due to a favorable electrical contact with the electrode and to the positive influence of the NW aspect ratio on charge extraction (clearly evidenced through a much lower series resistance for the AZO/ZnO-NW device with regard to the other AZO-based devices). These results clearly demonstrate the relevance of using bare AZO electrode to grow the NW, which, in addition, strongly simplify the overall fabrication protocol of the devices. This growing substrate has been chosen for the following studies. However, a low shunt resistance and fill factor suggest room for improvements as a function of NW properties. In the following sections, we will explore the influence of the precursor concentration and growing time of the NW related to the properties. These devices, made of ZnO NW grown on AZO substrates, will be discussed in terms of achieved PV performances.

3.2. *Influence of ZnO NWs' Growing Time*

Figure 1 shows the SEM images of samples prepared with different growing times from 2 h to 15 h (i.e., 2 h, 4 h, 6 h, and 15 h) with fixed precursor concentration ratio of $Zn(NO_3)_2$ to HMTA and PEI (($[Zn^{2+}]/[HMTA] = 1$), $[Zn^{2+}]/[PEI] = 300$). The average NWs' dimensions (length and diameter), the NWs' density, and the void fraction are summarized in Table 1.

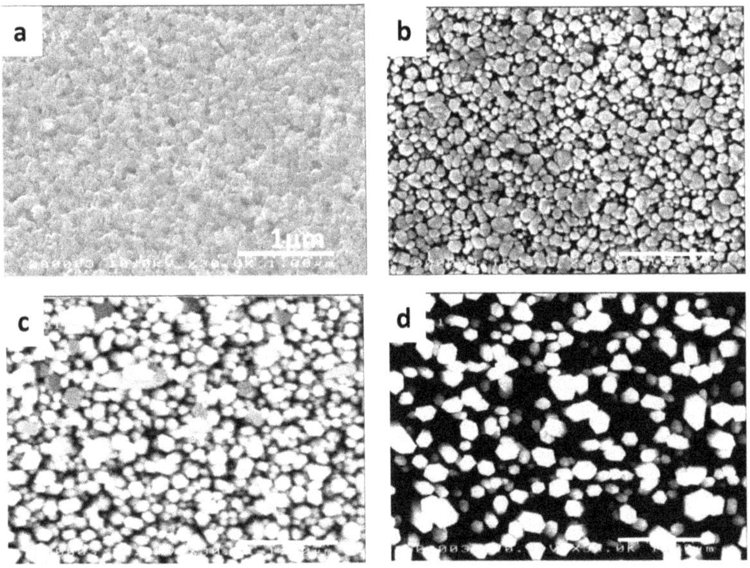

Figure 1. SEM images of ZnO NWs grown on AZO substrate during (**a**) 2 h, (**b**) 4 h, (**c**) 6 h, and (**d**) 15 h with fixed precursor concentration ratio of $Zn(NO_3)_2$ to HMTA and PEI ($[Zn^{2+}]/[HMTA] = 1$, $[Zn^{2+}]/[PEI] = 300$), respectively. All images are at the same scale (scale bar = 1 μm).

Table 1. Morphology of ZnO NW grown on AZO substrate depending on the growing time. In all cases, the concentration of $Zn(NO_3)_2$ precursor solution was 50 mM, and the $Zn(NO_3)_2$ to HMTA and PEI ratios were defined as $[Zn^{2+}]/[HMTA] = 1$ and $[Zn^{2+}]/[PEI] = 300$.

Growing Time	Length (L) (μm)	Average Diameter (d) (nm)	Density (D) (NW/μm^2)	Void Fraction (%)
2 h	0.1	-	-	-
4 h	0.3	-	-	-
6 h	0.4	127	53	~28
15 h	1.3	165	29	~58

Considering the length of the NWs, the growing time parameter has an important impact, and we can see the quite different morphological behavior, especially in the NWs' length, diameter, and void fraction values (Table 1). Indeed, the length increases from 0.1 to 1.3 μm by increasing the growing time from 2 h to 15 h. In addition, the growing time also has an influence on the NW density. For short growing times (i.e., 2 h and 4 h), very dense arrays are obtained, and the NWs are largely interconnected in the plane of the substrate (Figure 1a,b). As a result, no significant void fraction could be measured at such a short growing time. For a longer growing time (6 h and 15 h), a larger axial growth led to an increase in the void fraction from 28% to 58%, respectively. Consequently, a decrease of the NW density as a function of the growing time is clearly evidenced, while the average diameter of the NWs also increased from 127 nm after 6 h to 165 nm after 15 h, but at a much slower rate than the length, which is attributed to the presence of PEI hindering only the lateral growth of the ZnO NWs [63].

Considering the global morphology of the obtained ZnO NW arrays, we focused on samples grown for 6 h and 15 h to fabricate *n-i-p* perovskite solar cells of structure glass/AZO/ZnO NW/perovskite/spiro-OMeTAD/Au. Arrays grown at shorter times (i.e., 2 h and 4 h) did not enable a suitable perovskite deposition, making the investigations unworthy (not shown here). The UV–visible absorption and transmittance spectra of bare AZO, AZO/ZnO NW (6 h and 15 h), before and after perovskite deposition are presented as Supplementary Materials (Figure S2). A slight decrease in transmittance is observed after NW growth compared to bare AZO substrate, with transmission over 85% in the visible spectrum in all cases. After the perovskite deposition, a broad and intense absorption extending up to the near-infrared is observed, which is typical of the perovskite material used in this work. We note that a smaller absorption seems to be observed on the longer NW grown for 15h, which seems to indicate a difficulty for perovskite deposition in this case.

Figure 2 shows the forward and reverse scan (FS and RS, respectively) current–voltage characteristics under simulated solar illumination (AM1.5G, 100 mW cm^{-2}) of devices based on ZnO NW arrays as a function of growing time. We can clearly understand that it is challenging to get the consistent solar cell performance, even with the same batch of samples (see Figure S3 and Table S2), due to several factors such as interface defects, layer arrangement, perovskite infiltration, voids/pinholes, etc. Only champion cells are presented here, but several devices were characterized in each case (see Supplementary Materials, Figure S3 and Table S2), showing similar trends, and the corresponding photovoltaic parameters are presented in Table 2.

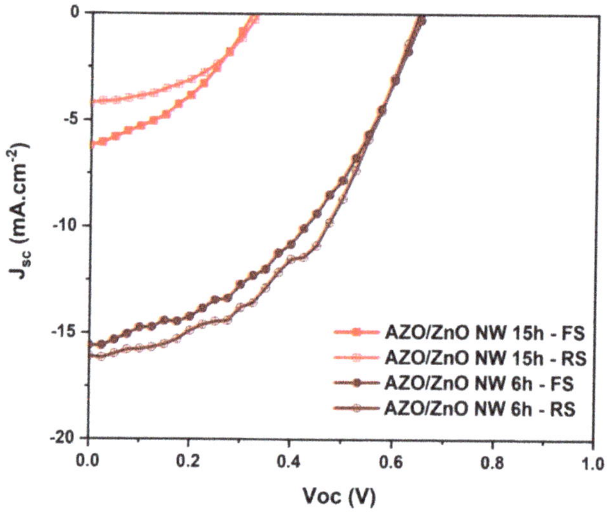

Figure 2. J–V characteristics under 1 sun of perovskite solar cells (best device) based on ZnO NW arrays grown on AZO for 6h and 15h, respectively.

Table 2. Photovoltaic parameters of perovskite solar cells fabricated as a function of ZnO NW array growing time. The concentration of $Zn(NO_3)_2$ precursor solution was 50 mM, and the $Zn(NO_3)_2$ to HMTA and PEI ratios were defined as $[Zn^{2+}]/[HMTA] = 1$ and $[Zn^{2+}]/[PEI] = 300$.

Growing Time	J_{sc} (mA cm^{-2})	Voc (V)	FF	PCE (%)	R_{sh} (Ω)	R_s (Ω)
6 h	16.1	0.64	0.47	4.9	1400	70
15 h	4.2	0.32	0.46	0.6	1200	130

Both cells are showing a clear photovoltaic effect with reduced hysteresis. Clearly, a longer growing time (15 h) is detrimental to device efficiency, mainly through reduced J_{sc} and V_{oc} values. Such observation seems consistent with the optical absorption measurements, as a reduced light-harvesting efficiency for the device based on the long NW was evidenced. It can suggest a more difficult crystallization for perovskite, which in turn leads to poor photovoltaic parameters (poor photocurrent). The SEM cross-section image of a device based on the short NW grown for 6h supports this assumption (see Supplementary Materials, Figure S4), as a well-defined sandwiched structure is observed, with the presence of compact perovskite grains on the ZnO NWs array. This performance variation between 6 h and 15 h devices might be due to the huge gap between nanowires without a proper alignment (Figure 1c,d), which possibly creates a direct contact between the perovskite and the AZO layers. And also, the 15 h-grown NWs length is higher (1.3 μm) than in the 6 h device, so the longest NWs can possibly penetrate the perovskite layer and make a direct connection with HTL (spiro-OMeTAD), leading to short-circuit current and recombination issues in the devices. Moreover, while having a higher NW diameter, it possibly does not offer a sufficient gap for the perovskite infiltration that restricts the charge extraction (electron), which causes the PV performance reduction. A much larger series resistance is also evidenced for longer NW, which can be associated with poorer charge transport properties, as reported in the literature for similar ZnO NWs [17,23,47]. In this present case, these preliminary observations indicate that a growing time of 6 h is in any case suitable as a starting point for further investigation of the different growth parameters governing the morphology and properties of the ZnO NWs.

3.3. *Optimization of the Hydrothermal Process*

In the following sections, we systematically investigate the effect of the main chemical parameters governing the growth of ZnO NWs, including the PEI, HMTA contents, and the growth solution concentration (related to Zn^{2+} quantity). Their influence on the length, density, and void fraction of the ZnO NWs will be discussed. The main objective is to optimize the precursor concentrations in order to achieve a suitable electrode morphology that favors the perovskite solar cell efficiency. In the following sections, the impact of the chemical parameters on the NW properties will be discussed. Then, the relation between the ZnO NW properties and the device performance will be presented.

3.3.1. Effect of PEI Content

The effect of the PEI content on the growth of ZnO NWs is already discussed in the literature, and the studies were devoted to NWs that are grown on the ZnO seed layer [46,64–66] or on the FTO layer [26]. They show that PEI addition significantly enhances the length of the NWs, while at the same time, it diminishes their aspect ratio [64,67,68]. In this work, we focus on the effect of the PEI content on the growth of NWs using AZO substrates as seed layers. Such a strategy aims to reduce the device fabrication complexity as no seed layer is required in this case. The ZnO NWs were synthesized at 90 °C for 6 h (temperature, growing time, and $Zn(NO_3)_2$ concentration 50 mM has been fixed) using various PEI contents ranging from 0 to 13 mM. The morphology characteristics deduced from SEM images (see Figure S5) are presented in Table 3. The PEI to HMTA ratio is also given as it obviously evolves as a function of PEI content.

Table 3. Evolution of the length, diameter, apparent density, and void fraction of ZnO NWs as a function of the [PEI]/[HMTA] ratio. The concentration of the $Zn(NO_3)_2$ solution is fixed at 50 mM, and the growing time and temperature are respectively 6 h and 90 °C.

[PEI] mM	[PEI]/[HMTA] Ratio	L (µm)	d (nm)	D NW/µm^2	Void Fraction%
–	–	0.3	131	59	~36
5	10	0.3	135	59	~37
7	7.15	0.4	147	49	~44
11	4.50	0.5	~125	58	~45
13	3.85	0.2	150	50	~28

For the reference sample without PEI, short ZnO NW with an average length of 0.3 µm is obtained. As the PEI content increases up to 11 mM, the length of the NWs increases to reach a maximum length of 0.5 µm. Adding over 11 mM (i.e., 13 mM), NW length decreases to 0.2 µm, reaching a shorter length than without using PEI. Additionally, the average void fraction of the NWs tends to increase with increasing PEI content (from 5 to 11 mM) with a maximum of 45%. Still, increasing PEI concentration to 13 mM implies a decrease of the parameters, and the void fraction decreases to 28%, and this trend is similar to that of previously published reports [64,68]. Also, the results are consistent with the fact that PEI mainly hinders the lateral growth of the wires and favors their axial growth in solution while maintaining a relatively high NW density. Therefore, considering the requirements suitable for perovskite photovoltaics, an appropriate void fraction is expected to favor the perovskite infiltration for an intimate electronic contact with the nanostructured ZnO [69]. In our study, the PEI concentration leading to the maximal void fraction is at 11 mM. In the next section, we will keep this parameter fixed to explore the influence of HMTA and zinc salt concentration on the NW morphology.

3.3.2. Effect of HMTA Content

Fixing the PEI concentration fixed at 11 mM, the $Zn(NO_3)_2$ to HMTA ratio has been tuned. SEM images (see Supplementary Materials, Figure S6) are used to extract the main morphologic features, which are summarized in Table 4 as a function of $Zn(NO_3)_2$ to HMTA ratio (from 1 to 2). This ratio was adjusted through the HMTA concentration from 50 to 25 mM. As previously mentioned, the concentration ratio $[Zn(NO_3)_2]/[PEI]$ is fixed to 4.5.

Table 4. Summary of the length, diameter, apparent density, and void fraction of ZnO NWs as a function of the $[Zn(NO_3)_2]/[HMTA]$ ratio. The concentration of the $Zn(NO_3)_2$ solution is 50 mM, and the growing time is fixed at 6 h (temperature at 90 °C).

$[Zn(NO_3)_2]/[HMTA]$ Ratio	[HMTA] mM	L (µm)	d (nm)	D (NW/µm^2)	Void fraction (%)	NWs Aspect Ratio L/d
1	50	0.5	124.6	58	~45	~4.01
1.33	37.5	0.5	125.0	58	~36	~4.01
2	25	0.3	101.3	76	~37	~2.66

The length of the obtained ZnO NWs decreases from about 0.5 to 0.3 µm as the $[Zn(NO_3)_2]/[HMTA]$ ratio increases from 1 to 2. The evolution of the diameter of the wires gives additional important insight into the role of HMTA in solution: the NW diameter decreases from ~125 to 101 nm as the HMTA content decreases. Using less HMTA, therefore, decreases both the NW diameter and length. Let us remember that Zn^{2+} ions are formed in solution following the solubilization of $Zn(NO_3)$. Then these ions react with the hydroxide OH^- ions (resulting from NH_3 protonation) to form $Zn(OH)_4{}^{2-}$ (Equation (3)). This latter species dehydrates, leading to the direct crystallization of ZnO, whereas OH^- ions and NH_3 play the role of ligands to form Zn(II) hydroxide and amine complexes. HMTA is initially hydrolyzed with heat, forming formaldehyde (HCHO) and NH_3. In precursor solution based on usual conditions, the produced HCHO does not take part in the growth, while in the presence of PEI, the HCHO molecules can react with $-NH_2$ groups, releasing Zn^{2+} ions chelated by PEI. –N=CH- or $-N=CH_2$ bonds are produced by the reaction between PEI and HCHO. The possible reaction between PEI and HCHO can be described in Scheme 1 according to the typical Mannich reaction between the –CH=O group and the $-NH_2$ or –NH- group [70].

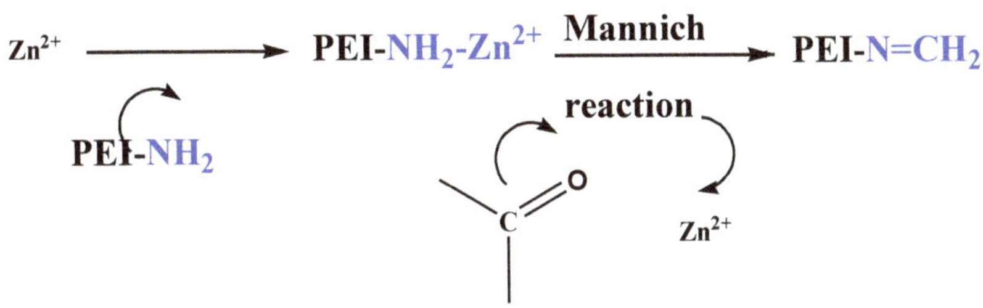

Scheme 1. Mannich reaction between PEI and HCHO.

This suggests that PEI addition has made Zn^{2+} the limiting species for the growth instead of NH_3. Meanwhile, increasing the HMTA concentration (which increases the HCHO concentration) leads to a faster Mannich reaction. This faster reaction brings more Zn^{2+} ions into the solution, increasing the growth rate of ZnO NW. Hence, when the $[Zn(NO_3)_2]/[HMTA]$ ratio is small, the limiting reactants for the axial growth of ZnO NWs are OH^- and Zn^{2+} ions, resulting in a fairly small axial and radial growth by inhibiting the development of their sidewalls [71,72].

This may explain our results, showing that, for $[Zn(NO_3)_2]/[HMTA]$ ratio equal to 1.33, ZnO NWs present a compromise between a high length and a small diameter, corresponding to a high aspect ratio (see Table 4). This morphology also results in a relevant void fraction compared to a $[Zn(NO_3)_2]/[HMTA]$ ratio of 1, with a constant density. Therefore using non-equimolar growth conditions with an excess of $Zn(NO_3)_2$ is required to reach an optimum morphology, especially with smooth surface without pinholes or voids. These conditions can be summarized as: $[Zn(NO_3)_2]$ = 50 mM, [PEI] = 11 mM, and [HMTA] = 37.5 mM, for growing time of 6 h at 90 °C, and 0.5 μm long ZnO NWs have been obtained in such conditions (see Figure S6). Such a length might appear too large for the application, considering the electrical shunt resistance that was pointed out in the previous section. Therefore, a study of the role of the zinc solution was conducted, and the aim was to try to control the NWs' length while keeping a void fraction over 35%.

3.3.3. Effect of $Zn(NO_3)_2$ Concentration

ZnO NWs were fabricated from various zinc salt concentrations $[Zn^{2+}]$ ranging from 20 to 50 mM, using fixed concentration ratios between $Zn(NO_3)_2$ and HMTA ($[Zn^{2+}]/[HMTA]$ = 1.33) and between $Zn(NO_3)_2$ and PEI ($[Zn^{2+}]/[PEI]$ = 4.5). SEM images were obtained (see Supplementary Materials, Figure S7) and analyzed to extract the main morphologic features of the samples (Table 5).

Table 5. Summary of the length, diameter, apparent density, and void fraction of ZnO NWs as a function of the concentration of zinc salt [$Zn(NO_3)_2$] ([Zn^{2+}]/[PEI] = 4.5; [Zn^{2+}]/[HMTA] = 1.33, growing time of 6 h). For the HMTA and PEI concentration values for different [$Zn(NO_3)_2$] concentrations, see Table S3.

[$Zn(NO_3)_2$] (mM)	L (μm)	d (nm)	D (NW/μm^2)	Void Fraction (%)	NWs Aspect Ratio L/d
20 mM	0.25	109	58	~43	2.28
35 mM	0.30	117	57	~46	2.56
50 mM	0.50	125	58	~36	4.00

As deduced from SEM pictures, the length of the NWs decreases with decreasing zinc nitrate concentration. For a concentration of 20 mM, the length is 0.25 μm (growing time 6 h) and increases up to 0.5 μm for a concentration of 50 mM. We emphasize that well-aligned NWs are obtained in this latter case (without any pinholes, see Supplementary Materials, Figure S7), leaving a comparable porosity of 36% (void fraction) compared to 20 mM and 35 mM of $Zn(NO_3)_2$ contents. Indeed, when the reagent concentration is increased, the transport rate of active ions increases too. It results in a relatively high nanowire growth rate in the axial direction, as seen with 50 mM of $Zn(NO_3)_2$ [64]. This study confirms that the length of the NWs increases with the concentration of the zinc salt.

Considering our initial assumption about the adverse influence of NWs' length on perovskite solar cell performance (refer to data from Section 0: influence of growing time), we now focus on devices based on NWs grown using different zinc salt concentrations. This simple approach aims to decipher the relation between PV performance and the morphological characteristics of the integrated ZnO NWs. Figure 3 shows the current density–voltage (J–V) characteristics under the illumination of perovskite solar cells based on ZnO NWs grown using various zinc salt concentrations, and the corresponding PV parameters are summarized in Table 6.

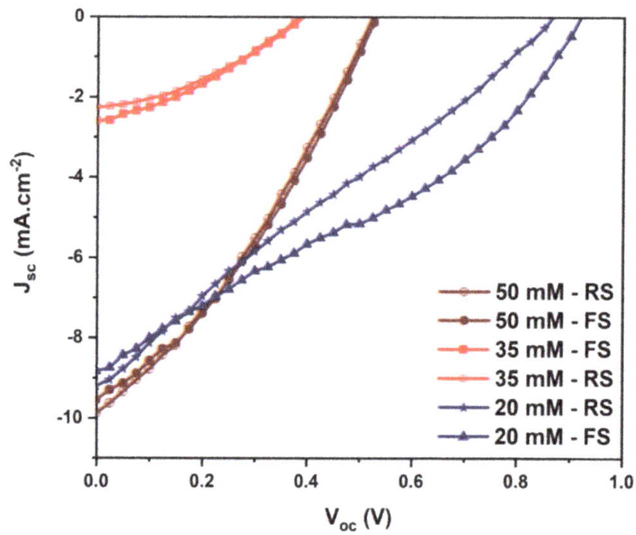

Figure 3. J–V characteristics of perovskite solar cells (best device) fabricated with the photoanodes of ZnO NWs depending on [$Zn(NO_3)_2$] concentrations.

Table 6. Performance of perovskite solar cells (reverse scan) fabricated with the photoanodes of ZnO NWs depending on the [$Zn(NO_3)_2$] concentration ([Zn^{2+}]/[PEI] = 4.5; [Zn^{2+}]/[HMTA] = 1.33, growing time of 6 h).

$ZnO(NO_3)_2$ Concentration	Jsc ($mA\ cm^{-2}$)	Voc (V)	FF	PCE (%)	R_{sh} (Ω)	R_s (Ω)
20 mM	8.9	0.92	0.33	2.7	600	200
35 mM	2.2	0.38	0.37	0.3	1250	470
50 mM	9.4	0.53	0.35	1.7	650	160

Clearly, by decreasing the zinc salt concentration down to 20 mM, better device performance is observed. A maximum power conversion efficiency of 2.7% was achieved for the PV cell corresponding to this zinc salt concentration (hence corresponding to the shorter NWs), mainly due to a combined high short-circuit current and open-circuit voltage. In general, the NWs' size and length should be controlled; otherwise, the perovskite cannot be well-deposited into the NW gaps or surface, which possibly creates unsmooth morphology, and increases chance of defects such as voids or pinholes in the layer [73]. This exploration of the main growth parameters enables us to point out suitable conditions for the synthesis of relatively short NW presenting a suitable void fraction, which is found to be critical to ensure suitable perovskite layer infiltration. This better infiltration may reduce the occurrence of direct shortcuts with the HTM. The main morphologic features of the ZnO NWs grown using all sets of parameters are summarized in Supplementary Materials Table S3. After careful adjustment of the main parameters of the ZnO NWs' growth, the final set ([$Zn(NO_3)_2$] = 20 mM, ([Zn^{2+}]/[PEI] = 4.5; [Zn^{2+}]/[HMTA] = 1.33, growing time of 6 h) is leading to a PV cell performance that is lower than the initial set ([$Zn(NO_3)_2$] = 50 mM, [Zn^{2+}]/[HMTA] = 1 and [Zn^{2+}]/[PEI] = 300, for 6 h growing time). The following Table 7 summarizes the corresponding PV cell characteristics.

Table 7. Performances of perovskite solar cells fabricated with different NW growing parameters (i.e., initial and final set).

Growing Time	J_{sc} ($mA\ cm^{-2}$)	Voc (V)	FF	PCE (%)	R_{sh} (Ω)	R_s (Ω)
Initial set	16.1	0.64	0.47	4.9	1400	70
Final set	8.9	0.92	0.33	2.7	600	200

It is worth mentioning here that while varying NW growing time, precursor concentrations (zinc nitrate hexahydrate, HMTA, and PEI), the obtained ZnO NWs' properties, especially NW length and diameter values, were associated and are consistent with the range of previously published reports (see Figure S8) [23,41–46,63,64,74]. The ZnO NWs' growing time should be limited to 6 h (0.4 μm long ZnO NWs) because longer NWs seem detrimental to the PV cell performance (Table 2). With our growth parameters, a PEI content of 11 mM seemed to be the optimal point to maximize the void content (here 45%) while keeping a limited length of the NWs (0.5 μm) (Table 3). In order to keep a good surface morphology with an appropriate void content, but also to decrease the ZnO NWs' length, the HMTA concentration effect was also studied. Apart from the NW length (0.5 μm), a smoother surface behavior was obtained for 37.5 mM than for other concentrations with a void fraction value of 36% (Table 4 and Figure S6). Finally, the $Zn(NO_3)_2$ zinc salt content was studied in order to fine-tune the NWs' length while trying to maintain a suitable void fraction (Table 5). This study showed that the void fraction decreases additionally to the length: for example, 20 mM zinc salt concentration led to 0.25 μm long NWs and a 43% void fraction compared to 35 mM, and this behavior is not same for the higher concentration (i.e., 50 mM). Therefore, the PV cells resulting from this last study seem to show that short ZnO NWs are prioritary to low dense NW arrays in order to maximize the PV cell efficiency.

In order to reinforce this conclusion and to get a deeper overview of the design choices, we decided to conduct a final study, by comparing performance of devices made with both initial and final sets of growth parameters and with two different HTLs. Consequently, in the following section we present a preliminary study over the performance of ZnO-NW arrays associated with an alternative HTL based on a carbazole moiety (Cz–Pyr).

3.4. Comparison between Spiro-OMeTAD and Cz–Pyr as HTL

As previously described, we succeeded to control the morphology parameters of ZnO NWs grown on AZO substrate. In order to discuss the relation between the morphology and the PV efficiencies, devices with ZnO NWs have been tested. The devices have been fabricated using two different HTG parameters: on one side the initial chemical process, and on the other side, the optimum recipe we have selected from the previous studies. Moreover, the devices have been fabricated using two different HTLs. Carbazole-based derivatives as HTL in perovskite solar cells have attracted attention because of their photochemical properties and their promising characteristics such as high mobilities, good charge transport, and high efficiencies [75–78]. In this context, reported Cz–Pyr [50] has been compared with the reference spiro-OMeTAD as HTL. In short, in this subsection, optimized NW growth conditions (i.e., for 20mM, $[Zn^{2+}]/[HMTA] = 1.33$ and $[Zn^{2+}]/[PEI] = 4.5$, for 6 h growing time) and initial NW growth conditions (i.e., for 50 mM $[Zn^{2+}]/[HMTA] = 1$ and $[Zn^{2+}]/[PEI] = 300$, for 6 h growing time) have been used to fabricate perovskite solar cells based on either spiro-OMeTAD or Cz–Pyr. Especially, two different $Zn(NO_3)_2$ concentrations, either $[Zn(NO_3)_2] = 20$ mM (NW referred in this case of ZnO—0.25 µm) or $[Zn(NO_3)_2] = 50$ mM (NW referred in this case as ZnO—0.40 µm), have been tested. We especially compare two NW lengths (i.e., 0.40 and 0.25 µm), and we also provide preliminary data regarding aging tests performed under continuous illumination in ambient conditions without encapsulation. Figure S9 shows the current–voltage characteristics of ZnO NW array-based perovskite solar cells for ZnO—0.25 µm and ZnO—0.40 **µm**, with the two different HTLs. The PV parameter results are presented in Table 8.

Table 8. Performances of perovskite solar cells fabricated with different NW-based ETLs (ZnO 0.40 µm and ZnO 0.25 µm) and HTLs (i.e., spiro-OMeTAD and Cz–Pyr).

			Jsc (mA.cm^{-2})	**Voc (V)**	**FF**	**PCE (%)**	**Rsh (Ω)**	**Rs (Ω)**
ZnO 0.40 µm/Spiro-OMeTAD	FS	Best	16.4	0.66	0.36	3.8	420	100
		Average	13.7	0.69	0.35	3.4	780	120
	RS	Best	16.1	0.64	0.47	4.9	1400	70
		Average	16.7	0.68	0.43	4.7	1350	80
ZnO 0.25 µm/Spiro-OMeTAD	FS	Best	9.1	0.86	0.25	2.0	530	330
		Average	8.2	0.85	0.25	1.8	560	380
	RS	Best	8.9	0.92	0.33	2.7	600	200
		Average	8.5	0.91	0.32	2.5	580	220
ZnO 0.40 µm/Cz–Pyr	FS	Best	7.3	0.82	0.35	2.1	1070	240
		Average	5.5	0.71	0.34	1.3	1350	400
	RS	Best	7.0	0.81	0.39	2.2	940	230
		Average	5.6	0.70	0.34	1.3	1100	400
ZnO 0.25 µm/Cz–Pyr	FS	Best	5.0	0.62	0.33	1.0	800	330
		Average	3.9	0.74	0.29	0.8	1170	970
	RS	Best	4.6	0.62	0.40	1.1	1840	280
		Average	3.9	0.73	0.30	0.8	1550	1030

It is clear that, while increasing the length of the NWs from 0.25 μm to 0.40 μm, the device performance significantly increased from 2.7% to 4.9% (using spiro-OMeTAD-HTL), and 1.1% to 2.2% (using Cz–Pyr-HTL), respectively. This trend confirms that there is a compromise for NW length to enable efficient charge extraction before series or shunt resistances finally reduce the device performance (see Section 3.2). Regarding the nature of the HTL, Cz–Pyr-based devices exhibit much lower J_{sc} values compared to devices based on spiro-OMeTAD. Such observation can be related to a larger hole mobility for spiro-OMeTAD (2.5×10^{-5} cm^2 V^{-1}s^{-1}) as measured using field-effect transistors in saturation regime, compared to Cz–Pyr (7.4×10^{-6} cm^2 V^{-1}s^{-1}) [50]. The incident photon-to-current conversion efficiency (IPCE) measurements show that the use of 0.40 μm long NWs instead of 0.25 μm enhances the photocurrent in the whole spectral region under study (300–800 nm), with both spiro-OMeTAD and Cz–Pyr (see Figure S10). Such behavior is clearly consistent with a better hole mobility of spiro-OMeTAD compared to Cz–Pyr, and points out the positive influence for NW of moderate length.

We finally conducted a preliminary aging test of the devices through continuous operation under solar illumination and without encapsulation for 20 min (Figure S11). Ambient conditions were used in this case, without thermal regulation at 25 °C, leading to a mean temperature of 55 °C and a moisture level of around 50% (conditions not finely controlled as in more adapted ISOS protocols) [79]. The Cz–Pyr-based device shows an interesting behavior compared to spiro-OMeTAD in these harsh conditions, with increasing performance mainly governed by J_{sc}, unlike the spiro-OMeTAD device, which sees its performance slowly decreasing with time. While many mechanisms have been discussed [80,81] to interpret the evolution of the perovskite active layer in these conditions, explaining common losses in V_{oc} and FF [82], this strong difference in device behavior suggests a specific role played by the carbazole HTL in device degradation. While Cz–Pyr shows a relatively lower glass transition temperature than spiro-OMeTAD [50], these results indicate a beneficial interfacial configuration with the perovskite active layer, which positively affects the performance evolution with time. The role of the dopants (*tert*-butyl pyridine and lithium salt), which are known to be a major source of performance drop in the case of spiro-OMeTAD [75], remains to be clarified in this case as well.

The comparison of published AZO/ZnO NWs/perovskite-based device PV performances with our proposed device results is demonstrated in Table 9. In our study, we have not annealed the AZO substrate before making ZnO NWs growth which might be another possible reason for the lower PV performance than the published report, especially with V. L. Ferrara et al.'s results, because their observation shows that the annealed AZO substrate at 150 °C evidently enhances the crystal quality compared to non-annealed sample that helps to achieve a good crystal nucleation sites [74]. Also, the present contribution provides novel device architecture (i.e., AZO/ZnO NWs/perovskite/Cz–Pyr/Au) with a significant efficiency of 2.2%, half of conventional spiro-OMeTAD-HTL-based devices (see Table 9).

Table 9. Published PV performance of AZO/ZnO NWs/perovskite architecture devices.

Device Structure	Jsc (mA cm^{-2})	Voc (V)	FF	PCE (%)	Ref
AZO/ZnO NRs/$MAPbI_3$/Spiro-OMeTAD/Au	16.00	0.80	0.53	7.00	[74]
AZO/ZnO NRs/$MAPbI_xCl_{3-x}$/Cu	14.87	0.86	0.28	3.62	[83]
AZO/ZnO NRs based DSSC	05.01	0.60	0.43	1.31	[84]
AZO/ZnO NWs/FACsPb(IBr)$_3$/Spiro-OMeTAD/Au	16.10	0.64	0.47	4.9	This work
AZO/ZnO NWs/FACsPb(IBr)$_3$/Cz–Pyr/Au	07.00	0.81	0.39	2.2	

4. Conclusions

This study focused on the ZnO NWs growth over the AZO substrate using the low-temperature hydrothermal growth (HTG) method to fabricate efficient perovskite solar cells based on AZO/ZnO NWs/perovskite/HTL (spiro-OMeTAD or Cz–Pyr)/Au device configuration. The effect of several parameters, especially HTG precursors concentrations such as zinc nitrate hexahydrate, HMTA, PEI, and growing time, were systematically studied to achieve dense and well-aligned ZnO NWs array with characterized void fraction in order to boost the PV device performance. Results highlight that a longer growing time (i.e., 15 h) and higher zinc salt concentrations (i.e., 35 mM and 50 mM) are unfavorable to the device performance. Also, AZO/ZnO NW-based devices demonstrate superior performance than AZO/ZnO or AZO/SnO_2 compact ETL-based devices. Finally, Cz–Pyr-HTL-based AZO/ZnO NWs device provides comparable performance to the conventional spiro-OMeTAD-HTL devices. Overall, the obtained solar cell efficiencies are lower than the published conventional perovskite device configurations, therefore, enhancing the ZnO NWs layer surface homogeneity, especially the NWs' alignment with an appropriate gap, will be a beneficial strategy to further increase the device efficiency in the near future. This current research observations suggest that using ZnO NW over AZO substrate is an effective approach for the perovskite solar cells in terms of performance and stability.

Supplementary Materials: The following supporting information can be downloaded at: https://www.mdpi.com/article/10.3390/nano12122093/s1. Figure S1: J–V characteristics of perovskite solar cells (best device) fabricated with different photoanodes of ZnO NWs depending on the seed layer. (a) Comparison of AZO/ZnO NW, FTO/ZnOc/ZnO NW, and FTO/SnO_2 devices; (b) comparison of AZO/ZnOc, AZO/SnO_2, and AZO/ZnO NW based devices, respectively. Figure S2: Recorded UV–visible absorbance (a) and transmittance (b) spectral response for ZnO NWs (6 h and 15 h) grown on AZO substrate. Figure S3: Comparison of several same batch device J–V characteristics of (a) AZO/ZnO NW (6h) and (b) AZO/ZnO NW (15 h) solar cells, respectively. Figure S4: SEM cross-section of a full device based on AZO/ZnO NW grown for 6 h. The image demonstrates a homogeneity of the sandwich structure, with a perovskite solution residing in the ZnO NW layer. Figure S5: SEM micrographs of ZnO NWs as a function of the [Zn(NO3)$_2$]/[PEI] ratio, (a) without PEI, (b) 10, (c) 7.15, (d) 4.5, and (e) 3.85, respectively. Figure S6: SEM micrographs of ZnO NWs as a function of the [$Zn(NO_3)_2$]/[HMTA] ratio: (a) 1; (b) 1.33; (c) 2. The concentration of the $Zn(NO_3)_2$ solution is kept constant to 50 mM in all cases, and the growing time is fixed at 6 h. Figure S7: SEM micrographs of ZnO NWs as a function of the [$Zn(NO_3)_2$] (a) 20 mM, (b) 35 mM, and (c) 50 mM, respectively. Figure S8: Published reports related to the ZnO NW properties, especially NW length and diameters. Comparison of NW diameters and lengths (a and b) based on growing time, (c and d) based on Zn–HMTA concentration, and (e and f) based on PEI concentration, respectively. Figure S9: J–V characteristics of perovskite solar cells (best device) fabricated with two different HTLs (spiro-OMeTAD and Cz–Pyr) using two different NWs, (a) ZnO 0.25 μm and ZnO 0.40 μm, respectively. Figure S10: IPCE spectra for the champion solar cells with two different NWs-based devices (ZnO 0.40 μm and ZnO 0.25 μm) using two different HTLs (spiro-OMeTAD and Cz–Pyr), respectively. Figure S11: Preliminary aging test of the ZnO NWs (ZnO 0.25 μm (a) and 0.40 μm (b)) based devices along with two different HTLs (spiro-OMeTAD and Cz–Pyr), respectively. Table S1: Performances of perovskite solar cells depending on the type of photoanodes. Table S2: Photovoltaic parameters of perovskite solar cells fabricated as a function of ZnO NW array growing time (i.e., 6 h and 15 h). Table S3: Summary table of the length, diameter, apparent density, and void fraction of ZnO NWs as a function of different conditions.

Author Contributions: Conceptualization, methodology, and writing—original draft preparation, R.N., K.S., K.N., J.B., B.S. and G.P.-V.; Formal analysis, R.N., K.S. and R.D.; writing—review and editing, K.S., K.N., J.B., B.S. and G.P.-V.; supervision, K.N., J.B., B.S. and G.P.-V. All authors have read and agreed to the published version of the manuscript.

Funding: The authors gratefully acknowledge the Regional Council Centre-Val de Loire who supported the CELEZ project (grant agreement 2016-00108356). This work has also received support under the CERTeM 5.0 Program, with the financial support of the Regional Council Centre-Val de Loire and Tours Val de Loire Metropolis (France). Similarly, this work is partially supported by the French National Research Agency (ANR-10-LABX-0074-01 Sigmalim). K.S. and J.B. specifically acknowledge the PLATINOM technology platform (University of Limoges), which hosted all device fabrication and characterization steps. Also, the CARMALIM common, FEDER (Europe), and Région Nouvelle Aquitaine are acknowledged for their contribution and indirect financial supports.

Institutional Review Board Statement: Not applicable.

Informed Consent Statement: Not applicable.

Data Availability Statement: The data presented in this study are available from the corresponding author upon request.

Acknowledgments: The authors acknowledge Taoufik Slimani Tlemcani and Camille Justeau for sharing their valuable experience.

Conflicts of Interest: The authors declare no conflict of interest.

References

1. Kazim, S.; Nazeeruddin, M.K.; Grätzel, M.; Ahmad, S. Perovskite as light harvester: A game changer in photovoltaics. *Angew. Chem. Int. Ed.* **2014**, *53*, 2812–2824. [CrossRef] [PubMed]
2. Lee, M.M.; Teuscher, J.; Miyasaka, T.; Murakami, T.N.; Snaith, H.J. Efficient hybrid solar cells based on meso-superstructured organometal halide perovskites. *Science* **2012**, *338*, 643–647. [CrossRef] [PubMed]
3. Kim, H.; Lee, C.; Im, J.; Lee, K.; Moehl, T.; Marchioro, A.; Moon, S.; Humphry-Baker, R.; Yum, J.; Moser, J.E.; et al. Lead Iodide Perovskite Sensitized All-Solid-State Submicron Thin Film Mesoscopic Solar Cell with Efficiency Exceeding 9%. *Science* **2012**, *2*, 1–7. [CrossRef] [PubMed]
4. Best Research-Cell Efficiency Chart Photovoltaic Research NREL. Available online: https://www.nrel.gov/pv/cell-efficiency.html (accessed on 30 March 2021).
5. Almora, O.; Baran, D.; Bazan, G.C.; Berger, C.; Cabrera, C.I.; Catchpole, K.R.; Erten-Ela, S.; Guo, F.; Hauch, J.; Ho-Baillie, A.W.Y.; et al. Device Performance of Emerging Photovoltaic Materials (Version 1). *Adv. Energy Mater.* **2020**, *11*, 2002774. [CrossRef]
6. Chen, L.-C.; Tseng, Z.-L. ZnO-Based Electron Transporting Layer for Perovskite Solar Cells. *Nanostruct. Sol. Cells* **2017**, *10*, 203–215. [CrossRef]
7. Grätzel, M. The light and shade of perovskite solar cells. *Nat. Mater.* **2014**, *13*, 838–842. [CrossRef]
8. Li, G.; Zhu, R.; Yang, Y. Polymer solar cells. *Nat. Photonics* **2012**, *6*, 153–161. [CrossRef]
9. Green, M.A.; Ho-Baillie, A.; Snaith, H.J. The emergence of perovskite solar cells. *Nat. Photonics* **2014**, *8*, 506–514. [CrossRef]
10. Han, G.S.; Chung, H.S.; Kim, B.J.; Kim, D.H.; Lee, J.W.; Swain, B.S.; Mahmood, K.; Yoo, J.S.; Park, N.G.; Lee, J.H.; et al. Retarding charge recombination in perovskite solar cells using ultrathin MgO-coated TiO2 nanoparticulate films. *J. Mater. Chem. A* **2015**, *3*, 9160–9164. [CrossRef]
11. Luo, J.; Wang, Y.; Zhang, Q. Progress in perovskite solar cells based on ZnO nanostructures. *Sol. Energy* **2018**, *163*, 289–306. [CrossRef]
12. Mahmood, K.; Swain, B.S.; Amassian, A. 16.1% Efficient Hysteresis-Free Mesostructured Perovskite Solar Cells Based on Synergistically Improved ZnO Nanorod Arrays. *Adv. Energy Mater.* **2015**, *5*, 1500568. [CrossRef]
13. Wang, P.; Zhao, J.; Liu, J.; Wei, L.; Liu, Z.; Guan, L.; Cao, G. Stabilization of organometal halide perovskite films by SnO_2 coating with inactive surface hydroxyl groups on ZnO nanorods. *J. Power Sources* **2017**, *339*, 51–60. [CrossRef]
14. Zhang, P.; Wu, J.; Zhang, T.; Wang, Y.; Liu, D.; Chen, H.; Ji, L.; Liu, C.; Ahmad, W.; Chen, Z.D.; et al. Perovskite Solar Cells with ZnO Electron-Transporting Materials. *Adv. Mater.* **2018**, *30*, 1703737. [CrossRef]
15. Mahmood, K.; Swain, B.S.; Jung, H.S. Controlling the surface nanostructure of ZnO and Al-doped ZnO thin films using electrostatic spraying for their application in 12% efficient perovskite solar cells. *Nanoscale* **2014**, *6*, 9127–9138. [CrossRef] [PubMed]
16. Son, D.Y.; Im, J.H.; Kim, H.S.; Park, N.G. 11% efficient perovskite solar cell based on ZnO nanorods: An effective charge collection system. *J. Phys. Chem. C* **2014**, *118*, 16567–16573. [CrossRef]
17. Dong, J.; Zhao, Y.; Shi, J.; Wei, H.; Xiao, J.; Xu, X.; Luo, J.; Xu, J.; Li, D.; Luo, Y.; et al. Impressive enhancement in the cell performance of ZnO nanorod-based perovskite solar cells with Al-doped ZnO interfacial modification. *Chem. Commun.* **2014**, *50*, 13381–13384. [CrossRef]
18. Mahmood, K.; Hameed, M.; Rehman, F.; Khalid, A.; Imran, M.; Mehran, M.T. A multifunctional blade-coated ZnO seed layer for high-efficiency perovskite solar cells. *Appl. Phys. A Mater. Sci. Process.* **2019**, *125*, 83. [CrossRef]
19. Wang, Z.L. Zinc oxide nanostructures: Growth, properties and applications. *J. Phys. Condens. Matter* **2004**, *16*, 829–858. [CrossRef]

20. Greene, L.E.; Yuhas, B.D.; Law, M.; Zitoun, D.; Yang, P. Solution-grown zinc oxide nanowires. *Inorg. Chem.* **2006**, *45*, 7535–7543. [CrossRef]
21. Özgür, Ü.; Alivov, Y.I.; Liu, C.; Teke, A.; Reshchikov, M.A.; Doğan, S.; Avrutin, V.; Cho, S.; Morkoç, H. A comprehensive review of ZnO materials and devices. *J. Appl. Phys. Appl. Phys.* **2005**, *98*, 041301. [CrossRef]
22. Zhou, H.; Shi, Y.; Wang, K.; Dong, Q.; Bai, X.; Xing, Y.; Du, Y.; Ma, T. Low-temperature processed and carbon-based ZnO/CH3NH3PbI3/C planar heterojunction perovskite solar cells. *J. Phys. Chem. C* **2015**, *119*, 4600–4605. [CrossRef]
23. Dymshits, A.; Iagher, L.; Etgar, L. Parameters influencing the growth of ZnO nanowires as efficient low temperature flexible perovskite-based solar cells. *Materials* **2016**, *9*, 60. [CrossRef] [PubMed]
24. Yang, J.; Siempelkamp, B.D.; Mosconi, E.; De Angelis, F.; Kelly, T.L. Origin of the Thermal Instability in CH3NH3PbI3 Thin Films Deposited on ZnO. *Chem. Mater.* **2015**, *27*, 4229–4236. [CrossRef]
25. Wang, Q.; Shao, Y.; Dong, Q.; Xiao, Z.; Yuan, Y.; Huang, J. Large fill-factor bilayer iodine perovskite solar cells fabricated by a low-temperature solution-process. *Energy Environ. Sci.* **2014**, *7*, 2359–2365. [CrossRef]
26. Law, M.; Greene, L.E.; Johnson, J.C.; Saykally, R.; Yang, P. Nanowire dye-sensitized solar cells. *Nat. Mater.* **2005**, *4*, 455–459. [CrossRef]
27. Léonard, F.; Talin, A.A. Electrical contacts to one- and two-dimensional nanomaterials. *Nat. Nanotechnol.* **2011**, *6*, 773–783. [CrossRef]
28. Susner, M.A.; Carnevale, S.D.; Kent, T.F.; Gerber, L.M.; Phillips, P.J.; Sumption, M.D.; Myers, R.C. Catalyst-free ZnO nanowires on silicon by pulsed laser deposition with tunable density and aspect ratio. *Phys. E Low-Dimens. Syst. Nanostruct.* **2014**, *62*, 95–103. [CrossRef]
29. Faraji, N.; Ulrich, C.; Wolff, N.; Kienle, L.; Adelung, R.; Mishra, Y.K.; Seidel, J. Visible-Light Driven Nanoscale Photoconductivity of Grain Boundaries in Self-Supported ZnO Nano- and Microstructured Platelets. *Adv. Electron. Mater.* **2016**, *2*, 1600138. [CrossRef]
30. Dahiya, A.S.; Opoku, C.; Alquier, D.; Poulin-Vittrant, G.; Cayrel, F.; Graton, O.; Huu Hue, L.P.T.; Camara, N. Controlled growth of 1D and 2D ZnO nanostructures on 4H-SiC using Au catalyst. *Nanoscale Res. Lett.* **2014**, *9*, 379. [CrossRef]
31. Boubenia, S.; Dahiya, A.S.; Morini, F.; Nadaud, K.; Alquier, D. A facile hydrothermal approach for the density tunable growth of ZnO nanowires and their electrical characterizations. *Sci. Rep.* **2017**, *7*, 15187. [CrossRef]
32. Dahiya, A.S.; Boubenia, S.; Franzo, G.; Poulin-Vittrant, G.; Mirabella, S.; Alquier, D. Photoluminescence Study of the Influence of Additive Ammonium Hydroxide in Hydrothermally Grown ZnO Nanowires. *Nanoscale Res. Lett.* **2018**, *13*, 249. [CrossRef] [PubMed]
33. Tlemcani, T.S.; Justeau, C.; Nadaud, K.; Poulin-vittrant, G.; Alquier, D. Deposition Time and Annealing Effects of ZnO Seed Layer on Enhancing Vertical Alignment of Piezoelectric ZnO Nanowires. *Chemosensors* **2019**, *7*, 7. [CrossRef]
34. Zheng, N.; Huang, Y.; Sun, W.; Du, X.; Liu, H. In-situ pull-off of ZnO nanowire from carbon fi ber and improvement of interlaminar toughness of hierarchical ZnO nanowire/carbon fi ber hydrid composite laminates. *Carbon N. Y.* **2016**, *110*, 69–78. [CrossRef]
35. Demes, T.; Ternon, C.; Morisot, F.; Riassetto, D.; Legallais, M.; Roussel, H.; Langlet, M. Mechanisms involved in the hydrothermal growth of ultra-thin and high aspect ratio ZnO nanowires. *Appl. Surf. Sci.* **2017**, *410*, 423–431. [CrossRef]
36. Alshehri, N.A.; Lewis, A.R.; Pleydell-Pearce, C.; Maffeis, T.G.G. Investigation of the growth parameters of hydrothermal ZnO nanowires for scale up applications. *J. Saudi Chem. Soc.* **2018**, *22*, 538–545. [CrossRef]
37. Yoo, K.; Lee, W.; Kang, K.; Kim, I.; Kang, D.; Oh, D.K.; Kim, M.C.; Choi, H.; Kim, K.; Kim, M.; et al. Low-temperature large-area fabrication of ZnO nanowires on flexible plastic substrates by solution-processible metal-seeded hydrothermal growth. *Nano Converg.* **2020**, *7*, 24. [CrossRef]
38. Yang, J.; Kelly, T.L. Decomposition and Cell Failure Mechanisms in Lead Halide Perovskite Solar Cells. *Inorg. Chem.* **2017**, *56*, 92. [CrossRef]
39. Cheng, Y.; Yang, Q.; Xiao, J.; Xue, Q. On the Decomposition of Organometal Halide Perovskite Films on Zinc Oxide Nanoparticles. *ACS Appl. Mater. Interfaces* **2015**, *7*, 19986. [CrossRef]
40. Schutt, K.; Nayak, P.K.; Ramadan, A.J.; Wenger, B.; Lin, Y.H.; Snaith, H.J. Overcoming Zinc Oxide Interface Instability with a Methylammonium-Free Perovskite for High-Performance Solar Cells. *Adv. Funct. Mater.* **2019**, *29*, 1900466. [CrossRef]
41. Nourdine, A.; Abdelli, M.; Charvin, N.; Flandin, L. Custom Synthesis of ZnO Nanowires for Efficient Ambient Air-Processed Solar Cells. *ACS Omega* **2021**, *6*, 32365–32378. [CrossRef]
42. Geng, Y.; Jeronimo, K.; Bin Che Mahzan, M.A.; Lomax, P.; Mastropaolo, E.; Cheung, R. Comparison of ZnO nanowires grown on e-beam evaporated Ag and ZnO seed layers. *Nanoscale Adv.* **2020**, *2*, 2814–2823. [CrossRef]
43. Maddah, M.; Unsworth, C.P.; Plank, N.O.V. Selective growth of ZnO nanowires with varied aspect ratios on an individual substrate. *Mater. Res. Express* **2018**, *6*, 015905. [CrossRef]
44. Tugral Arslan, H.; Arslan, C.; Baydogan, N. The effects of the curing parameters of the hydrothermal solution on the characteristic properties of ZnO nanowires. *J. Opt.* **2021**, *51*, 79–88. [CrossRef]
45. Eskandari, A.; Abdizadeh, H.; Pourshaban, E.; Golobostanfard, M.R. An investigation into the role of polyethyleneimine in chemical bath deposition of zinc oxide nanowires. *AIP Conf. Proc.* **2018**, *1920*, 020026. [CrossRef]
46. Zhitao, H.; Sisi, L.; Jinkui, C.; Yong, C. Controlled growth of well-aligned ZnO nanowire arrays using the improved hydrothermal method. *J. Semicond.* **2013**, *34*, 063002. [CrossRef]
47. Consonni, V.; Briscoe, J.; Kärber, E.; Li, X.; Cossuet, T. ZnO nanowires for solar cells: A comprehensive review. *Nanotechnology* **2019**, *30*, 362001. [CrossRef]

48. Justeau, C.; Tlemcani, T.S.; Poulin-Vittrant, G.; Nadaud, K.; Alquier, D. A comparative study on the effects of Au, ZnO and AZO seed layers on the performance of ZnO nanowire-based piezoelectric nanogenerators. *Materials* **2019**, *12*, 2511. [CrossRef]
49. Karthick, S.; Hawashin, H.; Parou, N.; Vedraine, S.; Velumani, S.; Bouclé, J. Copper and Bismuth incorporated mixed cation perovskite solar cells by one-step solution process. *Sol. Energy* **2021**, *218*, 226–236. [CrossRef]
50. Al-Zohbi, F.; Jouane, Y.; Benhattab, S.; Faure-Vincent, J.; Tran-Van, F.; Vedraine, S.; Bouclé, J.; Berton, N.; Schmaltz, B. Simple carbazole-based hole transporting materials with fused benzene ring substituents for efficient perovskite solar cells. *New J. Chem.* **2019**, *43*, 12211–12214. [CrossRef]
51. Li, Q.; Kumar, V.; Li, Y.; Zhang, H.; Marks, T.J.; Chang, R.P.H. Fabrication of ZnO Nanorods and Nanotubes in Aqueous Solutions. *Chem. Mater.* **2005**, *17*, 1001–1006. [CrossRef]
52. Qiu, J.; Li, X.; He, W.; Park, S.; Kim, H.; Hwang, Y.; Lee, J. The growth mechanism and optical properties of ultralong ZnO nanorod arrays with a high aspect ratio by a preheating hydrothermal method. *Nanotechnology* **2009**, *20*, 155603. [CrossRef] [PubMed]
53. Jiang, Q.; Zhang, X.; You, J. SnO2: A Wonderful Electron Transport Layer for Perovskite Solar Cells. *Small* **2018**, *14*, 1801154. [CrossRef] [PubMed]
54. Choi, H.; Kim, K.; Kim, M.; Kim, J.D.; Cho, I.; Kim, I.; Chae, H.; Han, I.; Kim, H.; Seo, J.H.; et al. Solution-Processable Ag-Mediated ZnO Nanowires for Scalable Low-Temperature Fabrication of Flexible Devices. *ACS Appl. Electron. Mater.* **2022**, *4*, 910–916. [CrossRef]
55. Lin, C.; Li, Q.; Guang, H.; An, M. Electrodeposited Zn: A promising alternative to ZnO seed layer for hydrothermal growth of ZnO nanowire array. *Mater. Lett.* **2022**, *314*, 131848. [CrossRef]
56. Hezam, M.; Algarni, A.; Ghaithan, H.; Alzahrani, H.; Alshehri, A.; Alfarhoud, A.; Labis, J.; Alduraibi, M. Hydrothermal growth optimization of vertically aligned ZnO nanowire arrays and their dye-sensitized solar cell performance under air/oxygen environments. *Mater. Res. Express* **2021**, *8*, 105501. [CrossRef]
57. Liu, L.; Cao, G.; Hong, K. Seed Free Growth of Aligned ZnO Nanowire Arrays on AZO Substrate. *J. Wuhan Univ.-Technol.-Mater. Sci. Ed.* **2018**, *33*, 1372–1375. [CrossRef]
58. Zhang, J.; Que, W. Preparation and characterization of solgel Al-doped ZnO thin films and ZnO nanowire arrays grown on Al-doped ZnO seed layer by hydrothermal method. *Sol. Energy Mater. Sol. Cells* **2010**, *94*, 2181–2186. [CrossRef]
59. Karthick, S.; Velumani, S.; Bouclé, J. Experimental and SCAPS simulated formamidinium perovskite solar cells: A comparison of device performance. *Sol. Energy* **2020**, *205*, 349–357. [CrossRef]
60. Lee, S.H.; Han, S.H.; Jung, H.S.; Shin, H.; Lee, J.; Noh, J.H.; Lee, S.; Cho, I.S.; Lee, J.K.; Kim, J.; et al. Al-doped ZnO thin film: A new transparent conducting layer for ZnO nanowire-based dye-sensitized solar cells. *J. Phys. Chem. C* **2010**, *114*, 7185–7189. [CrossRef]
61. Sun, J.; Li, N.; Dong, L.; Niu, X.; Zhao, M.; Xu, Z.; Zhou, H.; Shan, C.; Pan, C. Interfacial-engineering enhanced performance and stability of ZnO nanowire-based perovskite solar cells. *Nanotechnology* **2021**, *32*, 475204. [CrossRef]
62. Peña Alvarez, M.; Mayorga Burrezo, P.; Iwamoto, T.; Qiu, L.; Kertesz, M.; Taravillo, M.; Baonza, V.G.; López Navarrete, J.T.; Yamago, S.; Casado, J.; et al. Flexible, low-temperature, solution processed ZnO-based perovskite solid state solar cells. *Chem. Commun.* **2013**, *49*, 11089–11091. [CrossRef]
63. Weiguang, Y.; Yali, W.; Qiang, Z.; Weimin, S. Effect of growth time on morphology and photovoltaic properties of ZnO nanowire array films. *Rare Met.* **2011**, *30*, 676–680. [CrossRef]
64. Qiu, J.; Li, X.; Zhuge, F.; Gan, X.; Gao, X.; He, W.; Park, S.-J.; Kim, H.-K.; Hwang, Y.-H. Solution-derived 40 μm vertically aligned ZnO nanowire arrays as photoelectrodes in dye-sensitized solar cells. *Nanotechnology* **2010**, *21*, 195602. [CrossRef] [PubMed]
65. Huang, Q.; Fang, L.; Chen, X.; Saleem, M. Effect of polyethyleneimine on the growth of ZnO nanorod arrays and their application in dye-sensitized solar cells. *J. Alloys Compd.* **2011**, *509*, 9456–9459. [CrossRef]
66. Wu, W.; Hu, G.; Cui, S.; Zhou, Y.; Wu, H. Epitaxy of Vertical ZnO Nanorod Arrays on Highly (001) -Oriented ZnO Seed Monolayer by a Hydrothermal Route. *Cryst. Growth Des.* **2008**, *8*, 4014–4020. [CrossRef]
67. Tian, Z.R.; Voigt, J.A.; Liu, J.; Mckenzie, B.; Mcdermott, M.J.; Rodriguez, M.A.; Konishi, H.; Xu, H. Complex and oriented ZnO nanostructures. *Nat. Mater.* **2003**, *2*, 821–826. [CrossRef]
68. Parize, R.; Garnier, J.D.; Appert, E.; Chaix-Pluchery, O.; Consonni, V. Effects of Polyethylenimine and Its Molecular Weight on the Chemical Bath Deposition of ZnO Nanowires. *ACS Omega* **2018**, *3*, 12457–12464. [CrossRef]
69. Sami, A.; Ansari, A.; Idrees, M.D.; Alam, M.M.; Imtiaz, J. Solid-State Solar Cells Based on TiO_2 Nanowires and CH3NH3PbI3 Perovskite. *Coatings* **2021**, *11*, 404. [CrossRef]
70. Liu, K.; Wu, W.; Chen, B.; Chen, X.; Zhang, N. Continuous Growth and Improved PL Property of ZnO Nanoarrays with Assistance of Polyethylenimine. *Nanoscale* **2013**, *5*, 5986–5993. [CrossRef]
71. Parize, R.; Chaix-pluchery, O.; Verrier, C.; Appert, E.; Consonni, V. Effects of Hexamethylenetetramine on the Nucleation and Radial Growth of ZnO Nanowires by Chemical Bath Deposition. *J. Phys. Chem. C* **2016**, *120*, 5242–5250. [CrossRef]
72. Mcpeak, K.M.; Le, T.P.; Britton, N.G.; Nickolov, Z.S.; Elabd, Y.A.; Baxter, J.B. Chemical Bath Deposition of ZnO Nanowires at Near-Neutral pH Conditions without Hexamethylenetetramine (HMTA): Understanding the Role of HMTA in ZnO Nanowire Growth. *Langmuir* **2011**, *27*, 3672–3677. [CrossRef] [PubMed]
73. Mahmood, K.; Sarwar, S.; Mehran, M.T. Current status of electron transport layers in perovskite solar cells: Materials and properties. *RSC Adv.* **2017**, *7*, 17044–17062. [CrossRef]

materials combine the advantages of EDLCs and batteries since capacitive and charge-transfer processes coexist [13–15].

A proper device design development, involving economical and eco-friendly production processes, represents the most promising approach for succeeding in high-performance supercapacitors; in particular, nanomaterial-based electrodes play a crucial role in electrochemical energy storage [10] since the higher surface area can significantly contribute to improving the specific capacitance [16].

In recent years, transition metal oxides have been widely applied for electrochemical capacitors with high power density, taking advantage of their pseudocapacitive behavior. These oxides provide higher specific capacitance values because of their many oxidation states, characteristic of pseudocapacitive materials [17].

Among transition metal oxides, zinc oxide, an inorganic semiconductor relatively abundant in nature, has been extensively studied as an anode material candidate for lithium-ion batteries [18]. Owing to its eco-friendly nature and good electrochemical activity, ZnO also became a promising electrode material for supercapacitors [19–21]. Moreover, ZnO can be easily nanostructured in a multitude of morphologies by employing many different methods [22–26]. Solution routes for ZnO nanostructure synthesis have a multitude of advantages, such as cost containment, a simple laboratory setup, low-temperature processes, and fast kinetics growth [27].

It is worth mentioning that ZnO's properties are morphology dependent, which makes this material even more attractive for several scientific purposes [28,29], including energy storage applications.

Jayachandiran et al. reported ZnO nanoparticle fabrication obtained using a sol-gel approach, following a rGO/ZnO composite fabrication, obtained using ultrasonic-assisted solution-phase synthesis [30]. The composite, coated on nickel foil, showed pseudocapacitive behavior and a specific capacitance value of 312 F g^{-1} at 5 mV s^{-1}. Luo et al. synthesized ZnO tetrapods, prepared using vapor transport methods, coated onto Ni foam [31]. The material showed pseudocapacitive behavior, with a specific capacitance value of 160 F g^{-1} obtained at 1 A g^{-1}. Finally, Guo et al. obtained ZnO nanoparticles using a hydrothermal method, following a graphene–ZnO composite formation using microwave treatment [32]. Again, the material exhibited a pseudocapacitive behavior, with an obtained specific capacitance of 201 F g^{-1} at 1 A g^{-1}. All of these function in a neutral condition by using 1 M Na_2SO_4, where corrosion problems related to water splitting reactions are avoided [33]. Nonetheless, in these papers, the substrate's role was not considered nor evaluated, while it is known that C or Ni compounds act as energy storage materials. To optimize the study of the effective energy storage process in ZnO nanostructures, a proper substrate contribution evaluation is needed. This exercise is also useful in order to compare different ZnO nanostructure data in the literature, which usually appear as deposited on different substrates.

In this paper, a new nanostar-like ZnO nanostructure obtained at a very large yield by means of chemical bath deposition (CBD) is presented. These nanostars are composed of many crystalline wires, highly ordered within a common plane. The as-prepared nanostars were used as supercapacitor active material, and the energy storage mechanism was correlated with structural and morphological features.

2. Materials and Methods

2.1. Synthesis of ZnO Nanostars

ZnO nanostars (NSs) were synthesized by means of chemical bath deposition (CBD). Starting from an aqueous solution of zinc nitrate and hexamethylenetetramine (HMTA) [34], the ZnO NSs production was attained by adding ammonium fluoride to the bath [26]. Three solutions (50 mL each) were separately prepared with deionized water (MilliQ, 18 MΩ cm): (i) 25 mM zinc nitrate hexahydrate ($Zn(NO_3)_2 \cdot 6\,H_2O$, purum p.a., crystallized, ≥99.5%, Sigma Aldrich, Milan, Italy); (ii) 25 mM HMTA ($(CH_2)_6N_4 \geq 99.5\%$, Sigma Aldrich, Milan, Italy); and (iii) 16 mM ammonium fluoride (NH_4F, ≥99.99%, Sigma-Aldrich,

Milan, Italy). These solutions were placed in a bain-marie configuration to reach and maintain the desired temperature of 90 °C. Once the thermal equilibrium was reached, the three solutions were mixed in a large beaker where the 90 °C temperature was still maintained. $Zn(NO_3)_2$ and the HMTA solutions were mixed first, then after few seconds, the NH_4F solution was added. In order to study the growth kinetics, the synthesis solution was sampled at different growth times (0.5, 1, 3, 6, 10, 20, and 30 min). After 30 min, the solution was removed from the bain-marie configuration and left to reach room temperature.

The obtained solution was then washed with deionized water 4 times by means of decantation. The nanostar powders were finally dried in an oven in vacuum at 100 °C for 16 h (hereafter simply called NSs). A part of the product was then annealed in air onto a hot plate at 300 °C for one hour (hereafter called AnnNSs).

2.2. Characterization

The NS surface morphology was analyzed by using a Scanning Electron Microscope (Gemini field emission SEM Carl Zeiss SUPRA 25, Carl Zeiss Microscopy GmbH, Jena, Germany) equipped with an EDAX PV7715/89-ME energy-dispersive X-ray (EDX) spectrometer for elemental characterization. SEM images were analyzed by using ImageJ software to improve the brightness and contrast [35].

The ZnO NS crystal structure was examined by X-ray diffraction (XRD) using a Rigaku Smartlab diffractometer (Rigaku, Tokyo, Japan), equipped with a rotating anode of Cu K_α radiation operating at 45 kV and 200 mA in the grazing incidence mode (0.5°).

Photoluminescence (PL) measurements were performed by pumping with the 325 nm (3.81 eV) line of a He–Cd laser chopped through an acousto-optic modulator at a frequency of 55 Hz. The PL signal was analyzed using a single grating monochromator, detected with a Hamamatsu visible photomultiplier, and recorded with a lock-in amplifier using the acousto-optic modulator frequency as a reference [36].

Electrochemical measurements were carried out at room temperature using a VersaSTAT4 potentiostat (Princeton Applied Research, Oak Ridge, TN, USA), and a three-electrode setup was used with a platinum wire as a counter electrode, a saturated calomel electrode (SCE) as a reference electrode, and the ZnO NS samples on graphene paper (GP) substrates (240 μm thick, Sigma Aldrich, St. Louis, MO, USA) as a working electrode (WE). Solutions of 1 M Na_2SO_4, 1 M NaCl, and 1 M KCl (Sigma Aldrich, St. Louis, MO, USA, ≥85%) were used as supporting electrolytes. Cyclic voltammetry (CV) and galvanostatic charge–discharge (GCD) analyses were conducted in the potential range $0 \div 0.6$ V vs. SCE (Na_2SO_4), $-0.1 \div 0.5$ V (NaCl) and $-0.2 \div 0.4$ V (KCl). Electrochemical impedance spectroscopy (EIS) analyses were conducted at 0.3 V vs. SCE (in an Na_2SO_4 solution with a frequency range of 0.1–10,000 Hz and an amplitude of 10 mV RMS).

3. Results and Discussion

3.1. Material Characterization

The zinc oxide nanostructure morphology can be seen in Figure 1a–c, which show SEM images of NSs dropped on Si substrates after different CBD durations. Beyond rare single wires, most of the precipitate appeared in the form of nanostars, with each one made of six coplanar arms starting from a common center. Such arms are composed of a bunch of parallel wires, and each star draws six equally spaced angles. It is worth noting that the ZnO wurtzite phase has a hexagonal symmetry, which could explain the nanostructures' peculiar shape. As the CBD time increased, the arm length and thickness also increased.

A cross-linked center can be recognized, from which every wire starts with a lateral width of 70 nm, reaching a dimension of 20–30 nm on the tip (Figure S1). Once annealed, the morphology did not appreciably change (Figure S2).

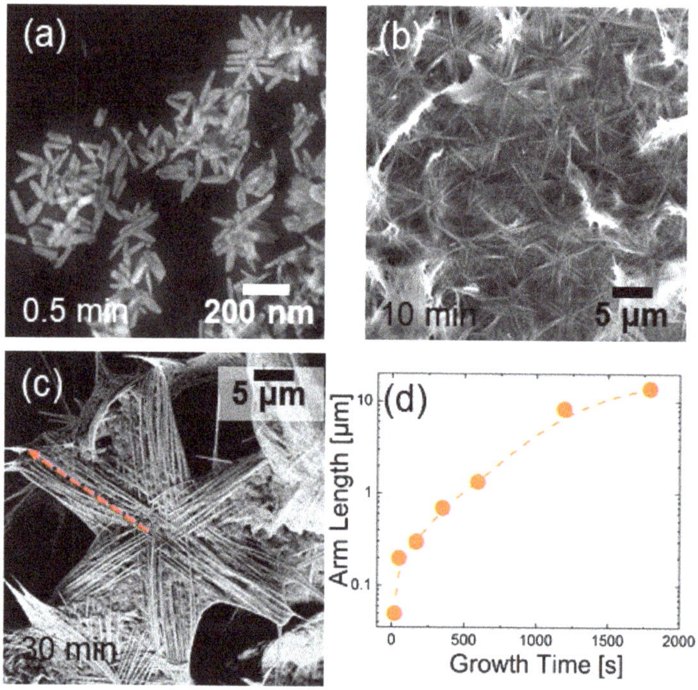

Figure 1. Kinetic study of ZnO NS growth. SEM images of nanostructures sampled after a growth of 0.5 (**a**), 10 (**b**), and 30 min (**c**); distribution of arm length as a function of time (**d**). The dashed red line in (**c**) indicates the arm length of an NS.

Figure 1a–c show the SEM micrographs of NSs grown for 0.5, 10, and 30 min, respectively. After just half a minute, the stars' central part was developed, while the arms clearly began to arrange themselves in preferential directions. After 30 min, the nanostars were very extended; their arms appeared dense with wires as long as 10–12 μm (dashed red line).

Arm length evolution as a function of time is plotted in Figure 1d. There was an initial explosive growth kinetics within 10 min, but the growth rate decreased at longer growth times.

Figure 2a shows the XRD patterns of 10-min-grown NSs before and after the annealing process. The NSs sample already shows many crystallographic peaks, which recall the hexagonal ZnO peaks at $2\theta = 31.77°$, $34.42°$, $36.25°$, $47.54°$, and $56.60°$ corresponding to (100), (002), (101), (102), and (110), respectively (see PDF Card No. 00-036-1451 in Figure 2a). In addition, there is a sharp peak at 20°, more intense than the other crystallographic peaks, due to the ZnOHF presence in the powders (PDF Card No. 74-1816 in Figure 2a) that is also responsible for (100) and (101)'s peculiar peaks splitting into two components [37]. Once annealed, this hydroxy fluoride species almost disappeared, producing pure zinc oxide powders. The 45° peak was related to the carbon tape used as substrate for AnnNSs [38]. The co-existence of the ZnO and ZnOHF phase in the NS powder was also confirmed by EDX analyses, as shown in Figures S3 and S4, revealing the presence of oxygen, zinc, and fluorine in the NSs. The PL spectra were acquired for both NSs and AnnNSs, as shown in Figure 2b. All the emission spectra were composed of a UV peak (2.7–3.5 eV) and a broad visible band (1.8–2.7 eV). Although the profile of the emission spectra is characteristic of ZnO [39], a small contribution of the ZnOHF phase in the near-UV–visible region [40,41] cannot be entirely excluded. At first glance, annealing in air induced a redshift and a visible band reduction. As the visible band is typically associated with defects [39], it can be concluded that thermal annealing reduces these defects. In order to better understand this aspect, the visible PL was analyzed in detail.

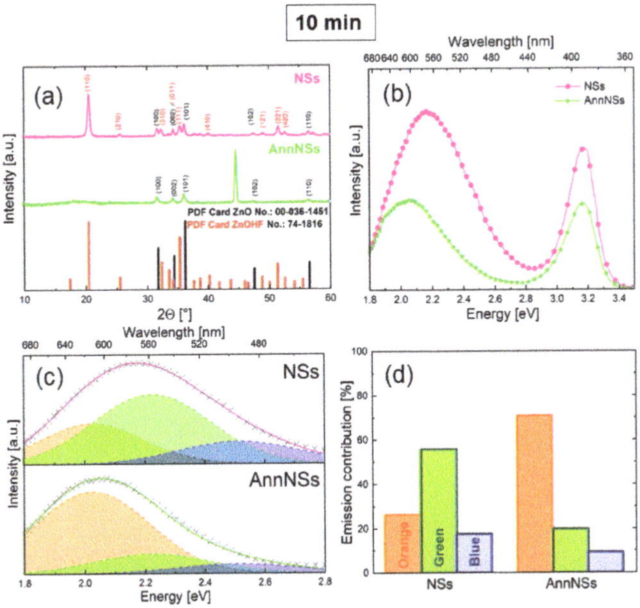

Figure 2. (**a**) XRD pattern of NSs and AnnNSs grown for 10 min; (**b**) room-temperature photoluminescence spectra; (**c**) visible emission band fitting with blue, green, and orange contributions, and (**d**) histogram of fit contributions for both NSs and AnnNSs.

The visible emission band can be described as a convolution of three Gaussian components attributed to the Blue (B, 2.52 eV), Green (G, 2.23 eV), and Orange (O, 2.03 eV) states [39].

In Figure 2c, the visible bands of the NSs and AnnNSs samples were fitted with the above Gaussian components, with FWHM fixed at 0.4 eV. The histograms in Figure 2d show the percent contribution of B, G, and O emissions for the NSs and AnnNSs. The B, G, and O emissions are related to radiation mediated by states into the ZnO bandgap [29,39]. The B emission is related to zinc vacancies (V_{Zn}^{2-}) states; the G emission is related to singly ionized oxygen vacancies (V_{O}^{+}) states at the surface; while the O emission is associated with oxygen vacancies at the ZnO core.

The annealing process led to a strong G band reduction and a great O contribution enhancement, as also confirmed by the clear redshift observed in Figure 2b. It is plausible to think that thermal annealing in air supplies oxygen atoms that recombine O vacancies at the surface, thus significantly reducing the G emission. Such surface defect modifications will have some effects on energy storage performance, as detailed below.

Grown nanostars are a ZnO and ZnOHF mixed phase, which converts to pure ZnO after annealing in air. A large amount of surface defects was present in the as-grown samples; such defects were recombined by air annealing at 300 °C. The ease of synthesis can be a great advantage for scalability; however, some nanostar yield estimations still needed to be conducted. The nanostar synthesis yield was evaluated by assuming the whole process to be limited by a Zn reactant and by considering a 1:1 molar relationship between zinc nitrate and zinc oxide, as expected. The NS yield was obtained by weighting the amount of NSs, grown for 10 min, after the annealing process (where only ZnO was present) in comparison to the utilized Zn nitrate amount. The ratio between ZnO and Zn nitrate moles indicated a nanostar yield as high as 40%. As reported in the literature [42], the yield of solution-derived ZnO nanopowders strongly depends on pH value, with higher yields achieved in highly basic media. NSs are synthesized in solutions with a pH value of 5.7; therefore, a 40% yield may be considered promising for high-volume production.

3.2. Electrochemical Measurements

To evaluate the ZnO NS electrochemical performances, CV and GCD analyses were performed on ZnO NSs-based electrodes, using a three-electrode setup. The WE was obtained by spin coating (300 rpm, 5 min) 40 µL of an NSs or AnnNSs aqueous solution (concentration of 2 mg mL^{-1}) on a 1 cm^2 graphene paper substrate, as shown in Figure 3a. The electrode was then dried on a hot plate at 60 °C in air, obtaining a mass of 0.2 mg, measured with a Mettler Toledo (Columbus, OH, USA) MX5 Microbalance (sensitivity: 0.01 mg). It should be noted that particular care was taken in order to have electrodes with the same mass so as to easily compare the electrochemical performances.

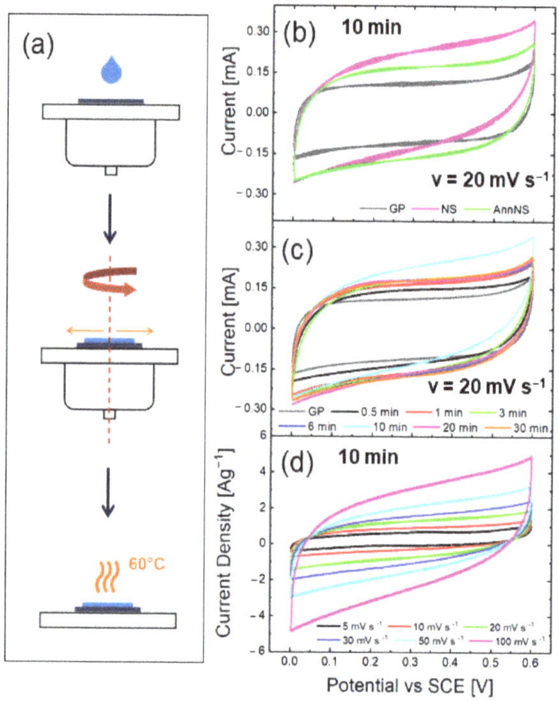

Figure 3. (**a**) Schematic of sample preparation for electrochemical characterization. CV curves in 1 M Na_2SO_4 of GP substrate (grey line); nanostars as prepared (magenta line) and after annealing (green line) at 20 mV/s (**b**); CV curves in 1 M Na_2SO_4 of NSs with 0.5, 1, 3, 6, 10, 20, and 30 min growth times and GP at 20 mV/s (**c**); CV curves in 1 M Na_2SO_4 of as-prepared NS at different scan rates (**d**).

Figure 3b shows the CVs of 10 min ZnO NSs as grown and after annealing, compared with bare graphene paper (GP). As expected, the bare substrate was active in the charge storage process with a CV typical of an electric double-layer capacitor (EDLC) [15]. Such activity of GP in charge storage was taken into account to investigate the ZnO NS net ability in storing charges.

When the GP was covered with as-grown ZnO NSs, a larger CV area was recorded, almost double in size, highlighting a significant pseudocapacitive storage mechanism attributable to ZnO NSs. It is worth noting that the substrate mass was some orders of magnitude greater than the ZnO NSs mass (0.2 mg); thus, a specific capacitance comparison is meaningless; however, for a solid evaluation of ZnO NS energy storage efficiency, the substrate contribution must be evaluated. The annealed ZnO NSs showed a smaller charge storage ability in comparison to the as-grown one. Indeed, the PL analysis evidenced that annealing induced a reduction of oxygen vacancy defects, which are expected to contribute to the storage mechanism through surface adsorption processes [29].

Moreover, the XRD results showed that the as-grown NSs partly contained a ZnOHF phase whose layered structure could facilitate charged ion intercalation [43].

In Figure 3c, CV curves for NSs with different growth times are presented, again compared to the GP curve, at a chosen scan rate of $\nu = 20$ mV s^{-1}. The bare GP substrate always showed a lower charge storage ability in comparison to that of GP covered with ZnO NSs, regardless of the ZnO NS growth times. By comparing GP covered with ZnO NSs at different growth times, what emerged is that, by increasing arm length, the stored charge increased up to 10 min, whereas for longer growth times, the charge storage started to decrease.

Figure 3d shows CV curves acquired for NSs grown for 10 min in 1M Na_2SO_4 at different scan rates, from 5 to 100 mV s^{-1}, which are useful to further deepen the charge storage mechanism.

The stored charge (Q_c, mC) can be determined from the CV curves as follows [44]:

$$Q_c = \frac{\int IdV}{v}$$

where V is the applied potential (V), I is the measured current (mA), and v is the scan rate voltage $\left(V\ s^{-1}\right)$. Consequently, the specific capacitance (C_s, F g^{-1}) can be determined from the CV curves as follows [45]:

$$C_s = \frac{\int IdV}{mv\Delta V}$$

where m is the active NS mass (mg), and ΔV is the potential range (V).

Figure 4a represents the stored charge from the CV as a function of the scan rate for GP and NSs on GP (total). The storage capacity for NSs (net) was extracted as the difference between these two. It should be noted that such a net storage capacity underestimates the ZnO NSs performances, as a non-negligible surface of the GP is covered by ZnO NSs, reducing the effective GP contribution to the charge storage in ZnO NSs-covered samples. The stored charge trend in the GP confirmed the EDLC mechanism due to its weak dependence on the scan rate (black spheres in Figure 4a). Instead, a strong dependence on the scan rate emerged for ZnO NS on GP (closed purples), with a stored charge of 17.2 mC at 5 mV s^{-1}, evidencing that ZnO NSs act as energy storage materials with a different mechanism in comparison to that of GP. In fact, by plotting the net stored charge in ZnO NSs (open magenta spheres), a clear decrease was observed with the scan rate, decreasing the stored charge at very high scan rates almost to zero.

In order to extract an effective ZnO NSs specific capacitance, we took into account the net stored charge. Figure 4b shows these effective C_s for all the growth times. In all the cases, a marked dependence on the scan rate was observed, as for the 10 min growth time discussed above. Indeed, the growth time significantly affected the C_s, with 6 and 10 min grown ZnO NSs exhibiting the largest values.

In order to study the effect of different electrolytes still at neutral pH, Figure 4c shows the effective C_s of the 10 min NSs as a function of the scan rate in 1 M Na_2SO_4 (magenta curve), 1 M KCl (green curve), and 1M NaCl (purple curve). While NaCl and KCl showed very similar results, Na_2SO_4 evidenced a larger specific capacitance at a lower scan rate.

Now, few considerations can be made in order to model the charge storage mechanism in ZnO NSs. By applying a positive voltage to WE, negative charges in the solution were attracted. KCl, NaCl, and Na_2SO_4 dissociate as follows:

$$NaCl \rightarrow Na^+ + Cl^-$$

$$KCl \rightarrow K^+ + Cl^-$$

$$Na_2SO_4 \rightarrow 2Na^+ + SO_4^{2-}$$

Among negative ions, SO_4^{2-} is larger and slower, so it requires a longer time to reach the WE surface (also taking into account the solvation shells) [46]. Hence, at low scan rates, the doubly charged sulfate ion allows a larger storage capacity since its lower diffusion plays a minor role with respect to chloride ions.

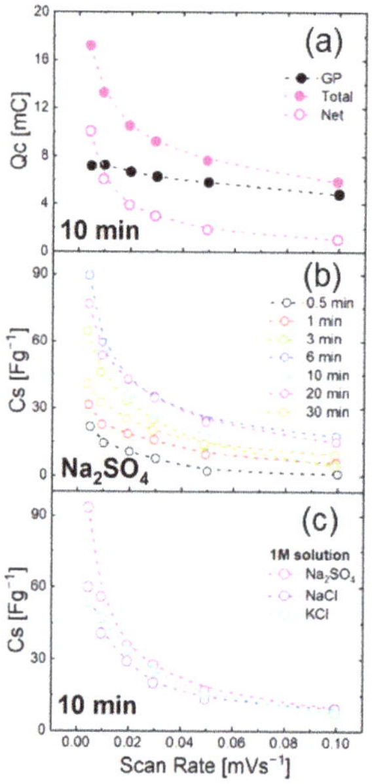

Figure 4. (**a**) Stored charge in GP (black symbols), 10 min ZnO NSs on GP (total, full magenta symbols) and their difference (net) (open magenta symbols); (**b**) specific capacitances extracted from CV for 0.5, 1, 2, 3, 6, 10, 20, and 30 min ZnO NSs in Na_2SO_4; and (**c**) specific capacitances of 10 min ZnO in NaCl, KCl, and Na_2SO_4 (purple, green, and magenta symbols respectively).

For a better understanding of the storage mechanism involved, the Dunn model was applied to the 10 min NSs CV curve in 1 M Na_2SO_4. The total charge stored can be split into three components: a faradic contribution from ion insertion; a faradic contribution from charge-transfer processes with surface atoms, referred to as pseudocapacitance; and a non-faradic contribution from the double-layer effect [47]. At nanoscale dimensions, both double-layer charging and pseudocapacitance can be substantially due to the high surface area.

These effects can be calculated by examining the cyclic voltammetry data at various scan rates according to:

$$I = av^b$$

where the measured current I obeys a power–law relationship with the scan rate v, and a and b are adjustable parameters. Specifically, b is determined from the log$I$ vs. log$v$ plot slope. There are two well-defined conditions since only diffusive or surface-limited phenomena can be present (b = 0.5 or 1, respectively) [47,48].

In Figure S5, the extrapolated b values for the 10 min NSs in 1 M Na_2SO_4 (magenta symbols) at different scan rates as a function of the potential values are presented. As previously explained, there were no peaks in the CV curves, so b never reached a value of 0.5, but swung between 0.75 and 0.9, indicating a dominant capacitive process with some pseudocapacitive contribution. The surface defects previously discussed in the PL analyses may play a fundamental role in WE surface charge-transfer phenomena.

To determine what happens at the electrode–electrolyte interface, EIS analysis was performed at a potential of 0.3 V versus SCE (see Bode plot in Figure S6) and compared with specific capacitance values. Figure 5a shows C_s (υ = 5 mV s^{-1}) as a function of growth time. The C_s values exhibited a clear bell-shaped trend. Figure 5b shows the impedance modulus and phases angle amplitudes at 1 Hz as a function of growth time. Focusing on the impedance modulus (blue), a funnel-shaped trend can be recognized. The 10 min growth point had the lowest impedance value (56 Ω), which is specular with the highest value of C_s (94 F g^{-1}) found for the same sample. The impedance module is inversely related to capacitance [49]. Hence, these two quantities being inversely proportional, lower $|Z|$ values mean higher capacitance values. It is unequivocal that the impedance modulus trend (magenta curve, Figure 5b) is the C_s bell trend's mirror image (red curve, Figure 5a).

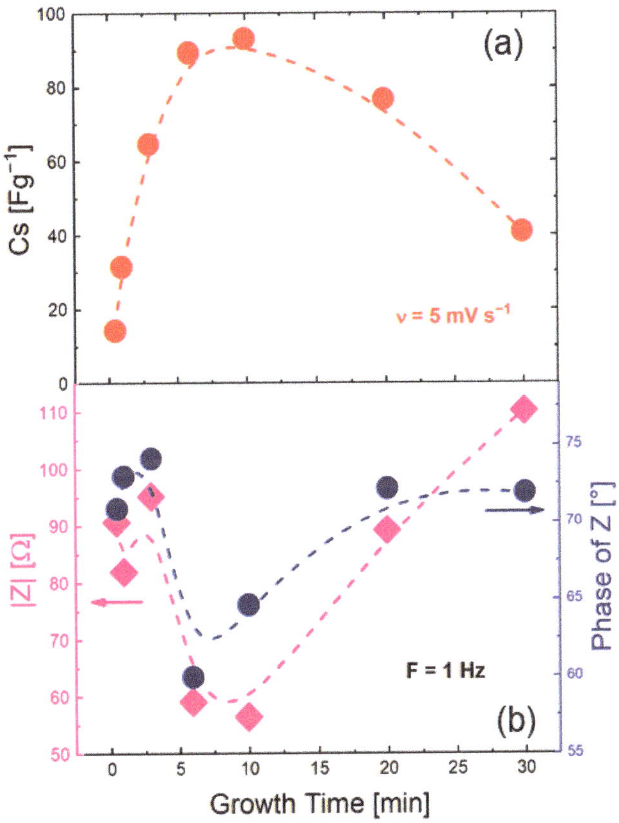

Figure 5. (**a**) C_s trend from CV curves acquired at 5 mV s^{-1} for all growth times analyzed and (**b**) impedance modulus and phase angle amplitude (magenta and blue symbols, respectively) trends (F = 1 Hz) as a function of growth time.

To counterproof the ZnO NSs charge storage process in real conditions, GCD curves were recorded. Figure 6a shows the GCD curves obtained at different current densities (from 0.5 to 10 A g^{-1}) in the same voltage range as that in the CV analyses. As expected, the NSs discharge time decreased with the increase in current density. The charge and discharge curves are symmetric, which indicates high reversibility, high columbic efficiency, and poor energy loss during the charge and discharge process. $C_{s,GCD}$ can be calculated from the GCD as follows [45]:

$$C_{s,GCD} = \frac{It_s}{m\Delta V}$$

where t_s is the discharge time (s), I is the applied current (mA), ΔV is the voltage window (V), and m is the active ZnO NSs mass (mg). Figure 6b shows $C_{s,GCD}$ as a function of the scan rate (CV, blue curve) and as a function of current density (GCD, red curve). The $C_{s,GCD}$ trend matched well with the values of the CV analyses, hence confirming again that all the data were consistent.

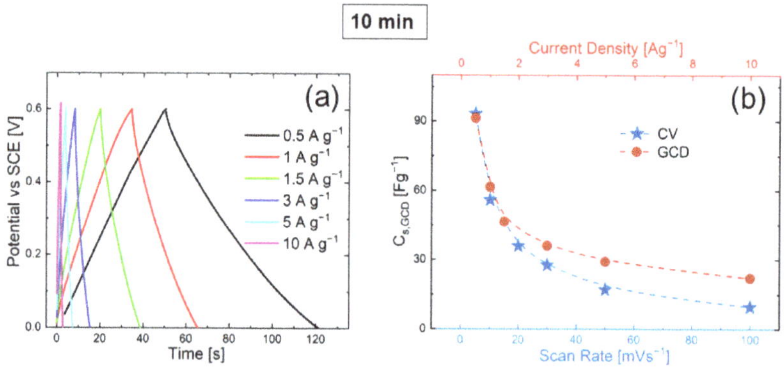

Figure 6. (**a**) GCD curves (0.5 A g^{-1} black, 1 A g^{-1} red, 1.5 A g^{-1} green, 3 A g^{-1} blue, 5 A g^{-1} light blue, and 10 A g^{-1} magenta lines) and (**b**) specific capacitance obtained by GCD (red dashed line and circles) and CV (light blue dashed line and stars) curves, trends of 10 min NS.

4. Conclusions

To conclude, we report novel ZnO nanostructures with a six-point star shape that have a good energy storage ability in neutral pH conditions. A large amount of ZnO/ZnOHF nanostars, highly ordered within the plane, were obtained via the CBD method. By varying the CBD duration, star arm elongation was observed, while annealing up to 300 °C induced a material purification towards the ZnO phase with concomitant surface oxygen vacancy defect removal. Energy storage was tested by covering a graphene paper substrate with ZnO nanostars and performing the electrochemical measurements. The ZnO nanostars showed a marked pseudocapacitive behavior at all the growth times, both as prepared and after annealing. The best performance was reached for the 10 min growth time without thermal treatments because of the presence of surface defects and low electrochemical impedance values. A careful measurement of the ZnO nanostars' net capacitance determined a maximum value of 94 F g^{-1}. The specific capacitance evaluation from the GCD was highly consistent with the values calculated from the CV analyses. All the results suggest that ZnO-based NSs can find promising application in efficient energy storage devices owing to the synergy between the double-layer capacitance and the pseudocapacitive effect.

Supplementary Materials: The following supporting information can be downloaded at: https://www.mdpi.com/article/10.3390/nano12152588/s1, Figure S1. (a) High magnification SEM of an NS center and (b) an NS tip. Figure S2. SEM image of AnnNSs. Figure S3. EDX spectrum of NSs taken in NSs grown for 10 min. Figure S4. SEM micrograph of (a) single grown NSs and (b–d) related SEM-EDX mapping for O, F, and Zn elements, respectively. EDX maps were acquired setting the operating voltage as 20 kV and a working distance of 8.5 mm. Figure S5. Extrapolated b values (magenta circles) for 10 min NSs in 1 M Na_2SO_4 at different scan rates as a function of potential values. Figure S6. Bode plot from the EIS analyses acquired at 0.3 V: (a) impedance modulus and (b) phase angle amplitudes for all growth times analyzed. Data for the GP substrate are also reported

(GP grey, 0.5 min black, 1 min red, 3 min green, 6 min blue, 10 min light blue, 20 min magenta, and 30 min Bordeaux lines and circles).

Author Contributions: Conceptualization, G.M.D.M., S.M. and E.B.; methodology, G.M.D.M., G.M., G.F., S.M., E.B. and V.S.; investigation, G.M.D.M., G.M., G.F., S.M., E.B. and V.S.; writing—original draft preparation, G.M.D.M.; writing—review and editing, G.M., G.F., S.M., E.B. and V.S.; supervision, S.M. and V.S. All authors have read and agreed to the published version of the manuscript.

Funding: This work was supported by programma ricerca di Ateneo UNICT 2020-22 linea 2 PIA.CE.RI "NaTI4Smart Sviluppo di NAnomateriali e Tecnologie Innovative per Smart detection", and by PRIN 2017 "CLEAN-Valorizing Sustainable Plastics through a CLEver use of NANoparticles" 20174FSRZS_003.

Institutional Review Board Statement: Not applicable.

Informed Consent Statement: Not applicable.

Data Availability Statement: The data presented in this study are available on request from the corresponding author.

Acknowledgments: The authors wish to thank G. Malandrino (University of Catania, Italy) for the useful discussions and experimental XRD contribution; the Bio-Nanotech Research and Innovation Tower (BRIT) laboratory of the University of Catania (grant no. PONa3_00136 financed by the MIUR) for the Smartlab diffractometer facility; G. Pantè, C. Percolla, and S. Tatì (CNR-IMM Catania, Italy) for the technical support; and L. Bruno for the useful discussion and scientific contribution.

Conflicts of Interest: The authors declare no conflict of interest.

References

1. IEA. Electricity. 2020. Available online: www.iea.org (accessed on 1 June 2022).
2. Abas, N.; Kalair, A.; Khan, N. Review of Fossil Fuels and Future Energy Technologies. *Futures* **2015**, *69*, 31–49. [CrossRef]
3. Electricity-Market-Report-January-2022. Available online: www.iea.org (accessed on 1 June 2022).
4. Schmidt, O.; Hawkes, A.; Gambhir, A.; Staffell, I. The Future Cost of Electrical Energy Storage Based on Experience Rates. *Nat. Energy* **2017**, *2*, 17110. [CrossRef]
5. Höök, M.; Tang, X. Depletion of Fossil Fuels and Anthropogenic Climate Change—A Review. *Energy Policy* **2013**, *52*, 797–809. [CrossRef]
6. Etacheri, V.; Marom, R.; Elazari, R.; Salitra, G.; Aurbach, D. Challenges in the Development of Advanced Li-Ion Batteries: A Review. *Energy Environ. Sci.* **2011**, *4*, 3243–3262. [CrossRef]
7. Kim, T.; Song, W.; Son, D.Y.; Ono, L.K.; Qi, Y. Lithium-Ion Batteries: Outlook on Present, Future, and Hybridized Technologies. *J. Mater. Chem. A* **2019**, *7*, 2942–2964. [CrossRef]
8. Vangari, M.; Pryor, T.; Jiang, L. Supercapacitors: Review of Materials and Fabrication Methods. *J. Energy Eng.* **2013**, *139*, 72–79. [CrossRef]
9. Sharma, P.; Bhatti, T.S. A Review on Electrochemical Double-Layer Capacitors. *Energy Convers. Manag.* **2010**, *51*, 2901–2912. [CrossRef]
10. Banerjee, S.; De, B.; Sinha, P.; Cherusseri, J.; Kar, K.K. *Applications of Supercapacitors*; Springer: Cham, Switzerland, 2020; Volume 300. [CrossRef]
11. Kim, B.K.; Sy, S.; Yu, A.; Zhang, J. Electrochemical Supercapacitors for Energy Storage and Conversion. In *Handbook of Clean Energy Systems*; Yan, J., Ed.; John Wiley & Sons: Hoboken, NJ, USA, 2015. [CrossRef]
12. Najib, S.; Erdem, E. Current Progress Achieved in Novel Materials for Supercapacitor Electrodes: Mini Review. *Nanoscale Adv.* **2019**, *1*, 2817–2827. [CrossRef]
13. Fleischmann, S.; Mitchell, J.B.; Wang, R.; Zhan, C.; Jiang, D.E.; Presser, V.; Augustyn, V. Pseudocapacitance: From Fundamental Understanding to High Power Energy Storage Materials. *Chem. Rev.* **2020**, *120*, 6738–6782. [CrossRef]
14. Dunn, B.; Kamath, H.; Tarascon, J.M. Electrical Energy Storage for the Grid: A Battery of Choices. *Science* **2011**, *334*, 928–935. [CrossRef]
15. Jiang, Y.; Liu, J. Definitions of Pseudocapacitive Materials: A Brief Review. *Energy Environ. Mater.* **2019**, *2*, 30–37. [CrossRef]
16. Yu, Z.; Tetard, L.; Zhai, L.; Thomas, J. Supercapacitor Electrode Materials: Nanostructures from 0 to 3 Dimensions. *Energy Environ. Sci.* **2015**, *8*, 702–730. [CrossRef]
17. Yi, C.-Q.; Zou, J.-P.; Yang, H.-Z.; Leng, X. Recent Advances in Pseudocapacitor Electrode Materials: Transition Metal Oxides and Nitrides. *Trans. Nonferrous Met. Soc. China* **2018**, *28*, 1980–2001. (In English) [CrossRef]
18. Wang, L.; Zhang, G.; Liu, Q.; Duan, H. Recent Progress in Zn-Based Anodes for Advanced Lithium Ion Batteries. *Mater. Chem. Front.* **2018**, *2*, 1414–1435. [CrossRef]

19. Ridhuan, N.S.; Abdul Razak, K.; Lockman, Z.; Abdul Aziz, A. Structural and Morphology of ZnO Nanorods Synthesized Using ZnO Seeded Growth Hydrothermal Method and Its Properties as UV Sensing. *PLoS ONE* **2012**, *7*, e50405. [CrossRef]
20. Özgür, Ü.; Alivov, Y.I.; Liu, C.; Teke, A.; Reshchikov, M.A.; Doğan, S.; Avrutin, V.; Cho, S.-J.; Morkc, H. A Comprehensive Review of ZnO Materials and Devices. *J. Appl. Phys.* **2005**, *98*, 041301. [CrossRef]
21. Klingshirn, C. ZnO: From Basics towards Applications. *Phys. Status Solidi Basic Res.* **2007**, *244*, 3027–3073. [CrossRef]
22. Kong, X.Y.; Wang, Z.L. Spontaneous Polarization-Induced Nanohelixes, Nanosprings, and Nanorings of Piezoelectric Nanobelts. *Nano Lett.* **2003**, *3*, 1625–1631. [CrossRef]
23. Chang, P.C.; Fan, Z.; Wang, D.; Tseng, W.Y.; Chiou, W.A.; Hong, J.; Lu, J.G. ZnO Nanowires Synthesized by Vapor Trapping CVD Method. *Chem. Mater.* **2004**, *16*, 5133–5137. [CrossRef]
24. Li, Y.; Cheng, G.S.; Zhang, L.D. Fabrication of Highly Ordered ZnO Nanowire Arrays in Anodic Alumina Membranes. *J. Mater. Res.* **2000**, *15*, 2305–2308. [CrossRef]
25. Kamalasanan, M.N.; Chandra, S. Sol-Gel Synthesis of ZnO Thin Films. *Thin Solid Films* **1996**, *288*, 112–115. [CrossRef]
26. Strano, V.; Greco, M.G.; Ciliberto, E.; Mirabella, S. Zno Microflowers Grown by Chemical Bath Deposition: A Low-Cost Approach for Massive Production of Functional Nanostructures. *Chemosensors* **2019**, *7*, 62. [CrossRef]
27. Hodes, G. Semiconductor and Ceramic Nanoparticle Films Deposited by Chemical Bath Deposition. *Phys. Chem. Chem. Phys.* **2007**, *9*, 2181–2196. [CrossRef] [PubMed]
28. Liao, F.; Huang, Y.; Ge, J.; Zheng, W.; Tedsree, K.; Collier, P.; Hong, X.; Tsang, S.C. Morphology-Dependent Interactions of ZnO with Cu Nanoparticles at the Materials' Interface in Selective Hydrogenation of CO_2 to CH_3OH. *Angew. Chem. Int. Ed.* **2011**, *50*, 2162–2165. [CrossRef] [PubMed]
29. Najib, S.; Bakan, F.; Abdullayeva, N.; Bahariqushchi, R.; Kasap, S.; Franzò, G.; Sankir, M.; Demirci Sankir, N.; Mirabella, S.; Erdem, E. Tailoring Morphology to Control Defect Structures in ZnO Electrodes for High-Performance Supercapacitor Devices. *Nanoscale* **2020**, *12*, 16162–16172. [CrossRef] [PubMed]
30. Jayachandiran, J.; Yesuraj, J.; Arivanandhan, M.; Raja, A.; Suthanthiraraj, S.A.; Jayavel, R.; Nedumaran, D. Synthesis and Electrochemical Studies of RGO/ZnO Nanocomposite for Supercapacitor Application. *J. Inorg. Organomet. Polym. Mater.* **2018**, *28*, 2046–2055. [CrossRef]
31. Luo, Q.; Xu, P.; Qiu, Y.; Cheng, Z.; Chang, X.; Fan, H. Synthesis of ZnO Tetrapods for High-Performance Supercapacitor Applications. *Mater. Lett.* **2017**, *198*, 192–195. [CrossRef]
32. Guo, Y.; Chang, B.; Wen, T.; Zhao, C.; Yin, H.; Zhou, Y.; Wang, Y.; Yang, B.; Zhang, S. One-Pot Synthesis of Graphene/Zinc Oxide by Microwave Irradiation with Enhanced Supercapacitor Performance. *RSC Adv.* **2016**, *6*, 19394–19403. [CrossRef]
33. Kim, J.H.; Lee, Y.S.; Sharma, A.K.; Liu, C.G. Polypyrrole/Carbon Composite Electrode for High-Power Electrochemical Capacitors. *Electrochim. Acta* **2006**, *52*, 1727–1732. [CrossRef]
34. Strano, V.; Smecca, E.; Depauw, V.; Trompoukis, C.; Alberti, A.; Reitano, R.; Crupi, I.; Gordon, I.; Mirabella, S. Low-Cost High-Haze Films Based on ZnO Nanorods for Light Scattering in Thin c-Si Solar Cells. *Appl. Phys. Lett.* **2015**, *106*, 3–9. [CrossRef]
35. Schneider, C.A.; Rasband, W.S.; Eliceiri, K.W. NIH Image to ImageJ: 25 Years of Image Analysis. *Nat. Methods* **2012**, *9*, 671–675. [CrossRef] [PubMed]
36. Bruno, L.; Strano, V.; Scuderi, M.; Franzò, G.; Priolo, F.; Mirabella, S. Localized Energy Band Bending in ZnO Nanorods Decorated with Au Nanoparticles. *Nanomaterials* **2021**, *11*, 2718. [CrossRef] [PubMed]
37. Yang, H.; Teng, F.; Gu, W.; Liu, Z.; Zhao, Y.; Zhang, A.; Liu, Z.; Teng, Y. A Simple Post-Synthesis Conversion Approach to Zn(OH)F and the Effects of Fluorine and Hydroxyl on the Photodegradation Properties of Dye Wastewater. *J. Hazard. Mater.* **2017**, *333*, 250–258. [CrossRef] [PubMed]
38. Seehra, M.S.; Geddam, U.K.; Schwegler-Berry, D.; Stefaniak, A.B. Detection and Quantification of 2H and 3R Phases in Commercial Graphene-Based Materials. *Carbon* **2015**, *95*, 818–823. [CrossRef] [PubMed]
39. Barbagiovanni, E.G.; Reitano, R.; Franzò, G.; Strano, V.; Terrasi, A.; Mirabella, S. Radiative Mechanism and Surface Modification of Four Visible Deep Level Defect States in ZnO Nanorods. *Nanoscale* **2016**, *8*, 995–1006. [CrossRef] [PubMed]
40. Dai, M.; Xu, F.; Lu, Y.; Lui, Y.; Xie, Y. Synthesis of Submicron Rhombic ZnO Rods via ZnOHF Intermediate using Electrodeposition Route. *Appl. Surf. Sci.* **2011**, *257*, 3586–3591. [CrossRef]
41. Tian, H.; Li, Y.; Zhang, J.; Ma, Y.; Wang, Y.; Wang, Y.; Li, Y. High Pressure Induced Phase Transformation through Continuous Topology Evolution in Zinc Hydroxyfluoride Synthesized via a Hydrothermal Strategy. *J. Alloys Compd.* **2017**, *726*, 132–138. [CrossRef]
42. Komarneri, S.; Bruno, M.; Mariani, E. Synthesis of ZnO with and without Microwaves. *Mater. Res. Bull.* **2000**, *35*, 1843–1847. [CrossRef]
43. Guo, Y.; Liu, N.; Sun, T.; Cui, H.; Wang, J.; Wang, M.; Wang, M.; Tang, Y. Rational Structural Design of ZnOHF Nanotube-Assembled Microsphere Adsorbents for High-Efficient Pb^{2+} removal. *CrystEngComm* **2020**, *22*, 7543–7548. [CrossRef]
44. Zubieta, L.; Bonert, R. Characterization of Double-Layer Capacitors for Power Electronics Applications. *IEEE Trans. Ind. Appl.* **2000**, *36*, 199–205. [CrossRef]
45. González, A.; Goikolea, E.; Barrena, J.A.; Mysyk, R. Review on Supercapacitors: Technologies and Materials. *Renew. Sustain. Energy Rev.* **2016**, *58*, 1189–1206. [CrossRef]
46. Tielrooij, K.J.; Garcia-Araez, N.; Bonn, M.; Bakker, H.J. Cooperativity in Ion Hydration. *Science* **2010**, *328*, 1006–1009. [CrossRef]

47. Wang, J.; Polleux, J.; Lim, J.; Dunn, B. Pseudocapacitive Contributions to Electrochemical Energy Storage in TiO_2 (Anatase) Nanoparticles. *J. Phys. Chem. C* **2007**, *111*, 14925–14931. [CrossRef]
48. Lindstrom, H.; So, S.; Solbrand, A.; Hjelm, J.; Hagfeldt, A.; Lindquist, S. Li Ion Insertion in TiO_2 (Anatase), Voltammetry on Nanoporous Films.Pdf. *J. Phys. Chem. B* **1997**, *2*, 7717–7722. [CrossRef]
49. Bard, A.J.; Faulkner, L.R.; White, H. *Electrochemical Methods: Fundamentals and Applications*; John Wiley: New York, NY, USA, 2004; Volume 677, pp. 368–375.

nanomaterials

MDPI

Article

Laser-Induced Au Catalyst Generation for Tailored ZnO Nanostructure Growth

Sebastien Durbach, Lars Schniedermeyer, Anna Marx and Norbert Hampp *

Department of Chemistry, University of Marburg, Hans-Meerwein Str. 4, 35032 Marburg, Germany
* Correspondence: hampp@uni-marburg.de

Abstract: ZnO nanostructures, semiconductors with attractive optical properties, are typically grown by thermal chemical vapor deposition for optimal growth control. Their growth is well investigated, but commonly results in the entire substrate being covered with identical ZnO nanostructures. At best a limited, binary growth control is achieved with masks or lithographic processes. We demonstrate nanosecond laser-induced Au catalyst generation on Si(100) wafers, resulting in controlled ZnO nanostructure growth. Scanning electron and atomic force microscopy measurements reveal the laser pulse's influence on the substrate's and catalyst's properties, e.g., nanoparticle size and distribution. The laser-induced formation of a thin SiO_2-layer on the catalysts plays a key role in the subsequent ZnO growth mechanism. By tuning the irradiation parameters, the width, density, and morphology of ZnO nanostructures, i.e., nanorods, nanowires, and nanobelts, were controlled. Our method allows for maskless ZnO nanostructure designs locally controlled on Si-wafers.

Keywords: ZnO-nanostructures; gradual ZnO growth manipulation; spatial-selective ZnO growth; laser-induced catalyst generation; thermal chemical vapor deposition

Citation: Durbach, S.; Schniedermeyer, L.; Marx, A.; Hampp, N. Laser-Induced Au Catalyst Generation for Tailored ZnO Nanostructure Growth. *Nanomaterials* **2023**, *13*, 1258. https://doi.org/10.3390/nano13071258

Academic Editor: Vincent Consonni

Received: 7 March 2023
Revised: 30 March 2023
Accepted: 31 March 2023
Published: 2 April 2023

1. Introduction

Regarding the ever-growing demand for new functional materials, semiconductor structures in the micro- and nanoscale remain of high interest in the research community. Due to their physical properties e.g., wide band gap [1] (~3.4 eV) and large exciton binding energy [2] of 60 meV, ZnO nanostructures are promising candidates for future electronic [3] or optoelectronic [4] devices as in UV-lasers [5], solar cells [6] or sensoric [7,8] applications. The generation of diverse ZnO nano-morphologies may be achieved by both wet [9,10] and dry [11–13] methods. In general dry approaches, e.g., thermal chemical vapor deposition (t-CVD), are more abundant and controllable. For ZnO nanostructure growth the vapor-solid- (VS) and the vapor-liquid-solid- (VLS) reactions are of relevance [11]. Control of which mechanism is dominating the growth process and hence guides the morphology expression is commonly controllable by adjusting the process variables, e.g., temperature [8,14] and reaction gas composition [15,16]. Most commonly a noble metal layer, e.g., gold, is applied before the growth process. Heating the thin film leads to thermal dewetting and nanoparticle formation. Upon introducing a semiconductor vapor source, e.g., through a carbothermal reaction [17], the nanoparticles act as catalysts (VLS) or nucleation points (VS) inducing the ZnO nanostructure growth. Spatial-selective control of the semiconductor growth is essential for the design of functional devices. The catalyst may be patterned by using a mask [18], by direct laser writing [19], or by lithographic methods [20], leading to binary growth regulation. Nanostructure growth is either fully suppressed or enabled.

In this work, we first analyze the ZnO morphology occurrence in dependence on underlying growth parameters and thus mechanisms. This allows the differentiation between ZnO modifications mediated by the growth parameters and the pulsed laser-induced dewetting (PLiD) of the catalyst layer (Au thin film). Particle analysis was conducted by electron microscopy images to investigate the type, size, and density of laser-generated catalysts. The growth of ZnO nanostructures can consequently be adjusted by the spatial-selective laser-induced catalyst generation, leading to the manifestation of various ZnO morphologies and sizes on a single substrate. Finally, the laser-induced formation of a passivating SiO_2 layer and its effect on the ZnO growth was investigated.

2. Materials and Methods

The silicon samples used in this study were mechano-chemically polished boron-doped (100) (Siegert Wafer, Aachen, Germany) single crystals with a native oxide layer of ca. 1.5 nm. The wafer fragments (ca. 1 × 1 cm) were coated with gold (0.6–5.4 nm) using a 108auto vacuum coater system (Cressington, Watford, UK). The thickness of thin films was measured with single-wavelength (633 nm) ellipsometry. ZnO growth was performed in a horizontal tube furnace inside of a quartz glass tube with an inner diameter of 40 mm and a length of 650 mm. The source material consisted of a mixture of ZnO (abcr, 99.99%) and graphite powder (99.9%) with a weight ratio of 1:1. The mixture (~1 g) was placed in an alumina boat at the center of the tube (Figure 1, $x = 0$ cm). The wafer fragments were placed on a quartz glass slab at a distance of 7 cm, 8 cm, and 9.5 cm downstream and at 2 cm upstream (edge of alumina boat). The sample placement and furnace setup are depicted in Figure 1a. Figure 1b shows the temperature gradient inside the tube furnace. The sample placements are mapped to their respective temperatures with colored indicators. The setup's heating curve is displayed in the supporting information (SI, Figure S1). The source was heated to a temperature of 1020 °C, initiating the generation of Zn-vapor according to the carbothermal reaction (1) and/or (2):

$$ZnO_{(s)} + C_{(s)} \rightarrow Zn_{(g)} + CO_{(g)} \quad (1)$$

$$ZnO_{(s)} + CO_{(g)} \rightarrow Zn_{(g)} + CO_{2(g)} \quad (2)$$

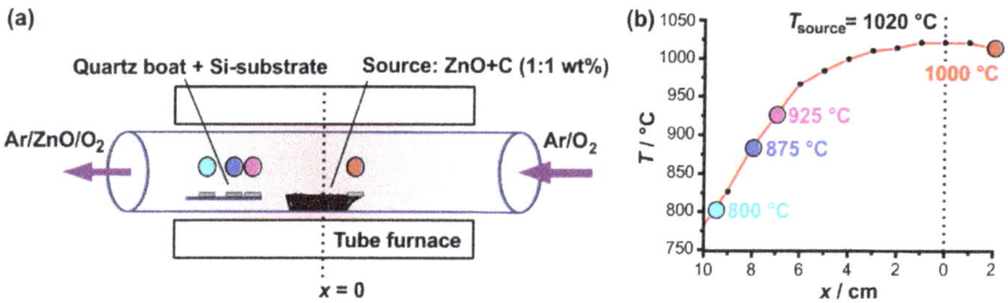

Figure 1. Schematic representation of the tube furnace system (**a**) and the corresponding temperature profile during the ZnO growth (**b**). The sample positions are mapped to their respective process temperatures by the colored indicators. The source is located at $x = 0$ cm and consists of a 1:1 mixture (weight) of ZnO and graphite powder.

Before each growth process, the tube was evacuated to a pressure of 2 mbar for several minutes. During the growth process, the pressure was kept at 10–15 mbar. As carrier gas pure argon (Nippon gases, Madrid, Spain, 99.999%, 1–1.5 bar) was optionally mixed with a mixture of argon and oxygen (Sanarc, 8% O_2). All experiments were conducted with a steady Ar (+O_2) gas flow of 30 sccm. Note that minimal leakage in the system has to be assumed, allowing the entry of O_2 in the ppm range, needed to initiate ZnO formation [16]. The quartz glass tube was cleaned with concentrated hydrochloric acid (HCl, 37%) before every growth process to ensure uniform growth parameters. This prevents a shift in the ratio of VS to VLS growth mediated by a variation of the carbon concentration [11].

Laser irradiation was performed with a frequency-doubled Nd:YVO_4 diode-pumped solid-state (DPSS) laser (explorer XP 532-5, Newport Corporation, Irvine, CA, USA) emitting ≈ 8 ns pulses (FWHM) of $\lambda_0 = 532$ nm wavelength in TEM_{00} ($M^2 < 1.1$) mode. The nanosecond-laser pulses were focused to a spot radius of 50 μm ($1/e^2$) at normal incidence to the substrate surface. The spot size was measured by fitting data from the knife-edge method at multiple z-heights. The beam was scanned over the substrate by a galvanometer scan head (SCANgine 14-532, Scanlab, Puchheim, Germany). More details on the laser setup used for irradiation are found in the SI (Figure S2). The laser power was measured with an LM-80V detector head (FieldMax II, Coherent, Santa Clara, CA, USA). The polarization state of the laser was measured with a polarimeter (Thorlabs TXP 5004 with the PAN5710VIS scan head). Scanning electron microscopy (SEM) was performed on a CrossBeam-Workstation (Gemini 2, Carl Zeiss, Oberkochen, Germany) with a silicon-drift EDX-detector (Ultim®Max. Oxford Instruments, Abingdon, UK). Cross-sections of the samples were achieved by splitting the substrates after ZnO growth. Atomic force microscopy (AFM, Nanoscope IV, Bruker, Billerica, MA, USA) was conducted in contact mode using sharp nitride layer (SNL) tips (f = 56 kHz, k = 0.24 N/m). Gwyddion, an open-source software, was used for the analysis of the particle's size and density, as well as for AFM image processing.

3. Results and Discussion

3.1. ZnO Growth Mechanism and Morphologies

As the ZnO-growth process is highly dependent on the complex combination of process parameters, e.g., system geometry (tube diameter) [21], temperature (T) [14], gas composition [15], and flow [15,22], pressure [21], type of substrate [23] and catalyst [12], the morphology expressions depend on the growth-setup. To differentiate between effects concerning the ZnO growth induced by the process parameters and those from laser irradiation of the catalytic gold thin film, ZnO-nanostructures were generated on unirradiated Au thin films with different layer heights (0.6–5.4 nm). Details about the thermal chemical vapor deposition (t-CVD) process are explained in detail in the Section 2. Figure 2a shows ZnO-nanostructures generated in a pure Ar–atmosphere (p = 12 mbar) at different temperatures T (hence different distances to the Zn-source, growth time t = 10 min). ZnO growth in pure Ar-atmosphere results in ZnO-nanorods (NRs) for all examined temperatures and gold layer heights. Analogue SEM micrographs for T = 1000 °C (upstream) are illustrated in the SI (Figure S3). The width and length of the NRs are dependent on the growth temperature. Lower temperatures lead to thicker, but shorter, NRs. Upon introducing 1% O_2 to the carrier gas, a mixture of different morphologies consisting of thin nanowires (NWs) and narrow nanobelts (NBs) are obtained at T = 800 °C (Figure 2b). The colors in Figure 2 match the temperatures and sample positions illustrated in Figure 1.

The vapor-liquid-solid (VLS) growth process is characterized by gold nanoparticles [12] forming liquid Au-Zn catalyst droplets, resulting in tip growth of nanostructures from precipitated ZnO [11]. For the liquid-solid (VS) mechanism, the Zn directly reacts to ZnO at the surface, resulting in ZnO bases (root growth). Although the presence of a noble metal catalyst is not inherently necessary, the metallic nanoparticles act as energetically favored nucleation sites for the Zn-vapor [24]. Both growth mechanisms are explained in detail in the SI (Section S2). In Figure 3 the three main types of ZnO-nanostructure morphologies grown in our setup, i.e., nanorods (NRs, 1D), nanowires (NWs, 1D) and nanobelts (NBs, 2D), are exemplarily displayed. ZnO-NRs (Figure 3a–c) grow from a ZnO base layer formed by the VS mechanism, as schematically illustrated in Figure 3g. As a consequence, the NRs tip consists of {0001}-facets (Figure 3c), typical for the preferential c-plane growth ([0001]-direction) of the strongly anisotropic wurtzite crystal structure of ZnO [11]. NWs (Figure 3d–f) grow by the VLS mechanism, thus a gold particle is observed at their tip, as highlighted in Figure 3f and schematically represented in Figure 3h. NBs (Figure 3d–f) represent a two-dimensional ZnO growth, induced by excess Zn atoms, not dissolved in gold nanoparticles, leading to a growth of the sides of NWs. This growth is often initiated at the junction of two NWs as illustrated in Figure 3i [14].

periodic arrays of relatively large, periodically arranged Au nanoparticles are formed at higher energy doses [27]. 2D-LIPSS nanoparticles are fully covered with a thin SiO_2 layer and partly sunken in the silicon. Before the formation of the highly arranged 2D-LIPSS, their predecessor structures are formed. These Au nanostructures are not periodically arranged but are covered with a thin SiO_2 passivation layer and are partly sunken into the silicon wafer too. At high laser energy doses, laser ablation of Au leads to the formation of a plasma. During the thermal relaxation following laser irradiation, the condensation of the aforementioned Au plasma leads to the formation of small nanoparticles. Unlike most other gold particles in this laser energy regime, these nanoparticles are not passivated by a SiO_2 layer and are located on the sample's surface. For all mentioned effects, the Au catalyst generation is schematically represented in Figure 4. PLiD, 2D-LIPSS generation, and the formation of small Au nanoparticles by condensation are extensively described in Durbach et al. [27]. The referenced work additionally illustrates the embedment of the various nanoparticles types in the substrate. Further explanations on the formation and properties of the passivating SiO_2 layer are presented in Section 3.4.

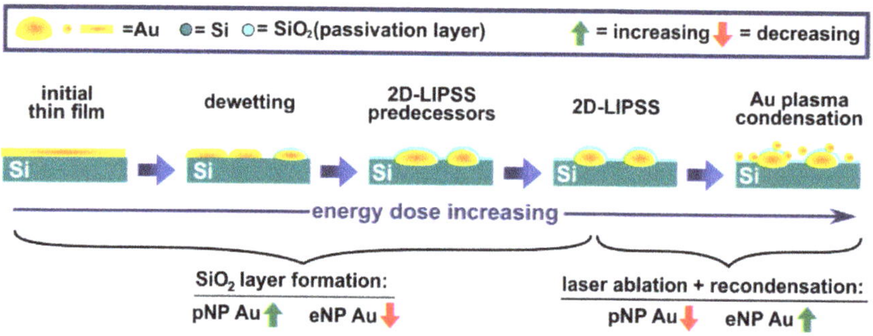

Figure 4. Schematic representation of the most important effects responsible for Au catalyst generation upon nanosecond laser irradiation. Displayed in order of their occurrence upon increased energy dose. Note that several effects may occur at the same time. Au nanoparticles (yellow) covered with a SiO_2-layer (cyan) are designated to be passivated hence called passivated nanoparticles (pNPs), in contrast to exposed nanoparticles (eNPs) as known from traditional thermal dewetting.

As the passivation of the Au nanostructures has significant implications on the ZnO growth, we classified the laser-generated nanoparticles into two categories. The first category consists of passivated nanoparticles, covered with a thin SiO_2 layer and partly sunken in the Si-wafer, e.g., 2D-LIPSS and their predecessor structures. They are further referred to as pNPs (passivated nanoparticles). The second category of nanoparticles consists of exposed nanoparticles (eNPs) and includes particles generated by PLiD, as well as particles formed by Au plasma condensation. The sum of all presented laser-induced effects results in an initially increasing proportion of passivated Au. Due to laser ablation and recondensation of Au nanoparticles, this trend is reversed for the highest laser energy doses, e.g., effective pulse number N, as illustrated in Figure 4.

The size and area density of both particle types were extracted from SEM images obtained by different detectors, as described in detail in the SI (Figure S5). To ensure comparability for uneven Au nanostructure shapes, the sizes are indicated by an equivalent to the radius of perfectly round particles possessing the same area. In Figure 5 the equivalent radius *r(eq.)* (black) and the area particle density ρ (particle count per area, red) are shown for particles generated by areal-irradiation using different effective pulse numbers N, a line-to-line distance (hatch) of h = 3.3 μm, a pulse fluence of φ_p = 127 mJ/cm^2 and a pulse frequency of f = 100 kHz. As ZnO-growth takes place at high temperatures, we examined the nanoparticle's ripening and dewetting mediated by the temperatures present during the heating phase of the ZnO-growth process. A blank growth cycle, analog to the regular ZnO growth process, but using only trace amounts of ZnO/C, was conducted. As such the process conditions relevant to dewetting and ripening of the nanoparticles were controlled. The ZnO nanostructure growth was consequently inhibited, while still allowing preliminary ZnO-nucleation and Au-Zn alloying. This method enables a detailed study of the active catalyst particles and the temperature-mediated changes to *r(eq.)* and the area density ρ are observed by comparing Figure 5a/b to Figure 5c/d. The dashed lines in Figure 5d represent the values concerning Au particles generated only by heat-induced dewetting (unirradiated: $N = 0$).

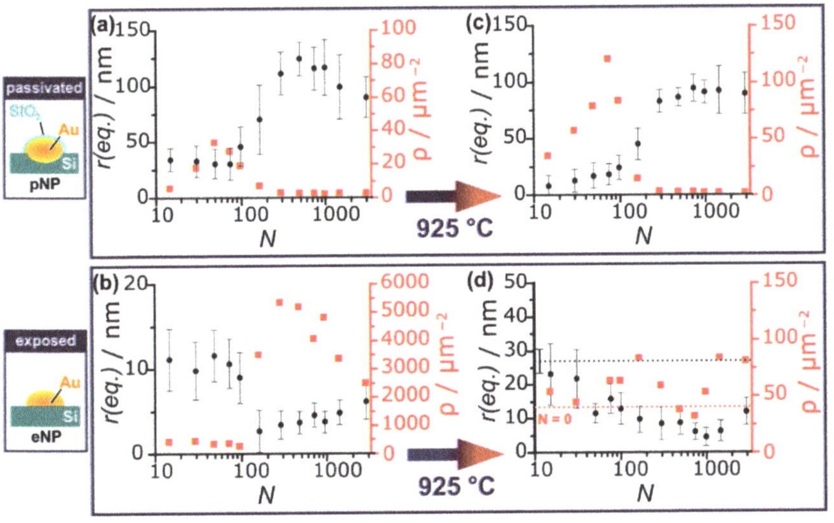

Figure 5. Equivalent radius *r(eq.)* (black) and area particle density ρ (red) of gold nanoparticles after laser irradiation (**a**,**b**) and after a subsequent heating process at a substrate temperature $T = 925$ °C (**c**,**d**) for both, passivated (pNPs, **a**,**c**) and exposed (eNPs, **b**,**d**) gold nanoparticles. The dashed lines represent the value for Au particles generated by the heating induced dewetting on unirradiated samples ($N = 0$).

Passivated nanoparticles (pNPs, Figure 5a) show a particle density peaking at about $N = 50$ ($\varphi_p = 127$ mJ/cm^2). For higher N the particle density converges to a value of about 3 particles/μm^2 as a result of the self-regulated 2D-LIPSS formation. Upon increasing N the size of 2D-LIPSS nanoparticles increases to a maximum of ~124 nm ($N = 498$) before slightly diminishing, as expected for LIPSS generation [28]. This decrease in the pNPs' size is explained by the partly ablation of the gold layer upon high energy doses, leading to the formation of small re-condensed Au-NP. At temperatures of 925 °C the pNPs undergo thermal dewetting, resulting in an increased count of smaller pNPs. Particles generated in the 2D-LIPSS generation parameter range, e.g., greater than $N = 100$ for $\varphi_p = 127$ mJ/cm^2, are relatively unaffected by the heating process, as they are well separated by the prior self-organization process. Exposed nanoparticles (eNPs, Figure 5b) show fairly constant particle sizes and densities for N smaller than 100, with an increase in the particle density by an order of magnitude and initial size drop for effective pulse numbers N greater than 100. This binary particle formation results from the aforementioned Au condensation from ns-laser laser ablation-generated plasma. Upon increasing N, the particle sizes (*r(eq.)*) decline constantly starting from a value of 27.1 nm for unirradiated samples. For the highest pulse numbers, a slight increase of *r(eq.)* is observed. After heating, the areal density *ρ(eNPs)* initially increases upon increasing N due to a dewetting process. After this initial increase, a dip in the particle density is observed. This laser parameter regime is optimal for 2D-LIPSS formation, resulting in an overall maximal proportion of Au residing in pNPs. Additionally, *ρ(eNPs)* the high particle density allows for additional particle fusing/ripening.

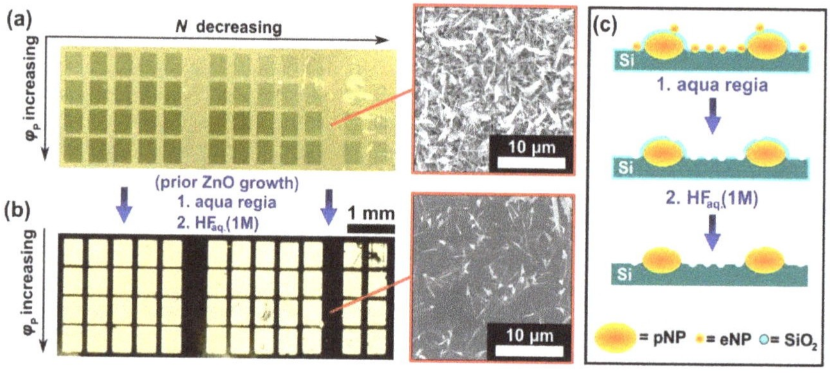

Figure 10. (**a**,**b**) Photography of an irradiated, initially gold-coated silicon wafer, overgrown with ZnO nanostructures. The wafer in (**a**) remained untreated after laser irradiation, whereas the wafer in (**b**) was consecutively treated with aqua regia and hydrofluoric acid before the ZnO growth. The visibly modified regions were irradiated with decreasing N (2984 to 15) from left to right and increasing pulse fluence φ_p (112 to 160 mJ/cm^2) from top to bottom (t = 10 min, T = 800 °C). (**c**) Schematic representation of the etching process including aqua regia and hydrofluoric acid, illustrating its effects on the gold nanoparticles and the silicon dioxide layer.

3.5. Overview of Laser-Irradiation's Effects on ZnO Growth

As both, the laser-induced generation of gold catalysts and the subsequent ZnO growth compromise a plethora of different effects, the laser irradiation's ability to affect ZnO growth should be recapitulated by highlighting the most important contributory factors. In general, the growth setup and the growth conditions determine the partial gas pressures of the reactive species in the vicinity of the substrate. Growth conditions are thus set for the entirety of the substrate, leading to predetermined ZnO nanostructure properties, e.g., morphology, height, width, and density. Laser irradiation of the Au surface before ZnO growth allows the generation of different Au catalysts. Depending on the laser parameters, e.g., pulse number N and pulse fluence φ_p, the size, density, and type of the laser-generated catalysts can be spatial-selectively controlled. For the system of Au@Si used in this work, two distinct types of Au-catalysts are generated. A part of the Au nanostructures is passivated with a thin SiO_2 layer and/or partly embedded in the substrate. As a consequence, the surface energy σ and predisposition for Au-Zn alloy formation, and therefore the VS and VLs growth mechanisms, are altered locally (Figure 9). Likewise, the density and size of the catalyst (Figures 4 and 5) have similar influences on the ZnO nanostructures' width or density (Figures 7 and 8). These parameters are a function of the applied laser energy dose. In Figure 11 the key aspects of the aforementioned interplay of ZnO growth conditions and the laser irradiation's effects are schematically summarized.

The spatial-selective and gradual change of the irradiated areas is evident by observing the different shades, corresponding to macroscopic optical properties, shown for a ZnO-covered Si-wafer in Figure 12. Complementary, Figure S6 (SI) shows an overview of the boundary of an irradiated and unirradiated ZnO-overgrown area, illustrating the effect of the irradiation on the ZnO-NRs' width and ZnO morphology-occurrence. The nanosecond laser-induced generation of Au catalysts shown in this work establishes a proof of principle for a maskless, controllable manipulation of ZnO nanostructures. By adjusting the irradiation conditions the presented principles are expected to be expandable to a variety of substrate materials. Likewise, a variation in the thickness of the underlying SiO_2-layer could lead to additional growth control. By allowing better control of the growth mechanism, e.g., by changing to a pure Zn-source, the laser-induced changes to the VLS growth should become even more apparent. The control of ZnO properties achieved in this work may be explored for complex ZnO nanostructure designs for opto-electrical devices.

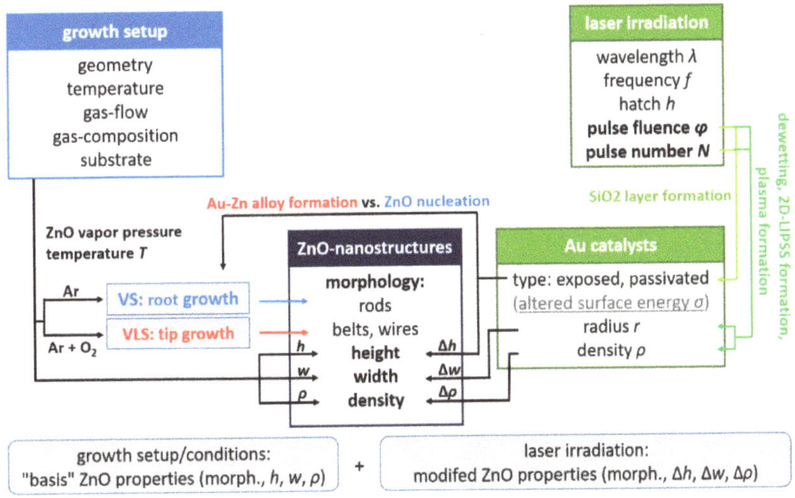

Figure 11. Schematic representation of the spatial-selective laser irradiation's effect on ZnO nanostructure growth as compared to the growth setup and conditions. Note that additional cross-relations, which are of less importance and thus not illustrated, may occur.

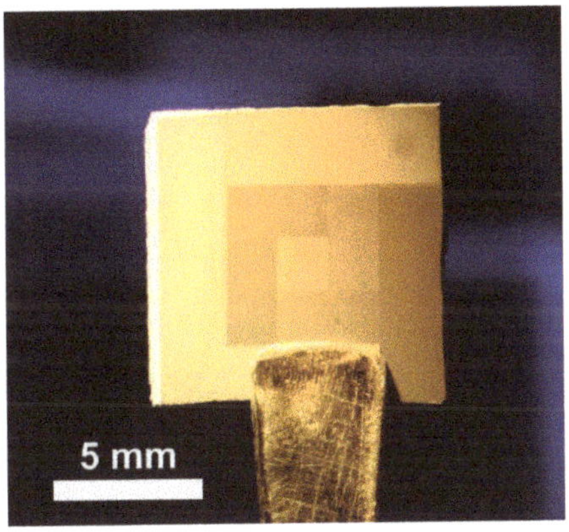

Figure 12. Photograph of a ZnO-nanostructure overgrown Si-wafer. The various shades correspond to different ZnO nanostructures grown on different laser-generated Au-catalysts.

4. Conclusions

While the current state-of-the-art ZnO growth mostly relies on binary growth control, we successfully demonstrated ZnO structures grown on ns-laser-modified gold catalysts, resulting in a plethora of different ZnO-nanostructures on a single Si-substrate in a maskless process. The catalyst modification involves pulsed laser-induced dewetting (PLiD) and 2D-LIPSS (and predecessor structures) formation, leading to two distinct gold particle types. Passivated gold nanoparticles (pNPs) are partly sunken into the wafer and covered by a thin SiO_2 layer leading to an active reduction in participating Au in the growth process. Exposed nanoparticles (eNPs) are playing a more active role in the ZnO growth process. Although their size and density ultimately change during the heating process through thermal dewetting and ripening, laser irradiation determines their final density and size. AFM measurements show additional modifications to the surface´s RMS value upon laser irradiation. Concerning ZnO nanostructures grown by a VS mechanism, we demonstrated the laser-induced alteration of the density and width of ZnO nanorods, as well as the height of an underlying accumulated ZnO layer. In the process-parameters regime favoring VLS growth, we demonstrated the laser-induced control of the morphology occurrence by gradually suppressing the growth of ZnO-nanowires and –belts. This spatial-selective control of ZnO nanostructures may allow the development of new optoelectronic nanodevices, with ZnO being a relevant but affordable semiconductor.

Supplementary Materials: The following supporting information can be downloaded at: https://www.mdpi.com/article/10.3390/nano13071258/s1, Supplementary information on the experimental setup (Figures S1 and S2), ZnO growth and characterizations (Figures S3 and S4), the particle analysis (Figure S5), the influence of the laser irradiation on ZnO growth (Figure S6), and a video of the growth-process (Video S1) are available in the supporting information [11,12,14,15,21–24,27,35].

Author Contributions: All authors contributed to the discussion and preparation of the manuscript. Conceptualization, S.D.; Methodology, S.D.; Validation, S.D. and N.H.; Formal Analysis, S.D. and L.S.; Investigation, S.D., A.M. and L.S.; Resources, N.H.; Data Curation, S.D.; Writing—Original Draft Preparation, S.D.; Writing—Review and Editing, N.H.; Visualization, S.D.; Supervision, N.H.; Project Administration, N.H. All authors have read and agreed to the published version of the manuscript.

Funding: This research received no external funding.

Data Availability Statement: I updated the data availability statement.The data that support the findings of this study are available from the corresponding author, N.H., upon reasonable request.

Acknowledgments: We thank Björn-Niclas Koch for his aid with temperature profile measurements.

Conflicts of Interest: The authors declare no conflict of interest.

References

1. Wang, T.; Jiao, Z.; Chen, T.; Li, Y.; Ren, W.; Lin, S.; Lu, G.; Ye, J.; Bi, Y. Vertically aligned ZnO nanowire arrays tip-grafted with silver nanoparticles for photoelectrochemical applications. *Nanoscale* **2013**, *5*, 7552–7557. [CrossRef]
2. Klingshirn, C.; Hauschild, R.; Fallert, J.; Kalt, H. Room-temperature stimulated emission of ZnO: Alternatives to excitonic lasing. *Phys. Rev. B* **2007**, *75*, 601. [CrossRef]
3. Kumar, A.; Navan, R.R.; Kushwaha, A.; Aslam, M.; Rao, V.R. Performance Enhancement of p-Type Organic Thin Film Transistors Using Zinc Oxide Nanostructures. *Int. J. Nanosci.* **2012**, *10*, 761–764. [CrossRef]
4. Galdámez-Martinez, A.; Santana, G.; Güell, F.; Martínez-Alanis, P.R.; Dutt, A. Photoluminescence of ZnO Nanowires: A Review. *Nanomaterials* **2020**, *10*, 857. [CrossRef]
5. Tong, C.; Jagadish, C. *Nanoscale Semiconductor Lasers*; Elsevier: Amsterdam, The Netherlands; Cambridge, MA, USA, 2019; ISBN 9780128141625.
6. Wibowo, A.; Marsudi, M.A.; Amal, M.I.; Ananda, M.B.; Stephanie, R.; Ardy, H.; Diguna, L.J. ZnO nanostructured materials for emerging solar cell applications. *RSC Adv.* **2020**, *10*, 42838–42859. [CrossRef] [PubMed]
7. Nahhas, A.M. Review of Recent Advances of ZnO Nanowires Based Sensors. *AJN* **2020**, *8*, 18–32. [CrossRef]
8. Zhao, L.; Gao, Z.; Zhang, J.; Lu, L.; Li, H. Effect of fabrication temperature on the manufacturability of lateral ZnO nanowire array UV sensor. *Sci. China Technol. Sci.* **2020**, *63*, 668–674. [CrossRef]
9. Fra, V.; Beccaria, M.; Milano, G.; Guastella, S.; Bianco, S.; Porro, S.; Laurenti, M.; Stassi, S.; Ricciardi, C. Hydrothermally grown ZnO nanowire array as an oxygen vacancies reservoir for improved resistive switching. *Nanotechnology* **2020**, *31*, 374001. [CrossRef]
10. Sakai, D.; Nagashima, K.; Yoshida, H.; Kanai, M.; He, Y.; Zhang, G.; Zhao, X.; Takahashi, T.; Yasui, T.; Hosomi, T.; et al. Substantial Narrowing on the Width of „Concentration Window" of Hydrothermal ZnO Nanowires via Ammonia Addition. *Sci. Rep.* **2019**, *9*, 14160. [CrossRef]
11. Ramgir, N.S.; Subannajui, K.; Yang, Y.; Grimm, R.; Michiels, R.; Zacharias, M. Reactive VLS and the Reversible Switching between VS and VLS Growth Modes for ZnO Nanowire Growth. *J. Phys. Chem. C* **2010**, *114*, 10323–10329. [CrossRef]
12. Zhang, Z.; Wang, S.J.; Yu, T.; Wu, T. Controlling the Growth Mechanism of ZnO Nanowires by Selecting Catalysts. *J. Phys. Chem. C* **2007**, *111*, 17500–17505. [CrossRef]
13. McIntyre, P.C.; Fontcuberta i Morral, A. Semiconductor nanowires: To grow or not to grow? *Mater. Today Nano* **2020**, *9*, 100058. [CrossRef]
14. Cheng, Q.; Ostrikov, K. Temperature-dependent growth mechanisms of low-dimensional ZnO nanostructures. *CrystEngComm* **2011**, *13*, 3455. [CrossRef]
15. Burshtein, G.; Lumelsky, V.; Lifshitz, Y. The Role of Reactive Gases in ZnO Nanowires Growth via the Carbothermal Reaction. *J. Phys. Chem. C* **2016**, *120*, 15424–15435. [CrossRef]
16. Park, J.-H.; Choi, Y.-J.; Park, J.-G. Synthesis of ZnO nanowires and nanosheets by an O2-assisted carbothermal reduction process. *J. Cryst. Growth* **2005**, *280*, 161–167. [CrossRef]

17. Shen, Y. Carbothermal synthesis of metal-functionalized nanostructures for energy and environmental applications. *J. Mater. Chem. A* **2015**, *3*, 13114–13188. [CrossRef]
18. Huang, M.H.; Wu, Y.; Feick, H.; Tran, N.; Weber, E.; Yang, P. Catalytic Growth of Zinc Oxide Nanowires by Vapor Transport. *Adv. Mater.* **2001**, *13*, 113–116. [CrossRef]
19. Sotillo, B.; Ariza, R.; Siegel, J.; Solis, J.; Fernández, P. Preferential Growth of ZnO Micro- and Nanostructure Assemblies on Fs-Laser-Induced Periodic Structures. *Nanomaterials* **2020**, *10*, 731. [CrossRef] [PubMed]
20. Zacharias, M.; Subannajui, K.; Menzel, A.; Yang, Y. ZnO nanowire arrays—Pattern generation, growth and applications. *Phys. Stat. Sol. B* **2010**, *247*, 2305–2314. [CrossRef]
21. Zhao, Y.; Li, C.; Chen, M.; Yu, X.; Chang, Y.; Chen, A.; Zhu, H.; Tang, Z. Growth of aligned ZnO nanowires via modified atmospheric pressure chemical vapor deposition. *Phys. Lett. A* **2016**, *380*, 3993–3997. [CrossRef]
22. Menzel, A.; Goldberg, R.; Burshtein, G.; Lumelsky, V.; Subannajui, K.; Zacharias, M.; Lifshitz, Y. Role of Carrier Gas Flow and Species Diffusion in Nanowire Growth from Thermal CVD. *J. Phys. Chem. C* **2012**, *116*, 5524–5530. [CrossRef]
23. Murillo, G.C.; Bojorge, C.D.; Heredia, E.A.; Reca, N.W.d. Study of the Substrate Influence in ZnO Nanowires Oriented Growth. *Procedia Mater. Sci.* **2015**, *8*, 630–634. [CrossRef]
24. Morris, R.J.H.; Dowsett, M.G.; Dalal, S.H.; Baptista, D.L.; Teo, K.B.K.; Milne, W.I. Spatial determination of gold catalyst residue used in the production of ZnO nanowires by SIMS depth profiling analysis. *Surf. Interface Anal.* **2007**, *39*, 898–901. [CrossRef]
25. Zhao, C.; Chen, A.; Ji, X.; Zhu, Y.; Gui, X.; Huang, F.; Tang, Z. Growth of vertically aligned ZnO nanowire arrays on ZnO single crystals. *Mater. Lett.* **2015**, *154*, 40–43. [CrossRef]
26. Fan, H.J.; Lee, W.; Scholz, R.; Dadgar, A.; Krost, A.; Nielsch, K.; Zacharias, M. Arrays of vertically aligned and hexagonally arranged ZnO nanowires: A new template-directed approach. *Nanotechnology* **2005**, *16*, 913–917. [CrossRef]
27. Durbach, S.; Hampp, N. Generation of 2D-arrays of anisotropically shaped nanoparticles by nanosecond laser-induced periodic surface patterning. *Appl. Surf. Sci.* **2021**, *556*, 149803. [CrossRef]
28. Dasbach, M.; Reinhardt, H.M.; Hampp, N.A. Formation of Highly Ordered Platinum Nanowire Arrays on Silicon via Laser-Induced Self-Organization. *Nanomaterials* **2019**, *9*, 1031. [CrossRef] [PubMed]
29. Chandrasekaran, H.; Sumanasekara, G.U.; Sunkara, M.K. Rationalization of nanowire synthesis using low-melting point metals. *J. Phys. Chem. B* **2006**, *110*, 18351–18357. [CrossRef] [PubMed]
30. Kolasinski, W.K. Catalytic growth of nanowires: Vapor–liquid–solid, vapor–solid–solid, solution–liquid–solid and solid–liquid–solid growth. *Curr. Opin. Solid State Mater. Sci.* **2006**, *10*, 182–191. [CrossRef]
31. Ho, S.-T.; Chen, K.-C.; Chen, H.-A.; Lin, H.-Y.; Cheng, C.-Y.; Lin, H.-N. Catalyst-Free Surface-Roughness-Assisted Growth of Large-Scale Vertically Aligned Zinc Oxide Nanowires by Thermal Evaporation. *Chem. Mater.* **2007**, *19*, 4083–4086. [CrossRef]
32. Dubrovskii, V.G.; Bol'shakov, A.D. Surface energy and modes of catalytic growth of semiconductor nanowhiskers. *Tech. Phys. Lett.* **2012**, *38*, 311–315. [CrossRef]
33. Nürnberger, P.; Reinhardt, H.M.; Rhinow, D.; Riedel, R.; Werner, S.; Hampp, N.A. Controlled growth of periodically aligned copper-silicide nanocrystal arrays on silicon directed by laser-induced periodic surface structures (LIPSS). *Appl. Surf. Sci.* **2017**, *420*, 70–76. [CrossRef]
34. Reinhardt, H.; Bücker, K.; Yang, F.; Nürnberger, P.; Hampp, N.A. Highly Dynamic Alloying and Dealloying in the Model System Gold–Silicon (AuSi). *J. Phys. Chem. C* **2015**, *119*, 5462–5466. [CrossRef]
35. Nguyen, P.; Ng, H.T.; Yamada, T.; Smith, M.K.; Li, J.; Han, J.; Meyyappan, M. Direct Integration of Metal Oxide Nanowire in Vertical Field-Effect Transistor. *Nano Lett.* **2004**, *4*, 651–657. [CrossRef]

MDPI
St. Alban-Anlage 66
4052 Basel
Switzerland
www.mdpi.com

Nanomaterials Editorial Office
E-mail: nanomaterials@mdpi.com
www.mdpi.com/journal/nanomaterials

MDPI